S0-AGV-396

# ESSENTIAL
# MATH
## FOR COLLEGE-BOUND STUDENTS

# ARCO

# ESSENTIAL MATH
## FOR COLLEGE-BOUND STUDENTS

**NORMAN LEVY**
**JOAN U. LEVY**

College
Preparation

MACMILLAN • USA

*For Joshua Seth and Jessica Dawn*
*two of our best reasons for writing this book*

Second Edition

Macmillan General Reference
A Simon & Schuster Macmillan Company
1633 Broadway
New York, NY 10019-6785

An Arco Book

Manufactured in the United States of America
10  9  8  7  6  5  4  3  2  1

**Library of Congress Cataloging-in-Publication Data**

Levy, Norman.
    Essential math for college-bound students  /  by Norman Levy, Joan U. Levy.
    p.  cm.
    ISBN 0-02-861313-9
    1.  Mathematics.  I.  Levy, Joan U. II. Title.
QA39.2.L47                96-27643
610—dc20    1996           CIP

# CONTENTS

# Introduction

This book, as the title suggests, is designed to review all the important components of high school mathematics oriented toward college-bound students. High school math curricula are in the process of change. Some school districts adhere to the traditional sequence of algebra, geometry, intermediate algebra and trigonometry. Other school districts have changed to the new integrated mathematics sequences I, II and III. Some topics have been deleted, some transposed and some added. Topics such as logic, probability and statistics are now included in these math sequences. Twelfth grade math, once known as advanced algebra or college algebra, is now called precalculus. This book encompasses both the old and the new spectrums and integrates them into unified topics.

This book is not a textbook. It makes no attempt to provide detailed analyses of the whys and wherefores of each process. Rather, it offers a comprehensive review of high school topics. It is based on the premise that mathematics is learned best by solving problems, not merely by reading about them. Thus, hundreds of author-directed examples and student problems are given, so that the college-bound student can review a subject thoroughly.

Each chapter is succinct. Important material is boxed. Every example is worked out in detail. For each example, the problem is given in the left column with sufficient space for the student to work out his own solution. In the right column is the author-provided solution as shown below.

---

**Example:**

Solve for $x$: $3x - 7 = 2x + 4$

**Solution:**

$$
\begin{array}{r}
3x - 7 = \phantom{-}2x + 4 \\
-2x \phantom{-7=} -2x \\
\hline
x - 7 = \phantom{-}4 \\
+7 \phantom{x-} +7 \\
\hline
x = \phantom{-}11
\end{array}
$$

---

The student should attempt to work out each example on his own before looking at the solution provided. At the end of each review section are multiple-choice problems similar to the problems you will face on college-entrance examinations. Detailed solutions to the practice problems are provided every four chapters.

*Essential Math for College-Bound Students* begins with a 50-question Pretest and a Diagnostic Chart indicating the subject area of each test question. Cumulative Review Tests of 20 questions each are provided every four chapters. These tests, too, are accompanied by a diagnostic analysis to help evaluate your progress. A 50-question Post-test at the end of the book serves as a final review of high school math. Throughout the text, over 1000 examples provide a "hands on" approach to mastering essential mathematical concepts.

This self-teaching text is designed to answer the common student lament: "If only I had a review sheet with all the important information." Here you have it—a capsule review of each topic covered in high school mathematics complete with practice questions to prepare you to score high on college entrance and college placement tests in math.

# Special Features of This Book

- Capsule reviews summarize the essential concepts of each area of high school math
- Key concepts are boxed for added emphasis
- Examples allow sufficient work space for students to try each problem before looking at the detailed solutions provided
- Cumulative review tests are offered every four chapters with extensive Diagnostic Charts for self-evaluation
- Diagnostic charts indicate subject area of each question, thus providing an instant analysis of individual trouble spots
- Author's Notes, indicated by the symbol on the right, provide additional clarification of important points
- Author's Warnings, indicated by the symbol on the right, point out common errors to be avoided
- Carefully worked through solutions are provided for every problem
- All levels of high school math are integrated by topic

**This book is designed to serve as:**
1. A review of high school math.
2. Preparation for standardized college entrance examinations.
3. Preparation for Math Level I and II Achievement tests.
4. Preparation for college placement tests in mathematics.

# Standardized Tests in Mathematics

Aside from the mathematics curriculum covered in the high school, college-bound students face a number of very important standardized tests during their high school years. These examinations are used by the colleges as part of their admissions criteria. Thus the standardized tests help to determine which students will gain admission to a given college. The importance of these tests should be stressed to every high school student. The standardized tests assist the college officials with a non-demographic, non-geographic comparison of skills.

The chart on page 9 summarizes the various tests.

## Preliminary Scholastic Assessment Test/National Merit Scholarship Qualifying Test (PSAT/NMSQT)

The Preliminary Scholastic Assessment Test/National Merit Scholarship Qualifying Test (PSAT/NMSQT) is taken by high school students in October of their junior year. The test is two hours long and consists of four 30-minute sections:

| | | |
|---|---|---|
| Section 1 | Verbal | 29 questions |
| Section 2 | Mathematics | 25 questions |
| Section 3 | Verbal | 29 questions |
| Section 4 | Mathematics | 25 questions |

The **Mathematics Section** of the PSAT/NMSQT has three types of questions:

Regular Multiple Choice:
A mathematical situation is presented involving algebra, arithmetic, geometry, spatial relationships or analytical reasoning. The student must select the correct solution from five choices.

Quantitative Comparison:
The student is presented with two relationships, one in column A and the other in column B. He is asked to evaluate the quantities under all allowed conditions of the problem and to determine whether:

- Column A is always greater than column B (answer choice A)
- Column B is always greater than column A (answer choice B)
- The two columns are always equal (answer choice C)
- An "always" relationship is not defined (answer choice D)

Student-Produced Response:
The student must solve each problem and grid the answer on the answer sheet. A grid response has a maximum of four characters.

**Scoring** for both Verbal and Mathematics sections is from 20 to 80. Each correct answer contributes upward to the score; each incorrect answer contributes fractionally downward from the score, and omitted questions and incorrectly answered Student-Response questions neither raise nor lower the score. The scoring is designed so that random guessing cannot significantly improve your score. It is generally better to review your completed work than to waste time guessing. However, if you are able to eliminate one or more choices, it may be advantageous to make an educated guess.

The **purpose** of the PSAT/NMSQT is multifold:

- First, it is, as its name implies, a preliminary measure of scholastic aptitude. It is meant to give the student some guidelines about his verbal and mathematical abilities prior to taking the Scholastic Assessment Test (SAT I).\* The PSAT score is *not* sent to colleges and does not affect college admission.
- It is an opportunity for the student to compare his abilities to those of other students nationwide who are competing for similar goals. The reported percentile scores are useful toward this end.
- Entrance to competition for National Merit scholarships and recognition is based on a selection index formulated by combining twice the verbal score plus the mathematics score. Thus the PSAT/NMSQT can have financial and/or prestige value for the college-bound student.

The PSAT/NMSQT is administered by the Educational Testing Service (ETS) on behalf of the College Board and National Merit Scholarship Corporation (NMSC). Further information and details are available from:

| | | |
|---|---|---|
| PSAT/NMSQT | NMSC | College Board |
| P.O. Box 24700 | 1 Rotary Center | PSAT/NMSQT |
| Oakland, CA 94623-1700 | 1560 Sherman Avenue | 45 Columbus Ave. |
| | Evanston, IL 60201 | New York, NY 10023 |

## Scholastic Assessment Test (SAT I)

The Scholastic Assessment Test (SAT I) is a three-hour, multiple-choice test offered by the Admissions Testing Program (ATP) of the College Board to help colleges compare the potential of students from different geographic and demographic circumstances. Since the SAT is one of the vehicles used by colleges and universities to help decide on admissions, the SAT assumes a great importance to the student.

The SAT I is a longer and more difficult version of the PSAT. It consists of seven sections of either 15 minutes or 30 minutes in length. They are:

Verbal—30 minutes  
Verbal—30 minutes  } 78 questions  
Verbal—15 minutes

Mathematics—30 minutes  
Mathematics—30 minutes  } 60 questions  
Mathematics—15 minutes

One Verbal or Mathematics  
Equating Section  —30 minutes

 The Equating Section does not contribute to the final SAT I score.

The three verbal sections are combined to yield a Verbal Score ranging from 200 to 800. Similarly, the three mathematics sections are combined to yield a Mathematics Score ranging

---

\*SAT, SAT I, SAT II, and Scholastic Assessment Test are registered trademarks that are the exclusive property of The College Entrance Examination Board.

from 200 to 800. Each correct answer contributes toward raising the score; each incorrect answer fractionally lowers the score, and omitted questions and incorrect Student-Produced responses neither raise nor lower the score. SAT scoring is designed so that random guessing will not significantly improve your score. It is generally better to review completed work for potential errors than to waste time on random guesses. However, if you are able to eliminate one or more choices, it may be advantageous to make an educated guess.

The **Mathematics Section** of the SAT I has three types of questions:

### Regular Multiple Choice:

A mathematical situation is presented involving analytic reasoning, algebra, geometry, spatial relationships or arithmetic processes. The student must select the correct solution to the given question from five choices. There are 40 such questions on the SAT. For example:

If $a$ apples cost $b$ cents, then how many apples can be bought for $b^2$ dollars?

A) $a^2$   B) $ab$   C) $100a^2$   D) $100ab$   E) $\dfrac{a^2}{100}$

The correct solution is D. Establish a proportion of apples to dollars:

$$\frac{\text{apples}}{\text{dollars}} : \frac{a}{\left(\dfrac{b}{100}\right)} = \frac{x}{b^2}$$

$$x = 100ab$$

### Quantitative Comparison:

Two relationships are presented, one in column A and the other in column B. The student is asked to evaluate the two quantities using any provided constraints and to determine whether:

- Column A is always greater than column B (answer choice A)
- Column B is always greater than column A (answer choice B)
- The two columns are always equal (answer choice C)
- An "always" relationship does not exist (answer choice D)

 Marking answer choice E is equivalent to omitting the question.

There are 15 such questions on the SAT. For example:

| *Column A* | *Column B* |
|:---:|:---:|
| $x \neq -4$ | |
| $x + 4$ | $\dfrac{x^2 - 16}{x - 4}$ |

The correct solution is C. The two quantities are always equal (when $x \neq -4$).

$$\frac{x^2 - 16}{x - 4} = \frac{(x + 4)(x - 4)}{(x - 4)} = x + 4$$

Student-Produced Response:
  A mathematical situation is presented. The student must grid a response within the four grid boxes provided on the answer sheet. There are 10 questions of this type. For example:

  What is the next integer following 8777 which contains three of the same digits?

  The answer is 8788. The test taker would grid the answer as follows:

| 8 | 7 | 8 | 8 |
|---|---|---|---|
|   | / | / |   |
| . | . | . | . |
|   | 0 | 0 | 0 |
| 1 | 1 | 1 | 1 |
| 2 | 2 | 2 | 2 |
| 3 | 3 | 3 | 3 |
| 4 | 4 | 4 | 4 |
| 5 | 5 | 5 | 5 |
| 6 | 6 | 6 | 6 |
| 7 | ● | 7 | 7 |
| ● | 8 | ● | ● |
| 9 | 9 | 9 | 9 |

  The major **purpose** of the SAT I is to provide college officials with a standardized basis of comparison of an individual's college potential. Colleges and universities use the results, together with grade point average, class rank, extracurricular activities, personal interview and application data, to evaluate candidates for admission to their respective schools.

  Further information and details are available from:

  College Board ATP
  CN 6200
  Princeton, NJ 08541-6200

# American College Testing (ACT) Program

  The American College Testing (ACT) Program is a standardized test that is used by colleges to help determine which students will be admitted to a particular college or university. The ACT test battery can be taken, in many cases, in lieu of the SAT I or SAT II.
  The ACT consists of four discrete sections:

  English Test—75 questions—45 minutes
  Mathematics Test—60 questions—60 minutes
  Reading Test—40 questions—35 minutes
  Science Reasoning Test—40 questions—35 minutes

The **Mathematics** test presents mathematics problems encompassing pre-algebra, intermediate algebra, plane geometry, coordinate geometry and trigonometry. The student must select the solution from five possible choices. For example:

If $a = b$ and $\dfrac{1}{c} = b$, then $c =$

A) $a$     B) $-a$     C) $b$     D) $\dfrac{1}{a}$     E) $-b$

The correct answer is D.

$$a = b = \frac{1}{c}$$

$$a = \frac{1}{c} \rightarrow ac = 1$$

$$c = \frac{1}{a}$$

The **scoring** on the ACT ranges from 1 to 36 for the individual test scores and the composite score and from 1 to 18 for the subscores. Unlike the SAT, no penalty is applied for incorrect answers. The student taking the ACT should always guess. No question should go unanswered.

The **purpose** of the ACT is to provide the college and university admissions officials with a standardized basis of comparison of an individual's college potential. These results, together with grade-point average, class rank, extracurricular activities, personal interview and application data, contribute to the admissions evaluation process.

Additional information and details are available from:

American College Testing Registration
P.O. Box 168
Iowa City, IA 52243

# SAT II—Subject Tests

Some colleges and universities require further standardized testing. They ask the student to take from one to three SAT II—Subject Tests in different subject areas. There are a total of seventeen tests given in fourteen subject areas. Each test is 60 minutes in duration.

**Scoring** for each test ranges from 200 to 800. As with the SAT I, each correct answer contributes toward raising the score; each incorrect answer fractionally lowers the score, and omitted questions neither raise nor lower the score. The scoring is designed so that random guessing will not significantly improve your score. However, if you are able to eliminate one or more choices, it may be advantageous to make an educated guess.

The SAT II—Subject Tests include:

| Languages | Sciences |
|---|---|
| French | Biology |
| German | Chemistry |
| Hebrew | Physics |
| Latin | |
| Spanish | |
| Italian | |
| Chinese | |
| Japanese | |

## Mathematics

Mathematics—Level IC

This test includes second-year algebra, geometry, trigonometry, elementary functions, reasoning, logic and elementary number theory. The IC test requires the use of a calculator.

Mathematics —Level IIC

This test includes third-year algebra, geometry, trigonometry, elementary functions, sequences, series, logic, probability, statistics and elementary number theory.

Mathematics—Level IC and Mathematics— Level IIC count as one Subject Test. Thus the student should choose between the two tests.

| **Social Studies** | **English** |
|---|---|
| American History and Social Studies | Writing |
| World History | Literature |

The **purpose** of the subject tests is defined by the college or university requesting that the tests be taken. They can serve either to aid the college officials in the admissions process or they can be used for placement in their respective subject areas for an entering freshman.

Further **information** can be obtained from:

College Board ATP
CN 6200
Princeton, NJ 08541-6200

# Mathematics Placement Test

Many colleges and universities require their students to take a placement test in mathematics prior to beginning their freshman year. The examination usually tests mathematical concepts and applications through precalculus. It is used for placement in beginning college mathematics courses. Further information can be obtained from your admitting college.

# Summary of Standardized Tests

| Test | Sections | Number of Questions | Minutes | Mathematics Level | For Further Information |
|---|---|---|---|---|---|
| Preliminary Scholastic Assessment Test/ National Merit Scholarship Qualifying Test (PSAT/NMSQT) | Verbal<br>Math<br>Verbal<br>Math | 23–30<br>25<br>25–30<br>25 | 30<br>30<br>30<br>30 | Through:<br>Arithmetic Reasoning,<br>Elementary Algebra<br>and Geometry | PSAT/NMSQT<br>P.O. Box 24700<br>Oakland, CA 94623-1700<br>*N.J.:* (609) 683-0449<br>*Calif.:* (415) 653-5595 |
| Scholastic Assessment Test I (SAT I) | Verbal<br>Verbal<br>Verbal<br>Math<br>Math<br>Math | 35<br>30<br>11–13<br>25<br>25<br>10 | 30<br>30<br>15<br>30<br>30<br>15 | Through:<br>Arithmetic Reasoning,<br>Elementary Algebra<br>and Geometry | College Board ATP<br>CN 6200<br>Princeton, NJ 08541-6200<br>*N.J.:* (609) 771-7588<br>*Calif.:* (415) 654-1200 |
| Equating | English or Math | Varies | 30 | | |
| American College Testing Program (ACT) | English<br>Mathematics<br>Reading<br>Science Reasoning | 75<br>60<br>40<br>40 | 45<br>60<br>35<br>35 | Through:<br>Second Yr. Algebra,<br>Geometry,<br>Number Concepts,<br>Arithmetic and Algebraic<br>Reasoning,<br>Trigonometry | American College Testing<br>Registration<br>P.O. Box 168<br>Iowa City, IA 52243<br>(319) 337-1270 |
| Scholastic Assessment Test II (SAT II: Subject Tests) | Chinese<br>French<br>German<br>Hebrew (Modern)<br>Italian<br>Japanese<br>Latin<br>Spanish<br>Biology<br>Chemistry<br>Physics<br>Mathematics—Level IC<br>Mathematics—Level IIC<br>American History and Social Studies<br>World History<br>Writing<br>Literature | 85<br>85<br>80–90<br>85<br>80–85<br>80–85<br>70–75<br>85<br>95<br>85<br>75<br>50<br>50<br>90–95<br>95<br>60 + Essay<br>55–65 | 60<br>60<br>60<br>60<br>60<br>60<br>60<br>60<br>60<br>60<br>60<br>60<br>60<br>60<br>60<br>60<br>60 | Level IC<br>Through:<br>Second Yr. Algebra,<br>Geometry,<br>Trigonometry,<br>Elementary Functions,<br>Reasoning, Logic and<br>Elementary Number Theory<br><br>Level IIC<br>Through:<br>Third Yr. Algebra,<br>Geometry<br>Trigonometry<br>Elementary Functions,<br>Sequences, Series<br>Logic, Probability, Statistics<br>and Elementary Number Theory | College Board ATP<br>CN 6200<br>Princeton, NJ 08541-6200<br>*N.J.:* (609) 771-7558<br>*Calif.:* (415) 654-1200 |
| Mathematics Placement Test | Math | | | Usually through Precalculus | Your Admitting College |

*The questions that follow represent each of the topics covered in this book. Solve each problem and select the correct answer from the five choices given. When you have completed the test, compare your answers with the solutions at the end of the test. Use the Diagnostic Chart on page 20 to evaluate your strengths and weaknesses.*

1. The $\sqrt{17}$ is a member of which set of numbers?

   A) rational    B) integers   C) irrational
   D) counting   E) imaginary

2. The inverse of $\sim p \rightarrow q$ is

   A) $p \wedge q$   B) $p \rightarrow \sim q$   C) $q \rightarrow \sim p$
   D) $q \rightarrow p$   E) $p \vee \sim q$

3. Factor $x^2 - 7x + 10$.

   A) $(x-10)(x+1)$   B) $(x+5)(x+2)$
   C) $(x-5)(x-2)$
   D) $(x+5)(x-2)$   E) $(x+10)(x+1)$

4. How many degrees are $\dfrac{2\pi}{15}$ radians?

   A) 12   B) 24   C) 36   D) 156   E) 180

5. If $\log 2 = A$ and $\log 3 = B$, then $\log 6$ is equal to

   A) $A+B$   B) $A-B$   C) $AB$   D) $\dfrac{A}{B}$   E) $A^B$

6. For which value of $c$ will the roots of the equation $f(x) = x^2 + 4x + c$ be real and equal?

   A) 1   B) 2   C) 3   D) 4   E) 5

7. Which is the fourth term in the expansion of $(x+3)^5$?

   A) $270x^2$   B) $135x^2$   C) $405x$
   D) $135x$   E) $-135x$

8. The circumference of a circle is $20\pi$. What is the area of the circle?

   A) $10\pi$   B) $20\pi$   C) $30\pi$   D) $100\pi$   E) $400\pi$

9. Perform the indicated operation and express in simplest form:

$$\frac{x^2 - 3x}{2x^2 + x - 6} \div \frac{x^2 - 5x + 6}{x^2 - 4}$$

   A) $\dfrac{x}{x+6}$   B) $\dfrac{x-3}{x+1}$   C) $\dfrac{x(x-3)}{(x-2)}$
   D) $\dfrac{x}{2x-3}$   E) $\dfrac{1}{2x-3}$

10. What are the coordinates of the center of a circle whose equation is $(x-1)^2 + (y+5)^2 = 7$?

    A) $(-1, 5)$   B) $(\sqrt{7}, 0)$   C) $(1, -5)$
    D) $(-5, 1)$   E) $(5, -1)$

11. The roots of the equation $x^2 - 6x - 2 = 0$ are

    A) $3 \pm \sqrt{7}$   B) $-3 \pm \sqrt{7}$   C) $3 \pm \sqrt{2}$
    D) $3 \pm \sqrt{11}$   E) $-3 \pm \sqrt{11}$

12. The expression $\dfrac{5}{2 - \sqrt{3}}$ is equivalent to

    A) $10 + 5\sqrt{3}$   B) $-2 - \sqrt{3}$   C) $-10 - 5\sqrt{3}$
    D) $2 + \sqrt{3}$   E) $2 - \sqrt{3}$

13. Find the mode of the following group of numbers: 8, 8, 9, 10, 11

    A) 8   B) 9   C) $8\frac{1}{2}$   D) $9\frac{1}{5}$   E) 11

14. What is the solution set of the equation

$$|3x + 2| = 5?$$

    A) $\{1\}$   B) $\{\frac{7}{3}\}$   C) $\{-\frac{7}{3}\}$
    D) $\{1, -\frac{7}{3}\}$   E) $\{-1, \frac{7}{3}\}$

15. In triangle $ABC$, if $a=2$, $b=4$ and $m \angle C = 60$, what is the length of side $c$?

    A) $2\sqrt{7}$  B) $2$  C) $2\sqrt{3}$  D) $\sqrt{21}$  E) $4\sqrt{7}$

16. If $m=8$, find the value of $m^{-2/3}$.

    A) $-\frac{16}{3}$  B) $-4$  C) $4$  D) $\frac{1}{4}$  E) $64$

17. The square of $(2-2i)$ is

    A) $0$  B) $-8i$  C) $4-4i$  D) $4$  E) $-4$

18. The amplitude of the graph of $y = 3 \cos 2x$ is

    A) $\pi$  B) $2$  C) $3$  D) $4\pi$  E) $\frac{2\pi}{3}$

19. What is the solution set for $(x+3)(x-2) > 0$?

    A) ←——○———  
             $-3$

    B) ——○———○——→  
             $-3$   $2$

    C) ———————○———→  
                 $2$

    D) ←——○———○——  
             $-3$   $2$

    E) ———●———●———  
             $-3$   $2$

20. How many different 6-player combinations are possible from a group of 10 students trying out for a team?

    A) 60  B) 105  C) 210  D) 360  E) 720

21. The locus of points in a plane that are a given distance $d$ from a point $P$ is

    A) one circle  B) two circles  
    C) one circle and one point  
    D) two parallel lines  E) a triangle

22. If 35% of number is 70, what is the number?

    A) 200  B) 20  C) 2000  D) 245  
    E) 24.5

23. If $f(x) = 2x - 1$ and $g(x) = 3x + 1$, then $g(f(x)) =$

    A) $2x+1$  B) $3x-1$  C) $6x-2$  D) $x$  
    E) $6x^2 - x - 1$

24. Find the lateral area of a right circular cone whose height is 10 and base radius is 2.

    A) $400\pi$  B) $20\pi$  C) $40\pi$  D) $2\sqrt{26}$  
    E) $4\pi\sqrt{26}$

25. Eliminate the parameter $n$ and identify the curve.

    $$x = 3 - \tfrac{1}{2}n^2, \; y = 1 + n$$

    A) line  B) circle  C) ellipse  
    D) parabola  E) hyperbola

26. What statement will this mini-program print?

    ```
    1  LET A=3
    2  LET B=4
    3  LET C=(B-A)²
    4  IF C>0 GO TO INSTRUCTION 7
    5  IF C=0 GO TO INSTRUCTION 9
    6  IF C<0 GO TO INSTRUCTION 11
    7  PRINT"A=3"
    8  STOP
    9  PRINT"B=4"
    10 STOP
    11 PRINT"B-A=1"
    12 STOP
    ```

    A) STOP  B) A=3  C) B=4  
    D) B−A=1  E) B−A=−1

27. $\displaystyle\prod_{k=1}^{3} (2k) =$

    A) $8$  B) $12$  C) $2k+1$  D) $6k-1$  
    E) $48$

28. $1+i$ is equivalent to

    A) $\sqrt{2}$  B) $\sqrt{2}(\cos 0° + i \sin 0°)$  
    C) $\sqrt{2}(\cos 30° + i \sin 30°)$  
    D) $\sqrt{2}(\cos 45° + i \sin 45°)$  
    E) $\sqrt{2}(\cos 60° + i \sin 60°)$

29. Evaluate $e^{-i\pi}$.

    A) $0$  B) $1$  C) $-1$  D) $\pi$  E) $-\pi$

30. If $\log 3.9 = .5911$ and $\log 1.97 = .2955$, find $\sqrt{390}$.

    A) 18.6  B) 19.3  C) 19.7  D) 19.9  E) 20.1

31. Sinh $x =$

    A) $\dfrac{1}{\sec x}$   B) $\dfrac{e^x - e^{-x}}{e^x + e^{-x}}$   C) $\dfrac{e^x + e^{-x}}{2}$

    D) $\dfrac{e^x - e^{-x}}{2}$   E) $1 - \cosh x$

32. $x$ varies jointly as $j$ and the $\sqrt{l}$. If $x$ is 3 when $j$ is 4 and $l$ is 4, find $x$ when $l$ is 9 and $j$ is 9.

    A) $\frac{81}{8}$   B) 72   C) $\frac{9}{8}$   D) $\frac{8}{3}$   E) 6

33. $y = \dfrac{(x+2)(x-3)(x-1)}{(x-7)(x+5)}$ has a vertical asymptote at

    A) $x = -2$   B) $x = 3$   C) $x = 1$
    D) $x = 7$   E) $x = 5$

34. Of 18 dogs in a show, 6 dogs have black spots, 7 dogs have brown spots and 3 dogs have both black and brown spots. How many dogs have neither black nor brown spots?

    A) 2   B) 4   C) 7   D) 8   E) 9

35. If $f(x) = x^2 - 2x + 1$, $f(x + 1) =$

    A) $x^3 - 2x^2 + 2$   B) $x^2 - 2x + 2$
    C) $x^2 - x + 2$   D) $x^2 - 2x + 3$   E) $x^2$

36. The expression $\log \sqrt{xy}$ is equivalent to

    A) $2 \log x + \log y$   B) $2(\log x + \log y)$
    C) $\frac{1}{2} \log x + \log y$   D) $\frac{1}{2}(\log x + \log y)$
    E) $2 \log x + \frac{1}{2} \log y$

37. Solve for $x$: $27^{2x} = 3^{3x-5}$.

    A) $-\frac{3}{5}$   B) $-\frac{5}{3}$   C) 5   D) $-5$
    E) 15

38. Find the value of $\sin 210°$.

    A) $-\dfrac{\sqrt{3}}{2}$   B) $\dfrac{\sqrt{3}}{2}$   C) $-\dfrac{1}{2}$   D) $\dfrac{1}{2}$

    E) $-\dfrac{7}{2}$

39. Express the sum of $(3 + 2\sqrt{-16})$ and $(7 - 3\sqrt{-81})$ in $a + bi$ form.

    A) $10 + \sqrt{-97}$   B) $10 + 5i$   C) $10 - 2i$
    D) $10 + 10i$   E) $10 - 19i$

40. Chords $AB$ and $CD$ of circle $O$ intersect at $E$. If $AE = 4$, $AB = 5$, $CE = 2$, find $ED$.

    A) 2   B) 3   C) 7   D) 8   E) 10

41. The greatest common monomial factor of $16x^5 + 6x^4 + 10x^3$ is

    A) 2   B) $4x$   C) $4x^2$   D) $2x^3$
    E) $x^4$

42. Express $.22222\overline{2}$ as a rational number.

    A) $.22$   B) $\frac{22}{100}$   C) $\frac{222}{1000}$   D) $\frac{2}{90}$   E) $\frac{2}{9}$

43. What is the length of the line segment joining the points $N(1, -1)$ and $L(-3, 3)$?

    A) $2\sqrt{5}$   B) $4\sqrt{2}$   C) $5\sqrt{2}$   D) 4
    E) $2\sqrt{2}$

44. What is the solution set for the following system of equations?

    $$2x - 3y = 4$$
    $$x - 2y = 3$$

    A) $\{2, -1\}$   B) $\{-1, -2\}$   C) $\{-1, 3\}$
    D) $\{3, -2\}$   E) $\{-2, -3\}$

45. What is the mean of the following data?

    | Score | Frequency |
    |-------|-----------|
    | 7     | 1         |
    | 11    | 2         |
    | 12    | 3         |

    A) 7   B) $10\frac{2}{3}$   C) 11   D) $10\frac{5}{6}$   E) 12

46. In triangle $JSL$, if $s = 10$, $l = 16$ and $m \angle J = 30$, the area of the triangle is

    A) 160   B) $80\sqrt{3}$   C) $40\sqrt{3}$   D) 40
    E) $40\sqrt{2}$

47. The product of the digits of a positive two digit number is 8. If the number is four times the sum of the digits, find the number.

    A) 42   B) 24   C) 81   D) 18   E) 16

48. What is the probability of exactly 2 heads out of 5 tosses of a fair coin?

    A) $_5C_2(\frac{1}{2})^2(\frac{1}{2})^3$  B) $_3C_2(\frac{1}{2})^2(\frac{1}{2})^3$  C) $_5C_5(\frac{1}{2})^2(\frac{1}{2})^3$
    D) $\frac{2}{5}$  E) $_5C_2(\frac{1}{2})$

49. $xy^2 = k$, where $k$ is a constant, is an example of what type of variation?

    A) direct  B) inverse  C) joint
    D) combined  E) sine wave

50. What is the sum of an arithmetic progression of 10 terms whose first term is 8 and whose common difference is $-3$?

    A) $-12$  B) $-47$  C) $-11$
    D) $-95$  E) $-55$

# Solutions to Pretest

1. **(C)** The $\sqrt{17}$ is an infinite non-repeating decimal. It is a member of the set of irrational numbers.

2. **(B)** The inverse is the statement with both the hypothesis and conclusion negated.

3. **(C)** $x^2 - 7x + 10 = (x-2)(x-5)$

4. **(B)** $\dfrac{2\cancel{\pi}}{15} \times \dfrac{180}{\cancel{\pi}} = \dfrac{2(\cancel{180}^{12})}{\cancel{15}} = 24°$

5. **(A)** $\log 6 = \log (2 \cdot 3) = \log 2 + \log 3$
$$= A + B$$

6. **(D)** If the discriminant $(b^2 - 4ac) = 0$, then the roots are real and equal.
$a = 1,\ b = 4,\ c = c$
$b^2 - 4ac = (4)^2 - 4(1)(c) = 0 \quad 16 = 4c$
$$c = 4$$

7. **(A)** kth term $= {}_nC_{k-1}(a)^{n-(k-1)}(b)^{k-1}$
$k = 4,\ n = 5,\ a = x,\ b = +3,\ k-1 = 3$
4th term $= {}_5C_3(x)^{5-3}(3)^3$
$$= \dfrac{5 \cdot 4 \cdot \cancel{3}}{\cancel{3} \cdot 2 \cdot 1} x^2(27)$$
$$= 270x^2$$

8. **(D)** $C = 2\pi r = 20\pi$
$$r = 10$$
$$A = \pi r^2 = \pi(10)^2 = 100\pi$$

9. **(D)** $\dfrac{x^2 - 3x}{2x^2 + x - 6} \div \dfrac{x^2 - 5x + 6}{x^2 - 4}$
$$= \dfrac{x^2 - 3x}{2x^2 + x - 6} \cdot \dfrac{x^2 - 4}{x^2 - 5x + 6}$$
$$= \dfrac{x\overset{(1)}{(x-3)}}{(2x-3)(x+2)} \cdot \dfrac{\overset{(1)}{(x+2)}\overset{(1)}{(x-2)}}{\underset{(1)}{(x-3)}\underset{(1)}{(x-2)}}$$
$$= \dfrac{x}{2x-3}$$

10. **(C)** $(x - h)^2 + (y-k)^2 = r^2$
Center: $(h, k)$, radius: $r$
$(x-1)^2 + (y+5)^2 = r^2$
Center: $(1, -5)$, radius: $\sqrt{7}$

11. **(D)** $x = \dfrac{-b \pm \sqrt{b^2 - 4ac}}{2a}$

$a = 1,\ b = -6,\ c = -2$

$x = \dfrac{-(-6) \pm \sqrt{(-6)^2 - 4(1)(-2)}}{2(1)}$

$= \dfrac{6 \pm \sqrt{36 + 8}}{2}$

$= \dfrac{6 \pm \sqrt{44}}{2}$

$= \dfrac{6 \pm 2\sqrt{11}}{2} = 3 \pm \sqrt{11}$

12. **(A)** $\dfrac{5}{2 - \sqrt{3}} \cdot \dfrac{2 + \sqrt{3}}{2 + \sqrt{3}} = \dfrac{5(2 + \sqrt{3})}{4 - 3} = 10 + 5\sqrt{3}$

13. **(A)** The mode is the number which occurs the most frequently. The mode $= 8$.

14. **(D)**
$$|3x + 2| = 5$$
$$\downarrow \qquad \downarrow$$
$$3x + 2 = -5 \quad 3x + 2 = 5$$
$$x = -\tfrac{7}{3} \qquad x = 1$$
check

| | |
|---|---|
| $\lvert 3x + 2 \rvert = 5$ | $\lvert(3x + 2)\rvert = 5$ |
| $\lvert 3(-\tfrac{7}{3}) + 2\rvert \overset{?}{=} 5$ | $\lvert 3(1) + 2\rvert = 5$ |
| $\lvert -7 + 2\rvert \overset{?}{=} 5$ | $\lvert 5\rvert = 5$ |
| $\lvert -5\rvert \overset{?}{=} 5$ | |
| $5 = 5$ | |

2 solutions $\{1, -\tfrac{7}{3}\}$

15. **(C)** $c^2 = a^2 + b^2 - 2ab \cos C$
$$c^2 = (2)^2 + (4)^2 - 2(2)(4) \cos 60°$$
$$c^2 = 4 + 16 - 16(\tfrac{1}{2})$$
$$c^2 = 12$$
$$c = 2\sqrt{3}$$

16. **(D)** $8^{-2/3}$. The 3 indicates cube root
The 2 represents squaring
The "—" indicates reciprocal
$$8^{-2/3} = \frac{1}{(\sqrt[3]{8})^2} = \frac{1}{4}$$

17. **(B)** $(2 - 2i)^2 = (2 - 2i)(2 - 2i)$
$$= 2(2 - 2i) - 2i(2 - 2i)$$
$$= 4 - 4i - 4i + 4i^2$$
$$= 4 - 8i + 4(-1)$$
$$= -8i$$

18. **(C)** The amplitude is 3.

19. **(D)** $(x + 3)(x - 2) = 0$
$$x = -3 \mid x = 2$$
The critical values are $-3$ and 2.
They are excluded due to the $>$ in the original problem (the values $-3$ and 2 are represented by hollow circles).

Test the interval $-\infty < x < -3$, e.g., $x = -10$
$(-10 + 3)(-10 - 2) = +84 > 0$        True

Test the interval $-3 < x < 2$, e.g., $x = 0$
$(0 + 3)(0 - 2) = -6 > 0$        False

Test the interval $x > 2$, e.g., $x = 3$
$(3 + 3)(3 - 2) = 6 > 0$        True

20. **(C)** $_{10}C_6 = {}_{10}C_4 = \dfrac{10 \cdot \overset{(3)}{9} \cdot 8 \cdot 7}{4 \cdot \underset{}{3} \cdot 2 \cdot 1} = 210$

21. **(A)**

A circle with center $P$ and radius d

22. **(A)** $\dfrac{35}{100} \cdot x = 70$
$$\frac{35x}{100} = 70$$
$$x = 200$$

23. **(C)** $g(f(x)) = g(2x - 1)$
$$= 3(2x - 1) + 1$$
$$= 6x - 3 + 1$$
$$= 6x - 2$$

24. **(E)** L.A. $= \pi r l$
$$= \pi(2)(2\sqrt{26})$$
$$= 4\pi\sqrt{26}$$

$l^2 = 10^2 + 2^2 = 104$
$l = \sqrt{104} = 2\sqrt{26}$

25. **(D)** $x = 3 - \tfrac{1}{2}n^2$
$y = 1 + n \Rightarrow n = y - 1$
$\therefore x = 3 - \tfrac{1}{2}(y - 1)^2$        Substitute for $n$
$x - 3 = -\tfrac{1}{2}(y - 1)^2$
This is the equation of a parabola.

26. **(B)**

| Value of A | Value of B | Value of $(B - A)^2 = C$ |
|---|---|---|
| 3 | 4 | $(4 - 3)^2 = 1$ |

Go to instruction 7
$A = 3$

27. **(E)** $\displaystyle\prod_{k=1}^{3} (2k) = [2(1)] \cdot [2(2)] \cdot [2(3)]$
$$2 \cdot 4 \cdot 6 = 48$$

28. **(D)** $x + yi = 1 + i \quad x = 1, y = 1$

$r = \sqrt{x^2 + y^2} = \sqrt{2}$

$\mathrm{Tan}\,\theta = \dfrac{y}{x} = \dfrac{1}{1} = 1, \theta = 45°$

$1 + i = \sqrt{2}(\cos 45° + i \sin 45°)$

29. **(C)** $e^{i\theta} = \cos\theta + i\sin\theta$

$e^{i(-\pi)} = \cos(-\pi) + i\sin(-\pi)$
$= -1 + i(0) = -1$

30. **(C)** $x = \sqrt{390}$

$\log x = \log\sqrt{390}$
$= \tfrac{1}{2}\log 390$
$= \tfrac{1}{2}(2.5911)$
$= 1.2955$
$x = \text{antilog}(1.2955)$
$x = 1.97 \times 10^1 = 19.7$

31. **(D)** $\mathrm{Sinh}\,x = \dfrac{e^x - e^{-x}}{2}$

32. **(A)** $\dfrac{x_1}{j_1\sqrt{l_1}} = \dfrac{x_2}{j_2\sqrt{l_2}}$

$\dfrac{3}{4\sqrt{4}} = \dfrac{x}{9\sqrt{9}}$

$\dfrac{3}{8} = \dfrac{x}{27}$

$x = \dfrac{81}{8}$

33. **(D)** Vertical asymptotes of $f(x) = \dfrac{g(x)}{h(x)}$ occur

when $h(x) = 0$
$(x-7)(x+5) = 0$
$x = 7 \,|\, x = -5$

34. **(D)**

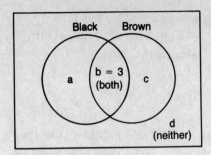

$b = 3$
$a + b = 6 \quad \therefore a = 3$
$b + c = 7 \quad \therefore c = 4$
$\text{Total} = a + b + c + d = 18$
$3 + 3 + 4 + d = 18$
$d = 8$

35. **(E)** $f(x) = x^2 - 2x + 1$

$f(x + 1) = (x + 1)^2 - 2(x + 1) + 1$
$= x^2 + 2x + 1 - 2x - 2 + 1$
$= x^2$

36. **(D)** $\log\sqrt{xy} = \log(xy)^{1/2}$

$= \tfrac{1}{2}\log xy = \tfrac{1}{2}(\log x + \log y)$

37. **(B)** $27^{2x} = 3^{3x-5}$

$(3^3)^{2x} = 3^{3x-5}$
$3^{6x} = 3^{3x-5}$
$6x = 3x - 5$
$x = \dfrac{-5}{3}$

38. **(C)**

$\sin 210° = -\sin 30°$

$\sin 30° = \dfrac{1}{2}$

$\therefore \sin 210° = -\sin 30° = -(\tfrac{1}{2}) = -\tfrac{1}{2}$

39. **(E)** $3 + 2\sqrt{-16} = 3 + 2(4i) = 3 + 8i$
$\qquad 7 - 3\sqrt{-81} = 7 - 3(9i) = 7 - 27i$
$\qquad (3 + 8i) + (7 - 27i) = 10 - 19i$

40. **(A)**

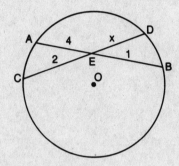

$\qquad (AE)(EB) = (CE)(ED)$
$\qquad\quad (4)(1) = (2)x$
$\qquad\qquad\quad x = 2$

41. **(D)**
1. The GCF $= 2x^3$
2. $\dfrac{16x^5 + 6x^4 + 10x^3}{2x^3} = 8x^2 + 3x + 5$
3. $16x^5 + 6x^4 + 10x^3 = 2x^3[8x^2 + 3x + 5]$

42. **(E)** $10x = 2.2222\overline{2}$
$\qquad\quad \dfrac{x = \phantom{0}.2222\overline{2}}{9x = 2}$
$\qquad\qquad x = \frac{2}{9}$

43. **(B)** $d = \sqrt{(x_2 - x_1)^2 + (y_2 - y_1)^2}$
$\qquad = \sqrt{(-3 - 1)^2 + (3 - (-1))^2}$
$\qquad = \sqrt{(-4)^2 + (4)^2}$
$\qquad = \sqrt{16 + 16} = \sqrt{32} = \sqrt{16 \cdot 2}$
$\qquad = 4\sqrt{2}$

44. **(B)**

$\qquad\qquad\quad 2x - 3y = 4$
$\qquad\qquad\quad (x - 2y = 3)(-2)$
$\qquad\qquad\quad \overline{\phantom{-}2x - 3y = 4}$
$\qquad\qquad\quad \underline{-2x + 4y = -6}$
$\qquad\qquad\qquad\qquad y = -2$

Substitute $y = -2$ into one of the original equations

$\qquad\qquad x - 2y = 3$
$\qquad\quad x - 2(-2) = 3$
$\qquad\qquad\qquad\quad x = -1$
$\qquad\qquad \{-1, -2\}$

45. **(D)** $\text{mean} = \text{average} = \dfrac{\Sigma f_i x_i}{\Sigma f_i}$

$\qquad\qquad = \dfrac{1(7) + 2(11) + 3(12)}{1 + 2 + 3}$

$\qquad\qquad = \dfrac{7 + 22 + 36}{6} = \dfrac{65}{6} = 10\frac{5}{6}$

46. **(D)**

$A = \frac{1}{2} sl \sin J$
$\quad = \frac{1}{2}(10)(16)\sin 30°$
$\quad = 80 \sin 30° = 80(\frac{1}{2}) = 40$

47. **(B)** Let $u = $ unit's digit
$\qquad\qquad t = $ ten's digit
$\qquad\quad (u)(t) = 8$
$\qquad\qquad 10t + u = 4(t + u) \Rightarrow 10t + u = 4t + 4u$
$\qquad\qquad\qquad\qquad 6t = 3u$
$\qquad\qquad\qquad\qquad 2t = u$

Substitute for $u$

$\qquad\quad u \cdot t = 8$
$\qquad\quad (2t)t = 8$
$\qquad\qquad t^2 = 4$
$\qquad\qquad t = \pm 2 \quad$ reject $t = -2$
$\qquad\qquad t = 2$
$\qquad\qquad u = 4$
The number is $10t + u = 24$

48. **(A)** $p = p(\text{success}) = \frac{1}{2}$, $\quad n = 5$ attempts
    $q = p(\text{failure}) = \frac{1}{2}$, $\quad r = 2$ successes
    $_5C_2(\frac{1}{2})^2(\frac{1}{2})^3$

49. **(B)** $x = \dfrac{k}{y^2}$ or $xy^2 = k$ is an example of inverse square variation.

50. **(E)** $S = \dfrac{n}{2}(2a + (n-1)d)$

$\qquad = \dfrac{10}{2}(2(8) + (10-1)(-3))$

$\qquad = 5(16 - 27) = 5(-11) = -55$

# How to Use the Diagnostic Chart (page 20)

The numbers at the left correspond to the question numbers in the test. The words and numbers at the top of the chart correspond to the chapters and chapter names. The darkened square in the row for each question indicates the subject area from which each question is drawn. Mark the chart with a $\sqrt{}$ for each question answered correctly and an $\times$ for each question answered incorrectly. You'll be able to see at a glance which topics need further study.

# Diagnostic Chart Pretest

*UNIT ONE*

UNIT ONE

# 1. Sets of Numbers

## 1.1 Fundamental Facts

The bracketed letters represent alpha notation frequently used to represent the respective set.

Irrational numbers are real numbers which cannot be expressed in $\frac{a}{b}$ form, where $a$ and $b$ are integers. These include non-terminating, non-repeating decimals and square roots of non-perfect squares.

 A bar over the last digit or last group of digits in a decimal number indicates infinite repetition of the sequence under the bar. For example $.13\overline{636} = .13636363636 \dots$

**Examples:**

1. Which sets of numbers comprise the set of complex numbers?

2. $7 + \sqrt{3}$ is a member of which lower order set?

**Solutions:**

1. Complex numbers are the union of the set of imaginary numbers and $R$, the set of real numbers.

2. $7 + \sqrt{3}$ is an irrational number. $7 + \sqrt{3}$ is an element of the set of irrational numbers.

## Practice Problems 1.1

1. Which statement is true?

   A) $R$ is a subset of $Q$
   B) $H$ is a subset of $W$
   C) $N$ is a subset of the set of fractions
   D) $J$ is a subset of $C$
   E) $R$ is a subset of $J$

2. $.166\overline{6}$ is a member of which set of numbers?

   A) $J$   B) $W$   C) $N$   D) $Q$
   E) Imaginary

3. The union of the set of irrational numbers and the set of rational numbers is the set of

   A) imaginary numbers   B) integers
   C) even numbers   D) real numbers
   E) improper fractions

4. $\frac{5}{7}$ belongs to all of the following sets except

   A) fractions   B) real   C) rational
   D) odd   E) complex

Solutions on page 44

# 2. Functions and Relations

## 2.1 Essentials

**Relation**

A **relation** is a set of ordered pairs $(x, y)$. The **domain** of the relation is the set of all first numbers of the relation; the **range** is the set of all resulting second numbers in the relation.

**Function**

A relation in which each value of the first number has a unique value for the second number.
(A given $x$ value cannot have more than one $y$ value.)

**Domain**

All acceptable values of the first number in a relation or function.

 $x$ values which create division by 0 are not acceptable values.
$x$ values which create imaginary values are not acceptable values.

**Range**

All values of the second number in a relation or function which can result from admissible values of the first number.

**$y = f(x)$**

Read as "$y$ is a function of $x$" or "$y$ equals $f$ of $x$".
Values of $x$ (the independent variable) represent numbers in the domain of the function. The resulting values of $y$ or $f(x)$ (the dependent variable) represent numbers in the range of the function.

 $f(x)$ is not read as $f$ times $x$.

**$y = f(a)$**

When a function $f(x)$ is defined by an equation, the value $f(a)$ is found by substituting the value '$a$' for each appearance of the variable '$x$'.

**Vertical Line Test**

The graphical technique to determine if a relation is a function. If a vertical line drawn anywhere on the coordinates can only cross the graph of the relation at one and only one intersection, then the relation is a function.

**Examples:**

1. If $f(x) = x^2 + 3x$, find $f(x + 2)$.

**Solutions:**

1.
$$f(x) = x^2 + 3x$$
$$f(x + 2) = (x + 2)^2 + 3(x + 2)$$
$$= x^2 + 4x + 4 + 3x + 6$$
$$= x^2 + 7x + 10$$

2. Does the graph of the relation shown below represent a function?

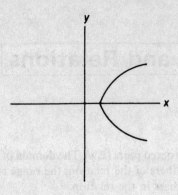

2. Solution: Use the Vertical Line Test. Since a vertical line *can* cross the graph at more than one point of intersection the relation is NOT a function.

3. If $f(x) = 3x - 2$, find $\dfrac{f(x + h) - f(x)}{h}$, $h \neq 0$.

3.             $f(x) = 3x - 2$

$$f(x + h) = 3(x + h) - 2 = 3x + 3h - 2$$

$$\frac{f(x + h) - f(x)}{h} = \frac{(3x + 3h - 2) - (3x - 2)}{h} = \frac{3h}{h} = 3$$

4. If $f(x) = x^2 + 4x - 2$, find the domain and range of the function.

4. Sometimes a simple sketch is very useful as an aid for determining domain and range.

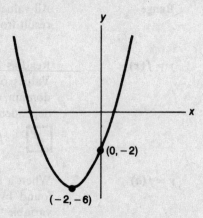

The domain: all real $x$ (all values of $x$ are good)
The range: $y \geq -6$ (there are no $y$ values below $y = -6$)

# Practice Problems 2.1

1. If $f(x) = 2x^2 - 2x + 1$, find $f(-3)$.
   A) $-23$   B) $-11$   C) $-3$   D) 13   E) 25

2. Which of the graphs shown on the next page does not represent a function?

A)

B)

C)

D)

E)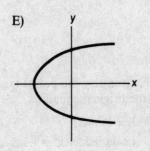

3. If $f(x) = 2x^2 - 2$, find $\dfrac{f(x+h) - f(x)}{h}$, $h \neq 0$.

 A) $4x + 2h$  B) $4x + 2h - 4$
 C) $4x^2 + 4x + 2h$  D) $4xh - 2h^2$  E) $2x + h$

4. Find the range of $f(x) = |x|$.

 A) All real numbers  B) $y \geqslant 0$  C) $y > 0$
 D) $y \geqslant x$  E) $y < 0$

5. If $f(x) = -x^2 - 3x - 2$, find $f(-3)$.

 A) $-2$  B) $-20$  C) 16
 D) $-17$  E) $-14$

6. Does the relation $x^2 + y^2 = 9$ represent a function?

 A) Never  B) Always
 C) Only when $x \geqslant 3$  D) Only when $y \geqslant 3$
 E) When $x$ and $y$ are $\geqslant 3$

Solutions on page 44

## 2.2 Symmetry

| Type of Symmetry | Conditions | | Example |
|---|---|---|---|
| y-axis | If $(x,y)$ is on the graph of $f(x)$, then $(-x,y)$ is also on the graph. <br> or <br> If $f(-x) = f(x)$ | also called an EVEN relation or function | |
| x-axis | If $(x,y)$ is on the graph of $f$, then $(x, -y)$ is also on the graph. <br><br> Note: these are relations not functions | | |

| Type of Symmetry | Conditions | | Example |
|---|---|---|---|
| Origin (0,0) | If $(x, y)$ is on the graph of $f(x)$, then $(-x, -y)$ is also on the graph. <br> or <br> If $f(-x) = -f(x)$ | also called an ODD relation or function | |
| The line x = y | If $(x, y)$ is on the graph of $f(x)$, then $(y, x)$ is also on the graph. | | |

**Examples:**

1. Is $x^2 + y^2 = 9$ symmetric to the $y$-axis?

2. Does the equation $xy = 8$ have origin symmetry?

3. Does the equation $x = y^2$ have $x$-axis symmetry?

4. Does $xy = 8$ have $y$-axis symmetry?

**Solutions:**

1. For $y$=axis symmetry the points $(x, y)$ and $(-x, y)$ both yield the original equation.

    *Test Points*
    $$x^2 + y^2 = 9 \quad \begin{array}{l} (x, y) \Rightarrow x^2 + y^2 = 9 \\ (-x, y) \Rightarrow (-x)^2 + y^2 = 9 \Rightarrow x^2 + y^2 = 9 \end{array}$$

    There is $y$-axis symmetry.

2. For origin symmetry the points $(x, y)$ and $(-x, -y)$ both yield the original equation.

    *Test Points*
    $$xy = 8 \quad \begin{array}{l} (x, y) \Rightarrow xy = 8 \\ (-x, -y) \Rightarrow (-x)(-y) = 8 \Rightarrow xy = 8 \end{array}$$

    There is origin symmetry.

3. For $x$=axis symmetry the points $(x, y)$ and $(x, -y)$ both yield the original equation.

    *Test Points*
    $$x = y^2 \quad \begin{array}{l} (x, y) \Rightarrow x = y^2 \\ (x, -y) \Rightarrow x = (-y)^2 = y^2 \end{array}$$

    There is $x$-axis symmetry.

4. For $y$=axis symmetry the points $(x, y)$ and $(-x, y)$ both yield the same equation.

    *Test Points*
    $$xy = 8 \quad \begin{array}{l} (x, y) \Rightarrow xy = 8 \\ (-x, y) \Rightarrow (-x)(y) = 8 \Rightarrow xy = -8 \end{array}$$

    There is *no* $y$-axis symmetry.

# Practice Problems 2.2

1. The graph of $y = |x| + 3$ has what type of symmetry?

   A) x-axis only   B) y-axis only
   C) Origin only   D) $x = y$ only
   E) x- and y-axes only

2. The graph of the equation $y = 7$ has what type of symmetry?

   A) x-axis only   B) y-axis only
   C) Origin only   D) $x = y$ only
   E) x- and y-axes only

3. The graph of the relation $x^2 - y^2 = 4$ has what type of symmetry?

   A) x-axis, origin and $x = y$ only
   B) y-axis, origin and $x = y$ only
   C) Origin only
   D) Origin, x-axis and y-axis only
   E) Origin and x-axis only

4. Classify the equation $y = \sin x$.

   A) Even function   B) Odd function
   C) Even relation but not a function
   D) Odd relation but not a function
   E) Neither even nor odd

5. The graph of the equation $xy = 8$ has what type of symmetry?

   A) Origin, $x = y$ and x-axis only
   B) Origin, $x = y$ and y-axis only
   C) Origin and $x = y$ only
   D) $x = y$ and x-axis only
   E) $x = y$ and y-axis only

6. The graph of $y = 2x - 1$ has what type of symmetry?

   A) None   B) $x = y$ only
   C) x-axis only   D) y-axis only
   E) Origin only

Solutions on page 45

## 2.3 The Algebra of Functions

| Given two functions $f(x)$ and $g(x)$ | |
|---|---|
| **Adding** | $(f + g)(x) = f(x) + g(x)$ |
| **Subtracting** | $(f - g)(x) = f(x) - g(x)$ |
| **Multiplying** | $(f \cdot g)(x) = f(x) \cdot g(x)$ |
| **Dividing** | $\left(\dfrac{f}{g}\right)(x) = \dfrac{f(x)}{g(x)}$ |
| **Composition** | $(f \circ g)(x) = f(g(x))$ |
| **Inverse** | $f^{-1}(x) =$ the inverse of $f(x)$ |

The **inverse** of a function denoted by $f^{-1}(x)$ represents the relation such that

$$f(f^{-1}(x)) = f^{-1}(f(x)) = x$$

Graphically, $f^{-1}$ of $x$ represents the reflection of $f(x)$ about the line $y = x$.

 $f^{-1}(x)$ does *not* mean $\dfrac{1}{f}$, nor $f^{-1}$ times $x$.

---

**Procedure for calculating the inverse of $f(x)$**

Example: Find the inverse of $f(x) = 2x - 3$

| *Step* | *Description* | *Example* |
|---|---|---|
| 1 | Replace $f(x)$ with $y$ | $y = 2x - 3$ |
| 2 | Swap every $x$ and $y$ | $x = 2y - 3$ |
| 3 | Solve for $y$ | $y = \dfrac{x + 3}{2}$ |
| 4 | Replace $y$ with $f^{-1}(x)$ | $f^{-1}(x) = \dfrac{x + 3}{2}$ |

---

☞ **$f(x)$, $f^{-1}(x)$ Notes:**

If $f(x) = 2x - 3$ and $f^{-1}(x) = \dfrac{x + 3}{2}$

$f(f^{-1}(x)) = 2\left(\dfrac{x + 3}{2}\right) - 3 = x$

$f^{-1}(f(x)) = \dfrac{(2x - 3) + 3}{2} = x$

NOTE:

$f(f^{-1}(x)) = x$
$f^{-1}(f(x)) = x$

NOTE: $f^{-1}(x)$ is the reflection of $f(x)$ about the line $x = y$.

**Examples:**

1. Find the inverse of $f(x) = \dfrac{x - 1}{x + 2}$.

**Solutions:**

1. Let $y = \dfrac{x - 1}{x + 2}$

Swap $x$ and $y$

$x = \dfrac{y - 1}{y + 2}$

Solve for $y$

$xy + 2x = y - 1$
$xy - y = -2x - 1$
$y(x - 1) = -2x - 1$
$y = \dfrac{2x + 1}{1 - x}$

replace $y$ with $f^{-1}(x)$

$f^{-1}(x) = \dfrac{2x + 1}{1 - x}$

2. If $f(x) = x^2 - 1$ and $g(x) = x + 1$, find

   A) $(f + g)(x)$  B) $(f - g)(x)$  C) $(f \cdot g)(x)$

   D) $\left(\dfrac{f}{g}\right)(x)$  E) $(f \circ g)(x)$  F) $(g \circ f)(x)$

   G) $(f \circ f)(x)$  H) $g^{-1}(x)$

2. A) $(f + g)(x) = f(x) + g(x) = (x^2 - 1) + (x + 1)$
$$= x^2 + x$$

   B) $(f - g)(x) = f(x) - g(x) = (x^2 - 1) - (x + 1)$
$$= x^2 - x - 2$$

   C) $(f \cdot g)(x) = f(x) \cdot g(x) = (x^2 - 1)(x + 1)$
$$= x^3 + x^2 - x - 1$$

   D) $\left(\dfrac{f}{g}\right)(x) = \dfrac{f(x)}{g(x)} = \dfrac{x^2 - 1}{x + 1} = \dfrac{(x + 1)(x - 1)}{(x + 1)} = x - 1$

   E) $(f \circ g)(x) = f(g(x)) = f(x + 1) = (x + 1)^2 - 1$
$$= x^2 + 2x$$

   F) $(g \circ f)(x) = g(f(x)) = g(x^2 - 1) = (x^2 - 1) + 1$
$$= x^2$$

   G) $(f \circ f)(x) = f(f(x)) = f(x^2 - 1) = (x^2 - 1)^2 - 1$
$$= x^4 - 2x^2$$

   H) $g^{-1}(x)$      $g(x) = x + 1$

| STEP | |
|---|---|
| 1 | $y = x + 1$ |
| 2 | $x = y + 1$ |
| 3 | $y = x - 1$ |
| 4 | $g^{-1}(x) = x - 1$ |

# Practice Problems 2.3

1. If $f(x) = 3x - 3$ and $g(x) = 2x + 1$, find $(f + g)(x)$.

   A) $x - 2$  B) $5x - 2$  C) $6x - 5$
   D) $6x^2 - 3x - 3$  E) $6x$

4. If $f(x) = 3x - 3$ and $g(x) = 2x + 1$, find $(g \circ f)(x)$.

   A) $x - 2$  B) $5x - 2$  C) $6x - 5$
   D) $6x^2 - 3x - 3$  E) $6x$

2. If $f(x) = 3x - 3$ and $g(x) = 2x + 1$, find $(f \cdot g)(x)$.

   A) $x - 2$  B) $5x - 2$  C) $6x - 5$
   D) $6x^2 - 3x - 3$  E) $6x$

5. If $f(x) = \dfrac{x - 1}{x + 1}$, find $f^{-1}(x)$.

   A) $\dfrac{x + 1}{x - 1}$  B) $1$  C) $\dfrac{x - 2}{x + 2}$

   D) $\dfrac{-(x + 1)}{x - 1}$  E) $x$

3. If $f(x) = 3x - 3$ and $g(x) = 2x + 1$, find $(f \circ g)(x)$.

   A) $x - 2$  B) $5x - 2$  C) $6x - 5$
   D) $6x^2 - 3x - 3$  E) $6x$

6. The graph which best represents the inverse of $f(x)$ is

C)

D)

A)

B)

E)

Solutions on page 46

## 2.4 Summary of Basic Function Types

| Type | Examples of Algebraic Representation | Examples of Graphic Representation |
|---|---|---|
| **Identity function** | $f(x) = x$ | |
| **Constant function** | $f(x) = a$ | |
| **Absolute Value function** | $f(x) = |x|$ | |
| **Linear function** | $f(x) = mx + b$<br>$m = $ slope<br>$b = y$ intercept | |
| **Quadratic function** | $f(x) = ax^2 + bx + c$ | |
| **Polynomial function** | $f(x) = a_n x^n + a_{n-1} x^{n-1} \ldots$ | |
| **Greatest Integer function (Step function)** | $f(x) = [x]$ $\begin{Bmatrix} [x] \text{ is the} \\ \text{greatest} \\ \text{integer less} \\ \text{than or equal} \\ \text{to x} \end{Bmatrix}$ | |
| **Rational function** | $f(x) = \dfrac{g(x)}{h(x)}$ | |
| **Inverse function** | $f^{-1}(x)$ | |
| **Trigonometric, Periodic, Circular or Wrapping functions** | $f(x) = \sin x$ | |
| **Exponential functions** | $f(x) = b^x$ | |
| **Logarithmic functions** | $f(x) = \log_b x$ | |
| **Hyperbolic functions** | $f(x) = \sinh x = \dfrac{e^x - e^{-x}}{2}$ | |

(Rows from Trigonometric through Hyperbolic are grouped under the vertical label **TRANSCENDENTOL FUNCTIONS**.)

NOTE: Circles and ellipses are relations not functions.

# 3. Absolute Value

## 3.1 Fundamental Facts

The symbol $|x|$ means the absolute value of $x$.

| | | | |
|---|---|---|---|
| $|x| = x \quad x \geq 0$ | $|x| = a$ | $\begin{array}{l} x = a \\ x = -a \end{array}$ | |
| | $|x| > a$ | $x > a$ or $x < -a$ | |
| | $|x| \geq a$ | $\begin{array}{l} x \geq a \text{ or } x \leq -a \\ (a \geq 0) \end{array}$ | |
| $|x| = x \quad x < 0$ | $|x| < a$ | $\begin{array}{l} x < a \text{ and } x > -a \\ (-a < x < a) \end{array}$ | |
| | $\begin{array}{l} |x| \leq a \\ (a > 0) \end{array}$ | $\begin{array}{l} x \leq a \text{ and } x \geq -a \\ (-a \leq x \leq a) \end{array}$ | |

**Examples:**

1. Evaluate $|5 - 3| - |3 - 5|$.

2. Evaluate $|0| + |-2| - |2| - |-2|$.

3. Graphically represent $x > 3$ or $x \leq -2$.

4. Graph the solution to $|x| = 5$.

**Solutions:**

1. $|5 - 3| = |2| = 2 \qquad |3 - 5| = |-2| = 2$
   $|5 - 3| - |3 - 5| = 2 - 2 = 0$

2. $|0| = 0, |-2| = 2, |2| = 2$
   $|0| + |-2| - |2| - |-2|$
   $= 0 + 2 - 2 - 2 = -2$

3. 
   $-2 \qquad 3$

4. $|x| = 5 \qquad x = 5$
   $\qquad\qquad x = -5$

34

# Absolute Value Equations (Graphs)

The absolute value function is generally in the shape of a "$\vee$" with the vertex at the value of $x$ which makes the sentence inside the absolute value sign equal to zero. This is the critical value of $x$ for the absolute value graph. To the left of the critical value the $|f(x)| = g(x)$ represents the equation $f(x) = -g(x)$, to the right of the critical value $|f(x)| = g(x)$ represents the equation $f(x) = g(x)$.

$$|f(x)| = g(x)$$

The vertex occurs at the value(s) such that $f(x) = 0$.

| This region represents the equation $f(x) = -g(x)$ | This region represents the equation $f(x) = g(x)$ |
|---|---|

Critical
value
of $x$

A chart can be developed to graph an absolute value function. Be sure to choose values above, below and at the critical value.

**Examples:**

5. Graph $|x - 1| = y$.

6. Graph $y = |x - 2| + 3$.

**Solutions:**

5. The critical value is $x = 1$.
   Using a chart

   | $x$ | $y$ |
   |-----|-----|
   | $-2$ | 3 |
   | $-1$ | 2 |
   | 0 | 1 |
   | 1 | 0 |
   | 2 | 1 |
   | 3 | 2 |
   | 4 | 3 |

6. The critical value is $x = 2$.
   Using a chart

   | $x$ | $y$ |
   |-----|-----|
   | $-1$ | 6 |
   | 0 | 5 |
   | 1 | 4 |
   | 2 | 3 |
   | 3 | 4 |
   | 4 | 5 |
   | 5 | 6 |

# Practice Problems 3.1

1. Evaluate $\left|\dfrac{-6}{-2}\right| + \dfrac{|-6|}{|-2|} + |6(-2)| + |6| \cdot |-2|$.

   A) 0   B) 6   C) 18   D) 30   E) 24

4. Graph $y = |x + 3| - 1$.

   A)

   B)

2. Graphically represent $|x| \leqslant 3$.

   A)

   B)

   C)

   D)

   E)

   C)

   D)

3. Given the equation $y = |2x - 6| + 5$, the equation which represents the left leg of the "**V**" is

   A) $y = -(2x - 6) + 5$   B) $y = (2x - 6) - 5$
   C) $y = (2x - 6) + 5$   D) $y = -(2x - 6) - 5$
   E) $y = (2x - 6)$

   E)

Solutions on page 46

# 4. Roots and Radicals

## 4.1 Basics and Beyond

To find the square root of a number means to find one of its two equal factors.

$\sqrt{\phantom{x}}$ is the symbol for square root and is called a radical sign.

$\sqrt[3]{\phantom{x}}$ is the notation for cube root and means to find one of its three equal factors.

 The $\sqrt{\phantom{x}}$ symbol indicates the principal square root. That is, the positive root only: $\sqrt{4} = 2$ not $-2$.

| Addition and Subtraction: | To add or subtract square roots, the value under the radical must be the same. Add or subtract the coefficients and retain the radical. $$m\sqrt{a} + n\sqrt{a} = (m + n)\sqrt{a}$$ $$m\sqrt{a} - n\sqrt{a} = (m - n)\sqrt{a}$$ |
|---|---|
| Multiply and Divide: | $$\sqrt{a} \cdot \sqrt{b} = \sqrt{ab}\ (a, b \geq 0), \quad m\sqrt{a} \cdot n\sqrt{b} = mn\sqrt{ab}$$ $$\frac{\sqrt{a}}{\sqrt{b}} = \sqrt{\frac{a}{b}}\left(\begin{matrix}a \geq 0\\ b > 0\end{matrix}\right), \quad \frac{m\sqrt{a}}{n\sqrt{b}} = \frac{m}{n}\sqrt{\frac{a}{b}}$$ |

 A square root should be expressed in simplest form: $2\sqrt{2}$, not $\sqrt{8}$.

 Common Error: $\sqrt{2} + \sqrt{3} = \sqrt{5}$. Should be $\sqrt{2} + \sqrt{3} = \sqrt{2} + \sqrt{3}$. These radicals cannot be added. To add radicals the values under the radical sign must be the same.

**Examples:**

1. Find $\sqrt{16}$.

2. Combine $3\sqrt{2} + 7\sqrt{2} - 5\sqrt{2}$.

**Solutions:**

1. $\sqrt{16} = 4$. Note: the $\sqrt{\phantom{x}}$ indicates the positive root only.

2. $3\sqrt{2} + 7\sqrt{2} - 5\sqrt{2} = (3 + 7 - 5)\sqrt{2} = 5\sqrt{2}$

 **To simplify a square root,** factor the value under the radical sign into two numbers, one of which is a perfect square. Then separate the original square root into the product of the square roots of the factors, and take the square root of the perfect square.

**Examples:**

3. Simplify $\sqrt{18}$.

4. Combine $\sqrt{125} + \sqrt{20} - \sqrt{500}$.

5. Multiply $(4\sqrt{3})(2\sqrt{6})$.

6. Find $(2\sqrt{5})^2$.

☞ For non-negative value of $x$: $\sqrt{x} \cdot \sqrt{x} = x$. For example, $\sqrt{1011} \cdot \sqrt{1011} = 1011$.

7. Divide $\dfrac{10\sqrt{20}}{2\sqrt{4}}$.

8. $(3 + \sqrt{2})(2 + \sqrt{10}) =$

9. $(2 + \sqrt{7})^2 =$

10. Simplify $\sqrt{a^4 b^5 c^3}$.

**Solutions:**

3. $\sqrt{18} = \sqrt{9 \cdot 2} = \sqrt{9}\sqrt{2}$
   $= 3\sqrt{2}$

4. First simplify each term:
   $\sqrt{25 \cdot 5} + \sqrt{4 \cdot 5} - \sqrt{100 \cdot 5}$
   $= \sqrt{25}\sqrt{5} + \sqrt{4}\sqrt{5} - \sqrt{100}\sqrt{5}$
   $= 5\sqrt{5} + 2\sqrt{5} - 10\sqrt{5}$
   $= -3\sqrt{5}$

5. $(4\sqrt{3})(2\sqrt{6}) = 4 \cdot 2\sqrt{3 \cdot 6} = 8\sqrt{18}$
   $= 8\sqrt{9 \cdot 2} = 8\sqrt{9}\sqrt{2}$
   $= 8(3)\sqrt{2}$
   $= 24\sqrt{2}$

6. $(2\sqrt{5})(2\sqrt{5}) = 2 \cdot 2\sqrt{5 \cdot 5}$
   $= 4\sqrt{25}$
   $= 4(5)$
   $= 20$

7. $\dfrac{10}{2}\sqrt{\dfrac{20}{4}} = 5\sqrt{5}$

8. $(3 + \sqrt{2})(2 + \sqrt{10}) =$
   $3(2) + 3\sqrt{10} + 2\sqrt{2} + \sqrt{2 \cdot 10}$
   $6 + 3\sqrt{10} + 2\sqrt{2} + \sqrt{20}$
   $6 + 3\sqrt{10} + 2\sqrt{2} + \sqrt{4 \cdot 5}$
   $6 + 3\sqrt{10} + 2\sqrt{2} + \sqrt{4}\sqrt{5}$
   $6 + 3\sqrt{10} + 2\sqrt{2} + 2\sqrt{5}$
   $6 + 2\sqrt{2} + 2\sqrt{5} + 3\sqrt{10}$

9. $(2 + \sqrt{7})(2 + \sqrt{7})$
   $= 2(2) + 2\sqrt{7} + 2\sqrt{7} + \sqrt{7}\sqrt{7}$
   $= 4 \quad + 4\sqrt{7} \qquad + 7$
   $= 11 + 4\sqrt{7}$

10. Note: $x^{\text{Positive and Even}}$ is a perfect square.
    $\sqrt{a^4 b^4 b^1 c^2 c^1}$
    $= \sqrt{a^4}\sqrt{b^4}\sqrt{b}\sqrt{c^2}\sqrt{c}$
    $= (a^2)(b^2)(\sqrt{b})(c)(\sqrt{c}) = a^2 b^2 c\sqrt{bc}$

# Practice Problems 4.1

1. Multiply and simplify $2\sqrt{18}\cdot 6\sqrt{2}$.

   A) 72   B) 48   C) $12\sqrt{6}$

   D) $8\sqrt{2}$   E) 36

2. Find $(3\sqrt{3})^3$.

   A) $27\sqrt{3}$   B) $81\sqrt{3}$   C) 81   D) $9\sqrt{3}$   E) 243

3. Combine $\sqrt{80}+\sqrt{45}-\sqrt{20}$.

   A) $9\sqrt{5}$   B) $5\sqrt{5}$   C) $-\sqrt{5}$   D) $3\sqrt{5}$
   E) $-2\sqrt{5}$

4. $(3+\sqrt{5})(3-\sqrt{5})=$

   A) 4   B) $4-6\sqrt{5}$   C) $-16$
   D) $-1$   E) 6

5. Simplify $\sqrt{20a^7b^3c^8}$.

   A) $10a^3bc^4\sqrt{ab}$   B) $2abc\sqrt{abc}$
   C) $5a^3bc^4\sqrt{2ab}$   D) $2abc^4\sqrt{5a^3b}$
   E) $2a^3bc^4\sqrt{5ab}$

6. $\dfrac{\sqrt{32b^3}}{\sqrt{8b}}=$

   A) $2\sqrt{b}$   B) $\sqrt{2b}$   C) $2b$
   D) $\sqrt{2b^2}$   E) $b\sqrt{2b}$

7. $(\frac{1}{2}\sqrt{2})(\sqrt{6}+\frac{1}{2}\sqrt{2})=$

   A) $\sqrt{3}+\frac{1}{2}$   B) $\frac{1}{2}\sqrt{3}$   C) $\sqrt{6}+1$
   D) $\sqrt{6}+\frac{1}{2}$   E) $\sqrt{6}+2$

8. Combine $\frac{1}{2}\sqrt{180}+\frac{1}{3}\sqrt{45}-\frac{2}{3}\sqrt{20}$.

   A) $3\sqrt{10}+\sqrt{15}+2\sqrt{2}$   B) $\frac{16}{5}\sqrt{5}$   C) $\sqrt{97}$
   D) $\frac{24}{3}\sqrt{5}$   E) 7

Solutions on page 47

## 4.2 Rationalizing the Denominator

To **rationalize** a fraction with a radical in the denominator is to write an equivalent fraction with the denominator being rational. This is done by multiplying by the identity element expressed in a useful form.

| Form of Denominator to be Rationalized | Form of Identity Element to be Used for Rationalizing |
|---|---|
| $\sqrt{x}$ | $\sqrt{x}$ |
| $a\sqrt{x}$ | $\sqrt{x}$ |
| $a\pm\sqrt{x}$ | $a\mp\sqrt{x}$ |
| $a\pm b\sqrt{x}$ | $a\mp b\sqrt{x}$ |
| $\sqrt{x}\pm\sqrt{y}$ | $\sqrt{x}\mp\sqrt{y}$ |

The expressions $a+b\sqrt{x}$ and $a-b\sqrt{x}$ or $a\sqrt{x}+b\sqrt{y}$ and $a\sqrt{x}-b\sqrt{y}$ are called conjugate pairs.

**Examples:**

1. Rationalize $\dfrac{1}{\sqrt{7}}$.

2. Rationalize $\dfrac{3}{4-\sqrt{5}}$.

3. Find the reciprocal of $2 + 2\sqrt{3}$.

4. Divide $\sqrt{5}$ by $\sqrt{7}$.

**Solutions:**

1. The denominator is of the form $\sqrt{x}$. Use $\sqrt{x}$ to rationalize.

$$\frac{1}{\sqrt{7}}\left(\frac{\sqrt{7}}{\sqrt{7}}\right)=\frac{\sqrt{7}}{7}$$

2. The denominator is of the form $a - \sqrt{x}$. Use $a + \sqrt{x}$ to rationalize.

$$\frac{3}{4-\sqrt{5}}\left(\frac{4+\sqrt{5}}{4+\sqrt{5}}\right)=\frac{3(4)+3\sqrt{5}}{4(4)+4\sqrt{5}-4\sqrt{5}-\sqrt{5}\sqrt{5}}$$

$$=\frac{12+3\sqrt{5}}{16-5}=\frac{12+3\sqrt{5}}{11}$$

3. The reciprocal is 1 over the number.

$$\text{Reciprocal}=\frac{1}{2+2\sqrt{3}}$$

Rationalizing:

$$\frac{1}{2+2\sqrt{3}}\left(\frac{2-2\sqrt{3}}{2-2\sqrt{3}}\right)=\frac{2-2\sqrt{3}}{4-4(3)}$$

$$=\frac{2-2\sqrt{3}}{4-12}=\frac{2(1-\sqrt{3})}{-8}=-\frac{(1-\sqrt{3})}{4}$$

4. $\dfrac{\sqrt{5}}{\sqrt{7}}\cdot\dfrac{(\sqrt{7})}{(\sqrt{7})}=\dfrac{\sqrt{35}}{7}$

# Practice Problems 4.2

1. Rationalize $\dfrac{2}{\sqrt{11}}$.

   A) $\dfrac{11\sqrt{11}}{2}$  B) $\dfrac{2\sqrt{11}}{11}$  C) $\dfrac{2+\sqrt{11}}{2}$

   D) $\dfrac{2-\sqrt{11}}{2}$  E) $\dfrac{4}{11}$

2. Rationalize $\dfrac{\sqrt{3}}{1+\sqrt{2}}$.

   A) $\dfrac{\sqrt{3}-\sqrt{6}}{-1}$  B) $\sqrt{6}-\sqrt{3}$  C) $-1$

   D) $1-\sqrt{2}$  E) $\sqrt{2}-1$

3. Find the reciprocal of $\dfrac{2\sqrt{3}}{3}$.

   A) $\dfrac{\sqrt{3}}{6}$  B) $9$  C) $\dfrac{3}{2}$

   D) $\dfrac{\sqrt{3}}{2}$  E) $\dfrac{\sqrt{6}}{2}$

4. The expression $\dfrac{1}{1+\sqrt{3}}$ is equivalent to

   A) $1$  B) $\dfrac{-\sqrt{3}}{2}$  C) $\dfrac{1-\sqrt{3}}{-2}$

   D) $1+\sqrt{3}$  E) $\dfrac{1+\sqrt{3}}{2}$

Solutions on page 48

# Cumulative Review

1. Find the numerical value of

   $|-3| + |7| - |-7|$.

   A) $-3$  B) 3  C) 11  D) $-11$  E) 17

2. Which is an irrational number?

   A) $\sqrt{9}$  B) $\sqrt{2} \cdot \sqrt{8}$  C) $\frac{2}{3}$  D) 0
   C) $\sqrt{2}$

3. The sum of $3\sqrt{3}$ and $\sqrt{12}$ is

   A) $5\sqrt{3}$  B) $7\sqrt{3}$  C) $3\sqrt{15}$
   D) $\sqrt{21}$  E) 18

4. If $h(x) = x^2 + 1$ and $k(x) = 3x$, then $(h - k)(x) =$

   A) $-2x + 1$  B) $x^2 - 2$  C) $x^2 - 3x + 1$
   D) $-x^2 + 3x - 1$  E) $9x^2 + 1$

5. The reciprocal of $\dfrac{\sqrt{3}}{2}$ is

   A) $\dfrac{3}{4}$  B) $\dfrac{3}{2}$  C) $\dfrac{2\sqrt{2}}{3}$

   D) $\dfrac{2\sqrt{3}}{3}$  E) $\dfrac{3\sqrt{2}}{2}$

6. If $f(x) = 3x + 2$, then $f^{-1}(x) =$

   A) $\dfrac{3}{x} + 2$  B) $2x + 3$  C) $x - 2$

   D) $\dfrac{x-2}{3}$  E) $\dfrac{x}{3} - 2$

7. Which of the following graphs represents a function?

   A)   B)   C)

   D)   E)

8. What type of symmetry does the equation $y = 2x^7 + 3x^5 - 2x$ have?

   A) $x$-axis  B) $y$-axis  C) Origin
   D) $x = y$  E) None

9. Given that $f(x) = x^2 - 1$ and $g(x) = x + 1$ find $(f \cdot g)(x)$.

   A) $x^2 + 2x$  B) $x^2$  C) $x^3 + x^2 - x - 1$
   D) $x^3 - 1$  E) $x^3 - x^2 + x - 1$

10. Which graph represents the equation $y = |x - 1| + 3$?

    A)   B)  C)

    D)   (−1,3)  E)   (1,3)

11. Rationalize $\dfrac{\sqrt{3}}{\sqrt{15} - 3}$.

    A) $\dfrac{\sqrt{15} + 3}{6}$  B) $\dfrac{\sqrt{15} + 3}{2}$  C) $\dfrac{\sqrt{5} + \sqrt{3}}{2}$

    D) $\dfrac{\sqrt{45} + \sqrt{9}}{6}$  E) $\dfrac{4\sqrt{3}}{3}$

41

12. If $f(x) = x^2 - 3$, find

$$\frac{f(x + h) - f(x)}{h}, \ h \neq 0.$$

   A) $2x + h$   B) $2x$   C) $2xh + h^2$

   D) $2x^2$   E) $\dfrac{2x^2 + 2xh + h^2 - 6}{h}$

13. Simplify $\sqrt{49a^3b} - \sqrt{a^3b} + 2a\sqrt{ab}$.

   A) $8a^2\sqrt{ab}$   B) $8a\sqrt{ab}$   C) $8a^3\sqrt{b}$

   D) $8a^3b$   E) $11a\sqrt{ab}$

14. The graphic representation of $|x + 1| < 3$ is:

   A) ⊸——⊸      B) ◄—⊸—⊸—►

   C) ●——●      D) ⊸——⊸

   C) ◄—⊸—⊸—►

15. If $f(x) = 3x^2 - 2$ and $g(x) = x + \sqrt{x}$, find $f(g(4))$.

   A) $46 + \sqrt{46}$   B) $10$   C) $22 + \sqrt{22}$

   D) $106$   E) $46$

16. Simplify $(2 - \sqrt{3})(2 + \sqrt{3})$.

   A) $4 - \sqrt{3}$   B) $4 + \sqrt{3}$   C) $1 + 4\sqrt{3}$

   D) $7$   E) $1$

17. If $f = \{(0, 1), (2, 7), (-1, 9), (-3, -3)\}$, find $f^{-1}$.

   A) $\{(0, -1), (2, -7), (-1, -9), (-3, 3)\}$

   B) $\{(0, 1), (-2, 7), (1, 9), (3, -3)\}$

   C) $\{(1, 0), (7, 2), (9, -1), (-3, -3)\}$

   D) $\{(-1, 0), (-7, 2), (-9, -1), (3, -3)\}$

   E) $\{(0, 1), (2, 9), (-1, 8), (-3, 0)\}$

18. Positive integers are synonymous with

   A) whole numbers   B) even numbers

   C) fractions   D) natural numbers

   E) rational numbers

19. An odd function is synonymous with

   A) $x$-axis symmetry   B) $y$-axis symmetry

   C) origin symmetry   D) $x = y$ symmetry

   E) a constant function

20. The number $\sqrt[3]{-\frac{1}{8}}$ is

   A) real and rational

   B) real and irrational

   C) imaginary   D) complex

   E) an improper fraction

Solutions on page 48

# Diagnostic Chart for Cumulative Review Test

# Solutions to Problems in Unit One

## Practice Problems 1.1

1. **(D)** $J$ is a subset of $C$. The integers are a subset of the complex numbers.

2. **(D)** $.166\overline{6}$ is a repeating non-terminating decimal and can be written in $\dfrac{a}{b}$ form. It is rational. In fact, $.166\overline{6} = \frac{1}{6}$. It is a member of set Q.

3. **(D)** The union of irrational and rational numbers yields the set of real numbers (**R**).

4. **(D)** $\frac{5}{7}$ is a proper fraction. It is not a member of the set of odd numbers (which is a subset of integers).

## Practice Problems 2.1

1. **(E)** 
$$f(x) = 2x^2 - 2x + 1$$
$$= 2(-3)^2 - 2(-3) + 1$$
$$= 2(9) + 6 + 1$$
$$= 18 + 6 + 1 = 25$$

2. **(E)** Using the vertical line test,

"E" is not a function

3. **(A)** 
$$f(x) = 2x^2 - 2$$
$$f(x + h) = 2(x + h)^2 - 2 = 2(x^2 + 2xh + h^2) - 2$$
$$= 2x^2 + 4xh + 2h^2 - 2$$
$$\frac{f(x+h) - f(x)}{h} = \frac{(2x^2 + 4xh + 2h^2 - 2) - (2x^2 - 2)}{h}$$

$$= \frac{4xh + 2h^2}{h} = 4x + 2h$$

44

**4. (B)** A simple sketch is very useful for determining domain and range.

Domain: All real numbers (all *x* are good values)
Range: $y \geq 0$ (only values of *y* greater than or equal to zero are possible)

**5. (A)** $f(x) = -x^2 - 3x - 2$
$= -(-3)^2 - 3(-3) - 2$
$= -9 + 9 - 2 = -2$

**6. (A)** A simple sketch is very useful. This is a circle with center (0,0) and radius = 3.

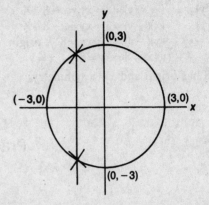

Using the vertical line test this *is not* a function.

# Practice Problems 2.2

**1. (B)** $y = |x| + 3$
    *Test Points*

| $(x, y)$ | $y = |x| + 3$ | |
|---|---|---|
| $(-x, y)$ | $y = |-x| + 3 = |x| + 3 \checkmark$ | *y*-axis |

| $(x, -y)$ | $-y = |x| + 3$ **X** |
|---|---|
| $(-x, -y)$ | $-y = |-x| + 3 = |x| + 3$ **X** |
| $(y, x)$ | $x = |y| + 3$ **X** |

Has *y*-axis symmetry only.

**2. (B)** $y = 7$
    *Test Points*

| $(x, y)$ | $y = 7$ | |
|---|---|---|
| $(-x, y)$ | $y = 7$ | *y*-axis $\checkmark$ |
| $(x, -y)$ | $-y = 7$ **X** | |
| $(-x, -y)$ | $-y = 7$ **X** | |
| $(y, x)$ | $x = 7$ **X** | |

Has *y*-axis symmetry only.

**3. (D)** $x^2 - y^2 = 4$
    *Test Points*

| $(x, y)$ | $x^2 - y^2 = 4$ | |
|---|---|---|
| $(-x, y)$ | $(-x)^2 - y^2 = 4 \Rightarrow x^2 - y^2 = 4 \checkmark$ | *y*-axis |

| $(x, -y)$ | $x^2 - (-y)^2 = 4 \Rightarrow x^2 - y^2 = 4 \checkmark$ | *x*-axis |
|---|---|---|
| $(-x, -y)$ | $(-x)^2 - (-y)^2 = 4 \Rightarrow x^2 - y^2 = 4 \checkmark$ | Origin |
| $(y, x)$ | $(y)^2 - (x)^2 = 4 \Rightarrow y^2 - x^2 = 4$ **X** | |

Has *x*, *y* and origin symmetry.

**4. (B)** $y = \sin x$
    *Test Points*

| $(x, y)$ | $y = \sin x$ | |
|---|---|---|
| $(-x, y)$ | $y = \sin(-x) = \sin x$ **X** | |
| $(-x, -y)$ | $-y = \sin(-x) = -\sin x \checkmark$ | Origin |

Origin symmetry is odd.

$y = \sin x$ is a function
$\therefore y = \sin x$ is an odd function

**5. (C)** $xy = 8$

*Test Points*

| $(x, y)$ | $xy = 8$ |
|---|---|
| $(-x, y)$ | $(-x)(y) = 8 \Rightarrow xy = -8$ **X** |
| $(x, -y)$ | $(x)(-y) = 8 \Rightarrow xy = -8$ **X** |
| $(-x, -y)$ | $(-x)(-y) = 8 \Rightarrow xy = 8$ √ origin |
| $(y, x)$ | $(y)(x) = 8 \Rightarrow xy = 8$ √ $x = y$ |

$xy = 8$ has origin and $x = y$ symmetry.

**6. (A)** $y = 2x - 1$

*Test Points*

| $(x, y)$ | $y = 2x - 1$ |
|---|---|
| $(-x, y)$ | $y = 2(-x) - 1 \Rightarrow -2x - 1$ **X** |
| $(x, -y)$ | $-y = 2x - 1$ **X** |
| $(-x, -y)$ | $-y = 2(-x) - 1 = -2x - 1$ **X** |
| $(y, x)$ | $x = 2y - 1$ **X** |

$y = 2x - 1$ has no symmetry.

# Practice Problems 2.3

**1. (B)**
$$(f + g)x = f(x) + g(x) = (3x - 3) + (2x + 1)$$
$$= 5x - 2$$

**2. (D)**
$$(f \cdot g)x = f(x) \cdot g(x) = (3x - 3)(2x + 1)$$
$$= 6x^2 - 3x - 3$$

**3. (E)**
$$(f \circ g)x = f(g(x)) = f(2x + 1) = 3(2x + 1) - 3$$
$$= 6x$$

**4. (C)**
$$(g \circ f)x = g(f(x)) = g(3x - 3) = 2(3x - 3) + 1$$
$$= 6x - 6 + 1 = 6x - 5$$

**5. (D)** $f(x) = \dfrac{x - 1}{x + 1}$

*Step 1*  $y = \dfrac{x - 1}{x + 1}$

*Step 2*  $x = \dfrac{y - 1}{y + 1}$

*Step 3*  $xy + x = y - 1$

$xy - y = -x - 1$

$y(x - 1) = -(x + 1)$

$y = \dfrac{-(x + 1)}{(x - 1)}$

*Step 4*  $f^{-1}(x) = \dfrac{-(x + 1)}{(x - 1)}$

**6. (C)** To be the graph of the inverse the relation should represent the reflection of $f(x)$ about the line $y = x$.

# Practice Problems 3.1

**1. (D)** $\left| \dfrac{-6}{-2} \right| = |3| = 3,$  $\dfrac{|-6|}{|-2|} = \dfrac{6}{2} = 3$

$|6(-2)| = |-12| = 12,$  $|6| \cdot |-2| = 6 \cdot 2 = 12$

$\left| \dfrac{-6}{-2} \right| + \dfrac{|-6|}{|-2|} + |6(-2)| + |6| \cdot |-2|$

$= 3 + 3 + 12 + 12 = 30$

**2. (B)** $|x| \leqslant 3 \Rightarrow x \leqslant 3$ and $-x \leqslant 3 \Rightarrow x \geqslant -3$

$-3 \leqslant x \leqslant 3$

**3. (A)** The critical value of $y = |2x - 6| + 5$ occurs when $2x - 6 = 0$ . . $x = 3$

| x | y |
|---|---|
| 0 | 11 |
| 1 | 9 |
| 2 | 7 |
| 3 | 5 |
| 4 | 7 |
| 5 | 9 |
| 6 | 11 |

$y = -(2x - 6) + 5$

$y = (2x - 6) + 5$

$(3,5)$

The answer is $y = -(2x - 6) + 5$

**4. (E)** The critical value is at $x + 3 = 0$
$\therefore x = -3$
$y = |x + 3| - 1$

| x | y |
|---|---|
| 0 | 2 |
| -1 | 1 |
| -2 | 0 |
| -3 | -1 |
| -4 | 0 |
| -5 | 1 |
| -6 | 2 |

$y = -(x + 3) - 1$  $(-3, -1)$  $y = (x + 3) - 1$

# Practice Problems 4.1

**1. (A)** $2\sqrt{18} \cdot 6\sqrt{2} = 2 \cdot 6\sqrt{18 \cdot 2}$
$$= 12\sqrt{36} = 12(6) = 72$$

**2. (B)** $(3\sqrt{3})^3 = (3\sqrt{3})(3\sqrt{3})(3\sqrt{3})$
$$= 27\sqrt{3}\sqrt{3}\sqrt{3}$$
$$= 27 \cdot 3 \cdot \sqrt{3}$$
$$= 81\sqrt{3}$$

**3. (B)** $\sqrt{80} + \sqrt{45} - \sqrt{20}$
$$= \sqrt{16 \cdot 5} + \sqrt{9 \cdot 5} - \sqrt{4 \cdot 5}$$
$$= \sqrt{16}\sqrt{5} + \sqrt{9}\sqrt{5} - \sqrt{4}\sqrt{5}$$
$$= 4\sqrt{5} + 3\sqrt{5} - 2\sqrt{5}$$
$$= 5\sqrt{5}$$

**4. (A)** $(3 + \sqrt{5})(3 - \sqrt{5})$
$$= 3(3) - 3\sqrt{5} + 3\sqrt{5} - \sqrt{5}\sqrt{5}$$
$$= 9 \qquad\qquad -5$$
$$= 4$$

**5. (E)** $\sqrt{20a^7b^3c^8}$
$$= \sqrt{4 \cdot 5a^6a^1b^2b^1c^8}$$
$$= \sqrt{4}\sqrt{5}\sqrt{a^6}\sqrt{a}\sqrt{b^2}\sqrt{b}\sqrt{c^8}$$
$$= 2\sqrt{5} \ a^3\sqrt{a} \ b \ \sqrt{b} \ c^4$$
$$= 2a^3bc^4\sqrt{5ab}$$

**6. (C)** $\dfrac{\sqrt{32b^3}}{\sqrt{8b}} = \sqrt{\dfrac{32b^3}{8b}} = \sqrt{4b^2}$
$$= \sqrt{4}\sqrt{b^2} = 2b$$

7. **(A)** $(\frac{1}{2}\sqrt{2})(\sqrt{6} + \frac{1}{2}\sqrt{2})$

$\quad = \frac{1}{2}\sqrt{2}\sqrt{6} + \frac{1}{2}\sqrt{2} \cdot \frac{1}{2}\sqrt{2}$

$\quad = \frac{1}{2}\sqrt{12} + \frac{1}{4}\sqrt{2}\sqrt{2}$

$\quad = \frac{1}{2}\sqrt{4 \cdot 3} + \frac{1}{4}(2)$

$\quad = \frac{1}{2}\sqrt{4}\sqrt{3} + \frac{2}{4}$

$\quad = \frac{1}{2}(2)\sqrt{3} + \frac{2}{4} = \sqrt{3} + \frac{1}{2}$

8. **(B)**

$\quad \frac{1}{2}\sqrt{180} + \frac{1}{3}\sqrt{45} - \frac{2}{5}\sqrt{20}$

$\quad = \frac{1}{2}\sqrt{36 \cdot 5} + \frac{1}{3}\sqrt{9 \cdot 5} - \frac{2}{5}\sqrt{4 \cdot 5}$

$\quad = \frac{1}{2}\sqrt{36}\sqrt{5} + \frac{1}{3}\sqrt{9}\sqrt{5} - \frac{2}{5}\sqrt{4}\sqrt{5}$

$\quad = \frac{1}{2}(6)\sqrt{5} + \frac{1}{3}(3)\sqrt{5} - \frac{2}{5}(2)\sqrt{5}$

$\quad = 3\sqrt{5} \quad + \sqrt{5} \quad - \frac{4}{5}\sqrt{5}$

$\quad = 4\sqrt{5} - \frac{4}{5}\sqrt{5}$

$\quad = \frac{16}{5}\sqrt{5}$

# Practice Problems 4.2

1. **(B)** $\dfrac{2}{\sqrt{11}}\left(\dfrac{\sqrt{11}}{\sqrt{11}}\right) = \dfrac{2\sqrt{11}}{11}$

2. **(A)**

$\dfrac{\sqrt{3}}{1 + \sqrt{2}}\left(\dfrac{1 - \sqrt{2}}{1 - \sqrt{2}}\right) = \dfrac{\sqrt{3} - \sqrt{3}\sqrt{2}}{1(1) - \sqrt{2} + \sqrt{2} - \sqrt{2}\sqrt{2}}$

$\quad = \dfrac{\sqrt{3} - \sqrt{6}}{1 - 2} = \dfrac{\sqrt{3} - \sqrt{6}}{-1}$

3. **(D)**

$\dfrac{1}{\frac{2\sqrt{3}}{3}} = \dfrac{3}{2\sqrt{3}}\left(\dfrac{\sqrt{3}}{\sqrt{3}}\right) = \dfrac{3\sqrt{3}}{2(3)} = \dfrac{\sqrt{3}}{2}$

4. **(C)**

$\dfrac{1}{1 + \sqrt{3}}\left(\dfrac{1 - \sqrt{3}}{1 - \sqrt{3}}\right) = \dfrac{1 - \sqrt{3}}{1 - \sqrt{3} + \sqrt{3} - 3}$

$\quad = \dfrac{1 - \sqrt{3}}{1 - 3} = \dfrac{1 - \sqrt{3}}{-2} \text{ or } \dfrac{\sqrt{3} - 1}{2}$

# Cumulative Review Test

1. **(B)** $|-3| = 3 \quad |7| = 7 \quad |-7| = 7$

$\quad |-3| + |7| - |-7|$

$\quad = 3 + 7 - 7 = 3$

2. **(E)** An irrational number cannot be written in $\dfrac{a}{b}$

form.

ans. $\sqrt{2}$

3. **(A)** $3\sqrt{3} + \sqrt{12}$

$\quad = 3\sqrt{3} + \sqrt{4 \cdot 3} = 3\sqrt{3} + \sqrt{4}\sqrt{3}$

$\quad = 3\sqrt{3} + 2\sqrt{3}$

$\quad = 5\sqrt{3}$

4. **(C)**

$(h - k)(x) = h(x) - k(x) = (x^2 + 1) - (3x)$

$\quad = x^2 - 3x + 1$

5. **(D)** $\dfrac{2}{\sqrt{3}}\dfrac{\sqrt{3}}{(\sqrt{3})} = \dfrac{2\sqrt{3}}{3}$

6. **(D)**

$$f(x) = 3x + 2$$
$$y = 3x + 2$$
$$x = 3y + 2 \quad \text{interchange } x \text{ and } y$$
$$y = \dfrac{x-2}{3} \quad \text{solve for } y$$
$$f^{-1}(x) = \dfrac{x-2}{3} \quad \text{replace } y \text{ with } f^{-1}(x)$$

7. **(E)** Using the vertical line test

8. **(C)** $y = 2x^7 + 3x^5 - 2x$

*Test Points*

$$(x, y) \Rightarrow \quad y = 2x^7 + 3x^5 - 2x$$
$$(-x, y) \Rightarrow \quad y = -2x^7 - 3x^5 + 2x$$
$$(x, -y) \Rightarrow -y = 2x^7 + 3x^5 - 2x$$
$$(-x, -y) \Rightarrow -y = -2x^7 - 3x^5 + 2x$$
$$\therefore \ y = 2x^7 + 3x^5 - 2x \ \checkmark$$

There is origin symmetry.

9. **(C)**

$$(f \cdot g)(x) = f(x) \cdot g(x) = (x^2 - 1)(x + 1)$$
$$= x^3 + x^2 - x - 1$$

10. **(E)** The critical value occurs when $x - 1 = 0$
$$\therefore x = 1.$$

When $x = 1 \quad y = 3$

$$y = -(x - 1) + 3 \qquad y = (x, - 1) + 3$$

11. **(C)**

$$\dfrac{\sqrt{3}}{\sqrt{15} - 3} \cdot \dfrac{\sqrt{15} + 3}{(\sqrt{15} + 3)} = \dfrac{\sqrt{3}(\sqrt{15} + 3)}{15 - 9}$$

*Note:*

$$\sqrt{45} = \sqrt{9}\sqrt{5}$$
$$= 3\sqrt{5}$$

$$= \dfrac{\sqrt{45} + 3\sqrt{3}}{6}$$

$$= \dfrac{3\sqrt{5} + 3\sqrt{3}}{6} = \dfrac{3(\sqrt{5} + \sqrt{3})}{6}$$

$$= \dfrac{\sqrt{5} + \sqrt{3}}{2}$$

12. **(A)**

$$f(x) = x^2 - 3$$
$$f(x + h) = (x + h)^2 - 3$$
$$= x^2 + 2xh + h^2 - 3$$
$$\dfrac{f(x + h) - f(x)}{h} = \dfrac{(x^2 + 2xh + h^2 - 3) - (x^2 - 3)}{h}$$
$$= \dfrac{2xh + h^2}{h} = 2x + h$$

13. **(B)**

$$\sqrt{49a^3b} = \sqrt{49a^2a^1b^1} = \sqrt{49}\sqrt{a^2}\sqrt{a}\sqrt{b}$$
$$= 7a\sqrt{a}\sqrt{b}$$
$$= 7a\sqrt{ab}$$
$$\sqrt{a^3b} = \sqrt{a^2a^1b^1} = \sqrt{a^2}\sqrt{a}\sqrt{b}$$
$$= a\sqrt{ab}$$

$$\sqrt{49a^3b} - \sqrt{a^3b} + 2a\sqrt{ab}$$
$$= 7a\sqrt{ab} - a\sqrt{ab} + 2a\sqrt{ab}$$
$$= 8a\sqrt{ab}$$

14. (A)

$$|x + 1| < 3$$

$$-(x + 1) < 3 \qquad x + 1 < 3$$
$$-x - 1 < 3 \qquad x < 2$$
$$-x < 4$$
$$x > -4$$

15. (D) $g(4) = 4 + \sqrt{4} = 6$

$f(6) = 3(6)^2 - 2 = 106$

16. (E)

$$(2 - \sqrt{3})(2 + \sqrt{3}) = 2(2) + 2\sqrt{3} - 2\sqrt{3} - 3$$
$$= 1$$

17. (C) The inverse involves interchange of the $x$ and $y$ coordinates:

$$f = \{(0, 1), (2, 7), (-1, 9), (-3, -3)\}$$
$$f^{-1} = \{(1, 0), (7, 2), (9, -1), (-3, -3)\}$$

18. (D) Positive integers are also called natural numbers.

19. (C) An odd function has origin symmetry.

20. (A) $\sqrt[3]{-\frac{1}{8}} = \dfrac{\sqrt[3]{-1}}{\sqrt[3]{8}} = \dfrac{-1}{2}$

This is real and rational.

# *UNIT TWO*

# 5. Exponents

## 5.1 Essentials

| $x^0 = 1 \ (x \neq 0)$ | $x^a \cdot x^b = x^{a+b}$ | $(xy)^a = x^a y^a$ |
|---|---|---|
| $x^1 = x$ | $\dfrac{x^a}{x^b} = x^{a-b} \quad (x \neq 0)$ | $\left(\dfrac{x}{y}\right)^a = \dfrac{x^a}{y^a}$ |
| $x^{-a} = \dfrac{1}{x^a} \quad (x \neq 0)$ | $(x^a)^b = x^{ab}$ | |
| $x^{1/a} = \sqrt[a]{x}$ | | If $x^a = x^b$, |
| $x^{b/a} = (\sqrt[a]{x})^b = \sqrt[a]{x^b}$ | | then $a = b \quad (x \neq 0, 1)$. |

 Common error: $3^2 \cdot 3^3 = 9^{2+3} = 9^5$. Should be: $3^2 \cdot 3^3 = 3^{2+3} = 3^5$. When multiplying exponential terms the base must be the same and remains the same; add the exponents.

**Examples:**

1. Evaluate $x^0 + x^{1/2} + x^{-2}$ when $x = 9$.

2. Simplify $\dfrac{3^7 \cdot 3^4}{3^5}$.

3. Simplify $\dfrac{n^{a+2} \cdot n^{2a-1}}{n^{2a+1}}$.

**Solutions:**

1. Substitute $x = 9$
$$9^0 + 9^{1/2} + 9^{-2}$$
$$= 1 + \sqrt{9} + \frac{1}{9^2}$$
$$= 1 + 3 + \frac{1}{81}$$
$$= 4\frac{1}{81}$$

2. $\dfrac{3^{7+4}}{3^5} = \dfrac{3^{11}}{3^5} = 3^{11-5} = 3^6$

3. $\dfrac{n^{(a+2)+(2a-1)}}{n^{2a+1}} = \dfrac{n^{3a+1}}{n^{2a+1}}$
$$= n^{(3a+1)-(2a+1)} = n^a$$

53

To **solve an equation which has the variable as an exponent**, express each side of the equation as a power of the same base, then set the exponents equal to each other.

**Examples:**                                               **Solutions:**

4. Solve $2^x = 32$.

    4. Write 32 as a power of 2:

$$2^x = 2^5$$

Since the bases are equal, equate the exponents:

$$x = 5$$

5. Solve for $x$: $3^{x+1} = 243$.

    5. $3^{x+1} = 3^5$
$$x + 1 = 5$$
$$x = 4$$

6. Solve for $y$: $2^y = 4^{y+1}$.

    6. Develop a common base:
$$2^y = (2^2)^{y+1}$$
$$2^y = 2^{2(y+1)} = 2^{2y+2}$$
$$y = 2y + 2$$
$$y = -2$$

7. Solve for $r$: $9^{r+3} = 27^{r-1}$.

    7. Convert to a common base:
$$(3^2)^{r+3} = (3^3)^{r-1}$$
$$3^{2(r+3)} = 3^{3(r-1)}$$
$$2(r + 3) = 3(r - 1)$$
$$2r + 6 = 3r - 3$$
$$r = 9$$

8. Solve for $x$: $125^x = \dfrac{1}{25}$.

    8. Write both sides of the equation in base 5:
$$(5^3)^x = \frac{1}{5^2} = 5^{-2}$$
$$5^{3x} = 5^{-2}$$
$$3x = -2$$
$$x = \frac{-2}{3}$$

9. Simplify $(x^2 y^3)^5$.

    9. $(x^2 y^3)^5 = x^{10} y^{15}$

10. Simplify $\dfrac{(x^2 y^7)^3}{x^4 y}$.

    10. $\dfrac{(x^2 y^7)^3}{x^4 y} = \dfrac{x^6 y^{21}}{x^4 y^1} = x^2 y^{20}$

To solve an equation of the form: $x^{-2/3} = 4$ or $2x^{1/3} = 3$, first rearrange the equation with only the variable term, with coefficient 1, on the left side, then raise each side of the equation to the reciprocal (up-side-down) power of the variable. The objective is to create the term $x^1 = x$.

**Examples:**

11. Solve $x^{1/2} = 8$.

12. Solve for $x$: $x^{-2/3} = 4$.

13. Solve for $x$: $2x^{1/3} = 3$.

**Solutions:**

11. $x^{1/2} = 8$

Raise each side of the equation to the $\frac{2}{1}$ power:

$$(x^{1/2})^{2/1} = 8^{2/1}$$
$$x^1 = 8^2 \implies x = 64$$

12. $(x^{-2/3})^{-3/2} = (4)^{-3/2}$

$$x^1 = 4^{-3/2} = \frac{1}{(\sqrt{4})^3} \implies x = \frac{1}{8}$$

13. $x^{1/3} = \frac{3}{2}$

$$(x^{1/3})^3 = \left(\frac{3}{2}\right)^3$$

$$x^1 = \frac{3^3}{2^3} \implies x = \frac{27}{8}$$

# Practice Problems 5.1

1. The value of $2^0 + 4^{1/2} + (\frac{1}{2})^{-1}$ is

   A) $1\frac{1}{2}$   B) $3\frac{1}{2}$   C) 4   D) 5   E) 8

2. Simplify $\dfrac{a^{x+1} \cdot a^{2x-1}}{a^{-x+1}}$.

   A) $a^{4x-1}$   B) $a^{2x+1}$   C) $a^{-x+1}$
   D) $a^{2x-1}$   E) $a^{4x+1}$

3. Solve for $m$: $4^{2m} = 2^{m+1}$.

   A) 4   B) 2   C) 1   D) $\frac{1}{3}$   E) $\frac{1}{15}$

4. Solve for $v$: $\frac{1}{16} = 2^{v-1}$.

   A) $-3$   B) $-1$   C) 0
   D) 3   E) 17

5. Solve for $x$: $x^{-2} = 16$.

   A) 18   B) $-8$   C) 4
   D) $-4$   E) $\pm\frac{1}{4}$

6. Simplify $\dfrac{2^3 \cdot 2^4}{(2^3)^2}$.

   A) $\frac{1}{2}$   B) 2   C) 4   D) $2^{7/6}$   E) $2^6$

7. Simplify $(5^{2a})(5)(5^{a+1})$.

   A) $5^{3a+1}$   B) $5^{3a+2}$   C) $15^{3a+1}$
   D) $25^{3a+1}$   E) $125^{3a+2}$

8. Evaluate $3x^0 + (3x)^0 + x^{-1} + 2x$
   when $x = \frac{1}{2}$.

   A) $-1$   B) $1\frac{1}{2}$   C) $6\frac{1}{2}$   D) 7   E) 10

Solutions on page 68

# 6. Scientific Notation and Logarithms

## 6.1 Scientific Notation

Very large or very small numbers are often expressed in the form:

$$N \times 10^a, \text{ where } 1 \leqslant N < 10 \text{ and } a \text{ is an integer.}$$

---

Procedure: **Converting a decimal number to scientific notation**

Step 1    Mark the existing decimal point.

Step 2    Write the number with the decimal point moved such that the value is now between 1 and 10. This is "$N$".

Step 3    Count the number of places the decimal point must move from step 2 to go back to step 1. This is "$a$".

Step 4    Note the direction of movement in step 3.    This is the sign of $a$.

Step 5    Write: $N \times 10^a$.

---

Scientific notation and the laws of exponents may be used together in calculations.

**Examples:**

1. Express 7123 in scientific notation.

2. Express .0048 in scientific notation.

**Solutions:**

1. Step 1    7123.

   Step 2    7.123    $N = 7.123$

   Step 3    7.123    $a = 3$

   Step 4    to the right, $a$ is "+"

   Step 5    $7.123 \times 10^{+3}$

2.    .0048    Step 1

   4.8    Step 2    $N = 4.8$

   004.8    Step 3    $a = 3$

   Step 4    $a$ is "−"

   $4.8 \times 10^{-3}$    step 5

3. If 6,000,000 is written as $6 \times 10^a$, find the value of $a$.

3. 6,000,000.

6.000,000   $N = 6$

6.000,000   $a = 6$
                 $a$ is "+"

$6,000,000 = 6 \times 10^{+6}$   $a = +6$

4. Using scientific notation, calculate $(250,000)(.000003)$.

4. $250,000 = 2.5 \times 10^5$

$.000003 = 3 \times 10^{-6}$

$(2.5 \times 10^5)(3 \times 10^{-6}) = (2.5)(3) \times 10^{5+(-6)}$

$= 7.5 \times 10^{-1} = .75$

# Practice Problems 6.1

1. Express 278300 in scientific notation.

A) $2783 \times 10^2$    B) $2.783 \times 10^5$
C) $.2783 \times 10^{-4}$   D) $2.783 \times 10^{-1}$
E) $27.83 \times 10^4$

2. If .007932 is written as $7.932 \times 10^a$, the value of $a$ is

A) 3  B) 2  C) 1  D) $-2$  E) $-3$

3. Calculate $\dfrac{12000}{.04}$ using scientific notation.

A) .00003  B) .003  C) .3
D) 300  E) 300,000

4. Write $\dfrac{1}{1000}$ in scientific notation.

A) $1 \times 10^3$  B) $1 \times 10^2$  C) $1 \times 10^{-2}$
D) $1 \times 10^{-3}$  E) $1 \times 10^{-4}$

Solutions on page 69

## 6.2  Logarithms

| | |
|---|---|
| $\log_b N = a \leftrightarrow b^a = N$ | $b$ = base (positive, real, $\neq 1$) <br> $N$ = number (positive, real) <br> $a$ = the logarithm, the exponent of $b$ to get $N$ |

$\log_b 1 = 0$ $\qquad$ $\log_b xy = \log_b x + \log_b y$

$\log_b b = 1$ $\qquad$ $\log_b\left(\dfrac{x}{y}\right) = \log_b x - \log_b y$

$\begin{cases} \log_b b^x = x \\ b^{\log_b x} = x \end{cases}$ $\qquad$ $\log_b x^y = y \log_b x$ $\quad$ *Note:* $\log_b \sqrt[y]{x} = \log_b x^{1/y}$

$\log_b \sqrt[y]{x} = \dfrac{\log_b x}{y} = \dfrac{1}{y} \log_b x$

$\log(N \times 10^a) = \log N + a$

 $\log_{10}$ is frequently written as log. "log" which is a logarithm to the base 10 is called a common logarithm.

☞ 
- $\log_e$ is frequently written as ln. "ln" which is a logarithm to the base $e$ is called a natural logarithm.
- All other bases must be indicated.
- $e \approx 2.718281828459 \ldots$

**Examples:**

1. Express $\log_7 49 = 2$ in exponential form.

2. Express $8^{-2} = \frac{1}{64}$ in logarithmic form.

3. Find the logarithm of $10^2$ in base 10.

4. Express $\log \dfrac{xy}{z}$ in expanded form.

5. Express $\log \dfrac{x^2 y^3 z}{\sqrt{m}}$ in expanded form.

6. If $\log 2.13 = t$, express $\log(2.13 \times 10^{-4})$ in terms of $t$.

7. If $\log 7.64 = m$, express the logarithm of 764 in terms of $m$.

8. If $\log_b 9 = 2$, find $b$.

9. If $\ln e^2 = a$, find $a$.

10. If $\log 2 = .3010$ and $\log 5 = .6990$, find $\log 40$.

**Solutions:**

1. $7^2 = 49$

2. $\log_8 \frac{1}{64} = -2$

3. $\log_{10} 10^2 = a$
   Change to exponential form: $10^a = 10^2$
   $a = 2$

4. $\log \dfrac{xy}{z} = \log x + \log y - \log z$

5. $\log \dfrac{x^2 y^3 z}{\sqrt{m}} = 2 \log x + 3 \log y + \log z$
   $- \dfrac{\log m}{2}$

6. $\log(N \times 10^a) = \log N + a$
   $\log(2.13 \times 10^{-4}) = \log 2.13 + (-4)$
   $= t - 4$

7. Express 764 in scientific notation:
   $\log 764 = \log (7.64 \times 10^2) = \log 7.64 + 2$
   $= m + 2$

8. $b^2 = 9 = 3^2$
   $b = 3$

☞ If $b^2 = 9$, $b = \pm 3$, however, in logarithms the base must be positive.

9. $\ln = \log_e$
   $\log_e e^2 = a$
   $e^a = e^2, \quad a = 2$

10. $\log 40 = \log(2^3 \cdot 5)$
    $= 3 \log 2 + \log 5$
    $= 3(.3010) + (.6990) = 1.6020$

11. If ln 7 = m,
   express ln $\frac{1}{7}$ in terms of m.

11. ln $\frac{1}{7}$ = ln 1 − ln 7
   ln 1 = 0   ($\log_b 1 = 0$)
   ln 7 = m
   ln 1 − ln 7 = 0 − m = −m

# Practice Problems 6.2

1. If $\log_m 125 = 3$, then m =
   A) 2  B) 3  C) 4  D) 5  E) 6

2. If $\log_{1/7} 49 = y$, then y =
   A) 7  B) 2  C) −2  D) $-\frac{1}{2}$
   E) 5

3. The expression $\ln \frac{a^2}{b^3}$ is equivalent to
   A) $2 \ln a - 3 \ln b$  B) $\frac{2}{3} \ln \frac{a}{b}$
   C) $\frac{(\ln a)^2}{(\ln b)^2}$  D) $\frac{2 \ln a}{3 \ln b}$
   E) $\frac{2}{3}(\ln a - \ln b)$

4. If log 3 = .4771 and log 7 = .8451, then $\log \sqrt{21}$ =
   A) 1.1499  B) 2.6444  C) .6611
   D) .2016  E) .6350

5. $2 \log a + \frac{\log b}{2}$ is equivalent to
   A) $\log\left(2a + \frac{b}{2}\right)$  B) $\log 2a + \log \frac{b}{2}$
   C) $\log \frac{a^2 b}{2}$  D) $\log a^2 \sqrt{b}$
   E) $\log \sqrt{a^2 b}$

6. If log 9.17 = t, express the log .00917 in terms of t.
   A) $3 - t$  B) $t - 3$  C) $3t$
   D) $2.17 + t$  E) $\frac{t}{3}$

7. Express $\log \sqrt{\frac{x^3}{zm}}$ in expanded form.
   A) $3 \log x - \log z + \log m$
   B) $\frac{1}{2}(3 \log x - \log z + \log m)$
   C) $\frac{1}{2}(3 \log x - \log z - \log m)$
   D) $\frac{3 \log x}{\log z + \log m}$
   E) $\frac{1}{2}\left(\frac{3 \log x}{\log z + \log m}\right)$

8. If $\log_8 x = \frac{1}{3}$, then x =
   A) $\frac{1}{8}$  B) 2  C) 24
   D) $\frac{8}{3}$  E) $3^8$

Solutions on page 69

# 7. Finding the Square Root of a Number

## 7.1 Basics and Beyond

There are two basic techniques for finding the square root of a number. One is iterative, the other logarithmic.

---

**Iterative Procedure:**

Numerical Example: $\sqrt{8}$

**Step 1**  Pair off the digits under the square root sign in each direction from the decimal point. Each pair of digits will generate one digit of the square root.

$$\sqrt{\overline{08}.\overline{00}\overline{00}}$$

**Step 2**  Bring up the decimal point.

$$\sqrt{08.0000}$$

**Step 3**  Find the largest whole number square root that is less than or equal to the first pair of numbers. Place this number on the top and the square of the number under the first pair of digits.

$$\begin{array}{r} 2. \\ \sqrt{08.0000} \\ 4 \end{array}$$

**Step 4**  Subtract and bring down the next pair of digits.

$$\begin{array}{r} 2. \\ \sqrt{08.0000} \\ \underline{4} \\ 400 \end{array}$$

**Step 5**  Double the top number and place it outside to the left. Put a dash after this number (this becomes the divisor) and a dash on top over the next pair of digits (this becomes the multiplier).

$$\begin{array}{r} 2.\_ \\ \sqrt{08.0000} \\ \underline{4} \\ 4\_\overline{)\,400} \end{array}$$

**Step 6**  Find the integer (0 through 9) which when placed on both dashes will provide the closest product, less than or equal to the required number. This is a trial and error process.

Try 7

$$\begin{array}{r} 2.7 \\ \sqrt{08.0000} \\ \underline{4} \\ 47\overline{)\,400} \\ 329 \leftarrow \boxed{\text{Too Small}} \end{array}$$

$7 \times 47 = 329$

---

Try 8

2.**8**

$\sqrt{08.0000}$

4

48 | 400

384 ← OK

$8 \times 48 = 384$

Try 9

2.**9**

$\sqrt{08.0000}$

4

49 | 400

441 ← Too Large

$9 \times 49 = 441$

Step 7    Write the digit that replaces the two dashes and show the product of the multiplication.

2.**82**

$\sqrt{08.0000}$

4

48 | 400

384

562 | 1600

1124

476

Step 8    Go back to step 4.

---

**Logarithmic Procedure:**

find the $\sqrt{8}$

Step 1A    Create an equation.    $x = \sqrt{8}$

Step 1B    Create a log equation.    $\log x = \log\sqrt{8} = \log(8^{1/2})$

Step 2    Use the rules of logarithms.    $\log x = \dfrac{\log 8}{2} = \dfrac{1}{2}\log 8$

Step 3    From the log tables, find log of the number.    $\log 8 = .9031$

Step 4    Substitute into the equation.    $\log x = \dfrac{.9031}{2}$

Step 5    Solve the equation for $\log x$.    $\log x = .45155$

Step 6    Use the table to find the antilog.    $x \approx 2.83$

---

**Examples:**

1. Find the $\sqrt{15}$ to the nearest tenth using the iterative procedure.

**Solutions:**

1. $\sqrt{15.0000}$

3.**87**

$\sqrt{15.0000}$

9

68 | 600

544

767 | 5600

5369

231

Try:

$68 \times 8 = 544$

$69 \times 9 = 621$    Too large

Try:

$767 \times 7 = 5369$

$768 \times 8 = 6144$    Too large

$\sqrt{15} = 3.9$ to the nearest tenth

2. Find the $\sqrt{15}$ to the nearest tenth using logarithms.
   Note: log 1.5 = .1761
         log 3.87 = .5877

2.     $x = \sqrt{15}$

$\log x = \log \sqrt{15} = \log(15^{1/2})$

$= \frac{1}{2} \log 15$

$= \frac{1}{2} \log(1.5 \times 10^1)$      $\log(1.5 \times 10^1)$

$= \frac{1}{2}(1.1761)$              $= \log 1.5 + \log 10^1$

$= .5881$                  $= .1761 + 1$

$x = \text{antilog } .5881$         $= 1.1761$

$x \approx 3.87$

$x = 3.9$ to the nearest tenth

# Practice Problems 7.1

1. Using the iterative procedure, find the $\sqrt{3}$ to the nearest tenth.

A) 1.6  B) 1.68  C) 1.7  D) 1.73  E) 1.732

2. Find $\sqrt{12}$ to the nearest tenth using the iterative procedure.

A) 3  B) 3.4  C) 3.46  D) 3.5  E) 3.464

3. Using logarithms, find the $\sqrt{3}$ to the nearest integer.

A) 1  B) 2  C) 1.7  D) 1.73  E) 1.732

4. Find the $\sqrt{12}$ to the nearest tenth using logarithms.

A) 3  B) 3.4  C) 3.46  D) 3.5  E) 3.464

Solutions on page 70

# 8. e, ln, Hyperbolics

## 8.1 Fundamental Facts

| e | an irrational number $\approx 2.718281828459\ldots$ |
|---|---|

| ln | natural logarithm $= \log_e$ |
|---|---|

$f(x) = e^x$ and $g(x) = ln\ x$ are inverse functions

| sinh $x$: | hyperbolic sin $x = \dfrac{e^x - e^{-x}}{2}$ |
|---|---|
| cosh $x$: | hyperbolic cos $x = \dfrac{e^x + e^{-x}}{2}$ |
| tanh $x$: | hyperbolic tan $x = \dfrac{e^x - e^{-x}}{e^x + e^{-x}}$ |

**Examples:**

1. Show that $\cosh^2 x - \sinh^2 x = 1$.

**Solutions:**

1. $\left(\dfrac{e^x + e^{-x}}{2}\right)^2 - \left(\dfrac{e^x - e^{-x}}{2}\right)^2 \overset{?}{=} 1$

$\left(\dfrac{(e^x + e^{-x})}{2} \cdot \dfrac{(e^x + e^{-x})}{2}\right) - \left(\dfrac{(e^x - e^{-x})}{2} \cdot \dfrac{(e^x - e^{-x})}{2}\right) \overset{?}{=} 1$

$\left(\dfrac{e^{2x} + 2e^0 + e^{-2x}}{4}\right) - \left(\dfrac{e^{2x} - 2e^0 + e^{-2x}}{4}\right) \overset{?}{=} 1$

$$\frac{2e^0 + 2e^0}{4} \stackrel{?}{=} 1$$

$$\frac{2+2}{4} \stackrel{?}{=} 1$$

$$1 = 1\surd$$

# Practice Problems 8.1

1. If ln 2 = .6931, find ln 16.

   A) $(.6931)^4$   B) $4(.6931)$   C) $2(.6931)$

   D) $(.6931)^2$   E) $\dfrac{1}{.6931}$

4. $2 \ln e^x =$

   A) $x$   B) $2x$   C) $x^2$
   D) $x + 2$   E) $x - 2$

2. Express ln 9.974 = 2.3 in exponential form.

   A) $e^{9.974} = 2.3$   B) $e^{.026} = 2.3$

   C) $e^{2.3} = 9.974$   D) $\dfrac{1}{e^{2.3}} = 9.974$

   E) $e^{-2.3} = 9.974$

3. In the graphs shown below, if the upper curve is $e^{2x}$, then the lower curve, which is the graph of its inverse, is given by what equation?

   A) $y = e^{-x}$   B) $y = e^{-2x}$   C) $y = \dfrac{1}{e^x}$

   D) $y = \ln 2$   E) $y = \ln x^{\frac{1}{2}}$

Solutions on page 70

# Cumulative Review Test

1. Solve for $x$: $4^{3x} = \dfrac{1}{8^{x-3}}$.

   A) 1  B) $\dfrac{-3}{2}$  C) $\dfrac{-3}{5}$

   D) $-3$  E) $-1$

2. Express .03 in scientific notation.

   A) 3  B) $\dfrac{3}{100}$  C) $3 \times 10^2$

   D) $3 \times 10^{-2}$  E) $3 \times 10^{-3}$

3. $\text{Log}(f^2 - g^2)$ equals

   A) $2 \log f - 2 \log g$
   B) $\log(f + g) + \log(f - g)$
   C) $\log 2f - \log 2g$  D) $\dfrac{2 \log f}{2 \log g}$
   E) $2 \log \dfrac{f}{g}$

4. The expression $\dfrac{2}{\sqrt{7} - 2}$ written as a fraction with a rational denominator is

   A) $2\sqrt{7} + 4$  B) $\dfrac{2\sqrt{7} + 4}{11}$  C) $\dfrac{4\sqrt{7}}{3}$

   D) $\dfrac{2\sqrt{7} + 4}{3}$  E) $\dfrac{2\sqrt{7} + 2}{3}$

5. Find $\sqrt{53}$ to the nearest tenth.

   A) 7.2  B) 7.28  C) 7.3
   D) 7.31  E) 7.4

6. Find the value of $2b^0 + b^{-1/2}$ when $b = 16$.

   A) $-4$  B) $-2$  C) $-8$
   D) $\dfrac{9}{4}$  E) $\dfrac{7}{4}$

7. The expression $r^n \div s^n$ is equivalent to

   A) $\left(\dfrac{r}{s}\right)^0$  B) $\dfrac{r}{s}$  C) $\dfrac{nr}{ns}$

   D) $\left(\dfrac{r}{s}\right)^n$  E) $\dfrac{r^n}{s^{-n}}$

8. The expression $\frac{1}{3}\log n - \frac{1}{2} \log l$ is equivalent to

   A) $\log(\sqrt[3]{n} - l^2)$  B) $\log(\sqrt[3]{n} - \sqrt{l})$

   C) $\log \dfrac{\sqrt[3]{n}}{\sqrt{l}}$  D) $\dfrac{1}{6} \log \dfrac{n}{l}$

   E) $\log \dfrac{n^3}{l^2}$

9. If $h(x) = \dfrac{x - 1}{x + 3}$, find $h(x - 3)$.

   A) $\dfrac{x - 4}{x}$  B) $\dfrac{x - 4}{x + 6}$  C) $\dfrac{x}{x - 4}$

   D) $\dfrac{x + 3}{x - 1}$  E) $2x + 2$

10. $.3 \times 10^{-4}$ is equivalent to

    A) .0003  B) $30 \times 10^3$
    C) $.03 \times 10^{-5}$  D) $3 \times 10^{-3}$
    E) $3 \times 10^{-5}$

11. The expression $-2(x)^3 + (-2x)^3 =$

    A) $4x^3$  B) $16x^3$  C) $16x^6$
    D) $-10x^3$  E) $-16x^3$

12. If $\log_8 x = \frac{2}{3}$, find the value of $x$.

    A) $\dfrac{16}{3}$  B) 4  C) $8\sqrt{2}$
    D) $\dfrac{4}{3}$  E) 16

13. Solve for $x$: $9^{3x} = 3^{x+5}$.

    A) $-2$  B) $-1$  C) 0
    D) 1  E) 2

14. Simplify $\dfrac{3^3 + 3^3 + 3^3}{3^4}$.

    A) $\frac{1}{3}$  B) $3^5$  C) 3  D) 1  E) 9

15. Express $\dfrac{3.2 \times 10^2}{.32 \times 10^{-2}}$ in scientific notation.

    A) 10  B) $1 \times 10^1$  C) $1 \times 10^4$
    D) $1 \times 10^5$  E) $1 \times 10^6$

16. log  4.1 = .6128,  log  6.40 = .8062 and
    log  6.41 = .8069. Find the $\sqrt{41}$ to the nearest
    tenth.

    A) 6.2  B) 6.3  C) 6.4
    D) 6.5  E) 6.6

17. Simplify $\left(\dfrac{x^{1/4}y^{1/2}}{z^{-1}}\right)^4$.

    A) $xy^2z^4$  B) $\dfrac{xy^2}{z^4}$  C) $xy^2z$

    D) $x^{17/4}y^{9/2}z^4$  E) $\dfrac{z}{xy^2}$

18. Evaluate $\dfrac{|-6|}{6} + \dfrac{|4|}{-4} + 2|-2| + (-2)|2|$.

    A) 8  B) 6  C) 0  D) 2  E) $-10$

19. $1 + \sinh^2 x =$

    A) $\cosh x$  B) $\dfrac{e^{2x} + e^{-2x}}{4}$

    C) $\dfrac{e^{2x} + 4 + e^{-2x}}{4}$  D) $\dfrac{e^x + e^{-x}}{2}$

    E) $\dfrac{e^{2x} + 2 + e^{-2x}}{4}$

20. $.35\overline{3535}$ is an example of a(n).

    A) imaginary number
    B) irrational number
    C) rational number
    D) improper fraction
    E) natural number

Solutions on page 71

# Diagnostic Chart for Cumulative Review Test

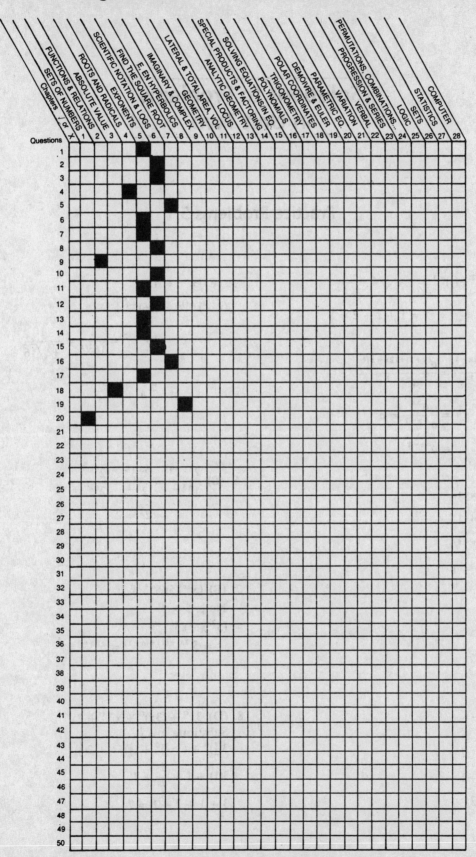

# Solutions to Problems in Unit Two

## Practice Problems 5.1

1. **(D)** $2^0 + 4^{1/2} + (\tfrac{1}{2})^{-1} = 1 + \sqrt{4} + \dfrac{1}{\tfrac{1}{2}}$

   $= 1 + 2 + 2 = 5$

2. **(A)** $\dfrac{a^{x+1} \cdot a^{2x-1}}{a^{-x+1}} = \dfrac{a^{(x+1)+(2x-1)}}{a^{-x+1}}$

   $= \dfrac{a^{3x}}{a^{-x+1}} = a^{3x-(-x+1)}$

   $= a^{4x-1}$

3. **(D)** $4^{2m} = 2^{m+1}$

   $(2^2)^{2m} = 2^{m+1}$

   $2^{4m} = 2^{m+1}$

   $4m = m + 1$

   $3m = 1$

   $m = \tfrac{1}{3}$

4. **(A)** $\dfrac{1}{16} = 2^{v-1}$

   $\dfrac{1}{2^4} = 2^{v-1}$

   $2^{-4} = 2^{v-1}$

   $-4 = v - 1$

   $v = -3$

5. **(E)** $x^{-2} = 16$

   $(x^{-2})^{-1/2} = (16)^{-1/2}$

   $x^1 = 16^{-1/2} = \dfrac{1}{16^{1/2}} = \dfrac{1}{\pm\sqrt{16}}$

   $= \pm \dfrac{1}{4}$

6. **(B)** $\dfrac{2^3 \cdot 2^4}{(2^3)^2} = \dfrac{2^{3+4}}{2^{3 \cdot 2}} = \dfrac{2^7}{2^6}$

   $= 2^1 = 2$

7. **(B)** $(5^{2a})(5)(5^{a+1})$

   Note: $5 = 5^1$

   $(5^{2a})(5^1)(5^{a+1})$

   $= 5^{(2a)+(1)+(a+1)} = 5^{3a+2}$

8. **(D)** $3x^0 + (3x)^0 + x^{-1} + 2x$

   Substitute $x = \tfrac{1}{2}$

   $3(\tfrac{1}{2})^0 + (3 \cdot \tfrac{1}{2})^0 + (\tfrac{1}{2})^{-1} + 2(\tfrac{1}{2})$

   $3(1) + 1 + \dfrac{1}{\tfrac{1}{2}} + 1$

   $3 + 1 + 2 + 1 = 7$

# Practice Problems 6.1

1. **(B)** 278300.

   2.78300   $N = 2.783$

   2.78300   $a = 5$

   $a = +$

   $2.783 \times 10^5$

2. **(E)** .007932

   7.932   $N = 7.932$

   007.932   $a = 3$

   $a$ is "$-$"

   $7.932 \times 10^{-3}$

   $a = -3$

3. **(E)**

   $12000 = 1.2 \times 10^4$

   $.04 = 4 \times 10^{-2}$

   $\dfrac{12000}{.04} = \dfrac{1.2 \times 10^4}{4 \times 10^{-2}} = \dfrac{1.2}{4} \times 10^{4-(-2)}$

   $= .3 \times 10^6$

   *Note:* .3 is *not* between 1 and 10. This is not in the correct format.

   $.3 \times 10^6 = 3. \times 10^5$

   $= 300,000$

4. **(D)** $\dfrac{1}{1000} = .001 = 1 \times 10^{-3}$

# Practice Problems 6.2

1. **(D)** $\log_m 125 = 3 \Rightarrow m^3 = 125 = 5^3$

   $m = 5$

2. **(C)** $\log_{1/7} 49 = y \Rightarrow (\tfrac{1}{7})^y = 49$

   $(7^{-1})^y = 7^2$

   $7^{-y} = 7^2$

   $y = -2$

3. **(A)** $\ln \dfrac{a^2}{b^3} = \ln a^2 - \ln b^3$

   $= 2 \ln a - 3 \ln b$

4. **(C)** $\log \sqrt{21} = \dfrac{\log 21}{2} = \dfrac{\log 7 \cdot 3}{2}$

   $= \dfrac{\log 7 + \log 3}{2}$

   $= \dfrac{(.8451) + (.4771)}{2} = .6611$

5. **(D)** $2 \log a + \dfrac{\log b}{2}$

   $= \log a^2 + \log \sqrt{b}$

   $= \log a^2 \sqrt{b}$

6. **(B)** $\log .00917 = \log(9.17 \times 10^{-3})$

   $= \log 9.17 - 3$

   $= t - 3$

7. **(C)**

   $\log \sqrt{\dfrac{x^3}{zm}} = \dfrac{1}{2}[\log x^3 - \log zm]$

   $= \tfrac{1}{2}[3 \log x - (\log z + \log m)]$

   $= \tfrac{1}{2}[3 \log x - \log z - \log m]$

8. **(B)** $\log_8 x = \tfrac{1}{3} \Rightarrow 8^{1/3} = x$

   $2 = x$

# Practice Problems 7.1

**1. (C)**

$$\sqrt{\underline{1.\ 7\ 3}\atop 03.0000}$$

$$\begin{array}{r} 1 \\ \hline \end{array}$$

$$27\overline{\smash{\big)}\ 200}$$
$$\phantom{27}\ 189$$

$$343\overline{\smash{\big)}\ 1100}$$

answer 1.73 = 1.7 to the nearest tenth

| | 27 | 28 | 342 | 343 |
|---|---|---|---|---|
| *Notes:* | ×7 | ×8 | ×2 | ×3 |
| | 189 | 224 | 684 | 1029 |

 If the square root is required to a particular place value (i.e., tenths), work the calculation one place beyond. If the digit in this place is 0, 1, 2, 3 or 4, retain the required place value as is. If the digit in this place is 5, 6, 7, 8 or 9, round the required place up by 1.

1.7̲3 = 1.7 to the nearest tenth
   less than 5

1.7̲6 = 1.8 to the nearest tenth
   5 or higher

**2. (D)**

$$\sqrt{\underline{3.\ 4\ 6}\atop 12.0000}$$

$$\phantom{000}9$$

$$64\overline{\smash{\big)}\ 300}$$
$$\phantom{64}\ 256$$

$$686\overline{\smash{\big)}\ 4400}$$
$$\phantom{686}\ 4116$$
$$\phantom{686}\ \overline{\ 284}$$

3.46 = 3.5 to the nearest tenth

**3. (B)**

$$x = \sqrt{3}$$      from the log table

$$\log x = \log \sqrt{3} = \log(3^{\frac{1}{2}})$$    $\log 3 = .4771$

$$= \tfrac{1}{2} \log 3$$

$$= \tfrac{1}{2}(.4771)$$

$$\log x = .23855$$     from the log table

$$x \approx 1.73$$     antilog .2380 = 1.73

1.7̲3 = 2. to the nearest integer

**4. (D)**

$$x = \sqrt{12}$$

$$\log x = \log \sqrt{12} = \log (12^{1/2})$$    $\log 12 = \log(1.2 \times 10^1)$

$$= \tfrac{1}{2} \log 12$$           $= \log 1.2 + \log 10^1$

$$= \tfrac{1}{2}(1.0792)$$

$$= .5396$$      from the log table:

$$x = \text{antilog } .5396$$     $\log 1.2 = .0792$

$$x \approx 3.46$$     $\therefore \log 12 = \log 1.2 + \log 10^1$

$$x = 3.5 \text{ to the nearest tenth}$$     $= .0792 + 1$

                                            $= 1.0792$

# Practice Problems 8.1

**1. (B)** $\ln 16 = \ln 2^4 = 4 \ln 2$

$$= 4(.6931)$$

$$= 2.7724$$

**2. (C)** $\log_a b = c \Rightarrow a^c = b$

$$\ln 9.974 = 2.3$$

$$= \log_e 9.974 = 2.3 \Rightarrow e^{2.3} = 9.974$$

**3. (E)** The inverse of $y = e^{2x}$

$$\text{is } x = e^{2y}$$

$$x = e^{2y} \Rightarrow \ln x = 2y$$

$$y = \tfrac{1}{2} \ln x$$

$$= \ln x^{1/2}$$

**4. (B)** Let $y = 2 \ln e^x$

$$y = 2 \log_e e^x$$

recall $2 \log a = \log a^2$

$$\therefore y = \log_e e^{2x}$$

changing to exponential form

$$e^y = e^{2x}$$

$$y = 2x$$

# Cumulative Review Test

**1. (A)** $4^{3x} = 8^{-(x-3)}$

$$4^{3x} = 8^{-x+3}$$

$$(2^2)^{3x} = (2^3)^{3-x}$$

$$2^{6x} = 2^{9-3x}$$

$$6x = 9 - 3x$$

$$x = 1$$

Since the hundredths place is greater than 5, the solution is 7.3 to the nearest tenth.

*Note:* 1445

$$\underline{\times 5}$$

$$7225$$

**2. (D)** .03    Step 1

3.0    Step 2  $N = 3$

03.    Step 3  $a = 2$

$\underset{\smile}{\phantom{03}}$  Step 4  $a$ is "$-$"

$3.0 \times 10^{-2}$

**6. (D)** $b^0 = 1 \quad b^{-1/2} = 16^{-1/2} = \tfrac{1}{4}$

$$2b^0 + b^{-1/2} = 2(1) + \tfrac{1}{4} = \tfrac{9}{4}$$

**7. (D)** $\left(\dfrac{a}{b}\right)^n = \dfrac{a^n}{b^n}$

$$\therefore \dfrac{r^n}{s^n} = \left(\dfrac{r}{s}\right)^n \quad (s \neq 0)$$

**3. (B)** $\log(f^2 - g^2) = \log[(f + g)(f - g)]$

$$= \log(f + g) + \log(f - g)$$

**8. (C)** $\tfrac{1}{3} \log n - \tfrac{1}{2} \log l$

$$= \log \sqrt[3]{n} - \log \sqrt{l}$$

$$= \log \dfrac{\sqrt[3]{n}}{\sqrt{l}}$$

**4. (D)**

$$\frac{2}{\sqrt{7} - 2}\left(\frac{\sqrt{7} + 2}{\sqrt{7} + 2}\right) = \frac{2(\sqrt{(7)} + 2)}{7 - 4} = \frac{2(\sqrt{7} + 2)}{3}$$

$$= \frac{2\sqrt{7} + 4}{3}$$

**9. (A)** $h(x) = \dfrac{x - 1}{x + 3}$

$$h(x - 3) = \frac{(x - 3) - 1}{(x - 3) + 3} = \frac{x - 4}{x}$$

**5. (C)**

$$\begin{array}{r} 7.\ 2\ \mathbf{5} \\ \sqrt{53.0000} \\ 49 \phantom{.0000} \\ \hline 142\ \overline{)\ 400} \\ 284 \\ \hline 1445\ \overline{)\ 11600} \end{array}$$

**10. (E)** $.3 \times 10^{-4} = .00003$

$$= 3 \times 10^{-5}$$

**11. (D)** $\underbrace{-2(x^3)} + (-2x)^3$

$$= -2x^3 + (-8x^3)$$

$$= -10x^3$$

12. **(B)** $\log_8 x = \frac{2}{3} \Rightarrow 8^{2/3} = x$

$$x = (\sqrt[3]{8})^2$$
$$= 4$$

13. **(D)** $9^{3x} = 3^{x+5}$

$$(3^2)^{3x} = 3^{x+5}$$
$$3^{6x} = 3^{x+5}$$
$$6x = x + 5$$
$$x = 1$$

14. **(D)** $\dfrac{3^3 + 3^3 + 3^3}{3^4} = \dfrac{3(3^3)}{3^4} = \dfrac{3^4}{3^4} = 1$

15. **(D)** $\dfrac{3.2}{.32} = 10 \qquad \dfrac{10^2}{10^{-2}} = 10^{2-(-2)} = 10^4$

$$\frac{3.2 \times 10^2}{.32 \times 10^{-2}} = 10 \times 10^4 = 1 \times 10^5$$

16. **(C)** $\log 4.1 = .6128$

$$\log 41 = 1.6128$$
$$x = 41^{1/2} \quad \log x = \tfrac{1}{2} \log 41$$
$$= \frac{1.6128}{2} = .8064$$
$$\log x = .8064$$
$$x = 6.4$$

17. **(A)** $\left( \dfrac{x^{1/4} y^{1/2}}{z^{-1}} \right)^4 = \dfrac{x^1 y^2}{z^{-4}} = xy^2 z^4$

18. **(C)** $|-6| = 6 \quad |4| = 4 \quad |-2| = 2 \quad |2| = 2$

$$\frac{|-6|}{6} + \frac{|4|}{-4} + 2|-2| + (-2)|2|$$

$$= \frac{6}{6} + \frac{4}{-4} + 2(2) + (-2)(2)$$

$$= 1 + (-1) + 4 + (-4) = 0$$

19. **(E)**

$$1 + \left( \frac{e^x - e^{-x}}{2} \right)^2 = 1 + \frac{e^{2x} - 2e^0 + e^{-2x}}{4}$$

$$= 1 + \frac{e^{2x} - 2 + e^{-2x}}{4}$$

$$= \frac{4 + e^{2x} - 2 + e^{-2x}}{4}$$

$$= \frac{e^{2x} + 2 + e^{-2x}}{4}$$

$$= \cosh^2 x$$

20. **(C)** $.35\overline{3535}$ is a *repeating* decimal. This is a rational number which can be expressed as a proper fraction.

# UNIT THREE

UNIT THREE

# 9. Imaginary and Complex Numbers

## 9.1 Fundamental Facts

$i$     The basic imaginary unit. It does not represent a variable. It equals $\sqrt{-1}$.

$a + bi$ or $x + yi$     The rectangular form of a complex number. The $a$ is the real part, the $bi$ is the imaginary part, where $a$ and $b$ are real numbers.

The successive non-negative integer powers of $i$ follow a repeated sequence of

$$1, i, -1, -i$$

☞
$$i^0 = 1$$
$$i^1 = i$$
$$i^2 = -1$$
$$i^3 = -i$$

$i^n$     The value of $i^n$, where $n$ is a non-negative integer $= i^r$, where $r$ is the remainder when $n$ is divided by 4. For example:

$$i^{27} = i^3 = -i \qquad \text{because } \tfrac{27}{4} = 6r3.$$

☞
- All integer powers of $i$ can and should be simplified to one of the four iterative values.
- Square roots of negative numbers should be put in $a + bi$ form prior to calculations.
- An expression with $i$ in the denominator should be rationalized by using its conjugate.
- Every complex number can and should be simplified to $a + bi$ form.

| VALUES OF $i$ | | | |
|---|---|---|---|
| $i^0 = 1$ | $i^4 = 1$ | $i^8 = 1$ | $i^n = i^r$ |
| $i^1 = i$ | $i^5 = i$ | $i^9 = i$ | |
| $i^2 = -1$ | $i^6 = -1$ | $\vdots$ | $n$ is a non-negative integer. |
| $i^3 = -i$ | $i^7 = -i$ | $\vdots$ | $r$ is the remainder when $n$ is divided by 4. |

75

| COMPLEX NUMBERS ( $=$ , $+$ , $-$ , $\cdot$ , $\div$ ) | | |
|---|---|---|
| **Equality** | $a + bi = c + di$; then $a = c$ and $b = d$ | Real terms are only equal to reals. Imaginary terms are only equal to imaginaries. |
| **Sum** | $(a + bi) + (c + di) = (a + c) + (b + d)i$ | Reals add to reals. Imaginaries add to imaginaries. |
| **Difference** | $(a + bi) - (c + di) = (a - c) + (b - d)i$ | Reals subtract from reals. Imaginaries subtract from imaginaries |
| **Product** | $(a + bi)(c + di) = (ac - bd) + (ad + bc)i$ | Use FOIL. Do regular binomial multiplication. Recall: $i^2 = -1$ |
| **Conjugate Product** | $(a + bi)(a - bi) = a^2 + b^2$ | Middle terms drop out. $i^2 = -1$ |
| **Quotient** | $\dfrac{(a + bi)}{(c + di)} = \dfrac{(a + bi)}{(c + di)} \dfrac{(c - di)}{(c - di)} = \dfrac{(ac + bd) + (bc - ad)i}{c^2 + d^2}$ | Rationalize the denominator |

**Examples:**

1. Add $\sqrt{-2} + \sqrt{-8}$.

2. Multiply $\sqrt{-2} \cdot \sqrt{-8}$.

3. Add $(3 + 2i) + (-2 - 6i)$.

4. Multiply $(2 + 3i)(2 - 3i)$.

**Solutions:**

1. 
$$\sqrt{-2} = \sqrt{2}\sqrt{-1} = \sqrt{2}i$$
$$\sqrt{-8} = \sqrt{8}\sqrt{-1} = 2\sqrt{2}i$$
$$\sqrt{-2} + \sqrt{-8} = \sqrt{2}i + 2\sqrt{2}i = 3\sqrt{2}i$$

2. 
$$\sqrt{-2} = \sqrt{2}i$$
$$\sqrt{-2}\sqrt{-8} = (\sqrt{2}i)(2\sqrt{2}i)$$
$$= 2(2)i^2$$
$$= 4i^2 = 4(-1) = -4$$

 $\sqrt{-2} \cdot \sqrt{-8} \neq \sqrt{16} = 4$
Use $i$ notation when working the square root of negative numbers.
$$(\sqrt{2}i)(\sqrt{8}i) = \sqrt{16}i^2 = 4(-1) = -4$$

3. $(3 + 2i) + (-2 - 6i) = (3 - 2) + (2i - 6i)$
$$= (3 - 2) + (2 - 6)i$$
$$= 1 - 4i$$

4. $(2 + 3i)(2 - 3i) = 4 - 6i + 6i - 9i^2$
$$= 4 - 9i^2 \quad i^2 = -1$$
$$= 4 - 9(-1)$$
$$= 13$$

5. Simplify $\dfrac{2+i}{1-3i}$.

5. $\dfrac{(2+i)}{(1+3i)}\dfrac{(1+3i)}{(1+3i)} = \dfrac{2+6i+i+3i^2}{1+3i-3i-9i^2} = \dfrac{2+7i+3i^2}{1-9i^2}$

$\qquad\qquad = \dfrac{2+7i+3(-1)}{1-9(-1)} = \dfrac{-1+7i}{10}$

6. Find the reciprocal of $i$.

6. $\dfrac{1}{i}\cdot\dfrac{i}{i} = \dfrac{i}{i^2} = \dfrac{i}{-1} = -i$

7. Subtract $3 + 2i$ from $2 - i$.

7. $(2-i)-(3+2i) = 2-i-3-2i$

$\qquad\qquad\qquad = -1-3i$

☞ The $\sqrt{-8}$ can be written as $2i\sqrt{2}$ or $2\sqrt{2}i$. The former, $2i\sqrt{2}$, is used to make clear that the $i$ is not under the radical sign. The latter, $2\sqrt{2}i$, is used because of its association to $a + bi$ form. Both notations are used depending on the author.

## Practice Problems 9.1

1. Find the reciprocal of $\dfrac{4+3i}{25}$.

A) $4 - 3i$   B) $3 - 4i$   C) $\dfrac{25(4-3i)}{7}$

D) $\dfrac{4-3i}{25}$   E) $4 + 25i$

2. Find the simplified value of $i^{25}$.

A) 0   B) 1   C) $i$
D) $-1$   E) $-i$

3. Add $(3 + 2i) + (3 + 6i) + (-4 - 4i)$.

A) 2   B) $4i$   C) $8i$   D) $2 + 4i$
E) $4 + 2i$

4. Express $\sqrt{-3} + 3\sqrt{-27} - \sqrt{-12}$ in $a + bi$ form.

A) $8\sqrt{3} + i$   B) $0 + 8\sqrt{3}i$   C) $8\sqrt{3} + 0i$
D) $1 + 8\sqrt{3}i$   E) $8\sqrt{3} + 1i$

5. Simplify $\dfrac{3+2i}{i}$.

A) $2 - 3i$   B) $2 + 3i$   C) $3 - 2i$
D) $3 + 2i$   E) 5

6. Multiply $(3 + \sqrt{2}i)(3 - \sqrt{2}i)$.

A) $11 - 6\sqrt{2}i$   B) $11 + 6\sqrt{2}i$   C) 9
D) $9 - 6\sqrt{2}i$   E) 11

7. If $a + 3i = 7 - bi$ then $b =$

A) 7   B) $-3$   C) $-7$
D) 3   E) $-4$

8. $(3 + 7i)(2 - i) =$

A) $13 + 11i$   B) $-1 + 11i$   C) $13 + 9i$
D) $-1 + 9i$   E) $3i$

Solutions on page 127

## 9.2 Graphing of Complex Numbers

$a + bi$ can be represented as an ordered pair $(a, b)$ on rectangular coordinates whose abscissa is the real axis and whose ordinate is the imaginary axis.

To add two complex numbers graphically, find the ordered pair associated with the diagonal of the parallelogram obtained using each complex number as one of the sides. For example, add $(2 + 3i) + (1 - i)$ graphically.

To subtract two complex numbers graphically, find the ordered pair associated with the diagonal of the parallelogram obtained using the first complex number and the negative of the complex number being subtracted. For example:

$$(2 + 3i) - (1 - i) \Rightarrow (2 + 3i) + (-1 + i)$$

# Practice Problems 9.2

1. Add $(-2 + 3i) + (3 - 2i)$ graphically.

2. Subtract $(-2 + 3i)$ from $(6 + 2i)$ graphically.

Solutions on page 128

# 10. Geometry

## 10.1 Angles

| Angle | Acute | Right | Obtuse | Straight | Reflex |
|-------|-------|-------|--------|----------|--------|
| $\angle$ is the symbol for angle. | $\angle NJL$ is acute. $0 < x < 90$ | $\angle NJL$ is a right angle. $x = 90$ ⌐ is the symbol for a right angle. | $\angle NJL$ is obtuse. $90 < x < 180$ | $\angle NJL$ is a straight angle. $x = 180$ | $\angle NJL$ is a reflex angle. $180 < x < 360$ |

  Within this text:
$m\angle A$ indicates the measure of angle $A$ as measured in degrees. Hence it is not necessary to use the word 'degrees' or the symbol (°) when referring to the measure of an angle.

**Complementary Angles**

Two angles whose sum is 90°
$m\angle a + m\angle b = 90$

**Supplementary Angles**

Two angles whose sum is 180°
$m\angle a + m\angle b = 180$

**Adjacent Angles**

Angles that share a common side
$\angle a$ is adjacent to $\angle b$.

**Vertical Angles**

The opposite angles formed when two lines intersect
Vertical angles are equal
$m\angle a = m\angle c$
$m\angle b = m\angle d$

| Perpendicular Lines | Parallel Lines |
|---|---|
| <br>Lines which intersect at right ∠s.<br>The symbol ⊥ indicates perpendicular.<br>$n \perp l$ | <br>Lines which do not meet.<br>The symbol ∥ indicates parallel.<br>$n \parallel l$ |
| **Transversal** | **Concurrent** |
| <br>A line which intersects parallel lines.<br>Line $p$ is a transversal. | Lines are concurrent if they<br>share a common intersection.<br>Line $j$, $s$ and $l$ are concurrent. |

# Parallel Lines

interior: the space between the parallel lines.

Transversal

When a transversal cuts two parallel lines a series of relationships exists among the eight angles formed.

## Adjacent Angles are Supplementary

$m\angle 1 + m\angle 2 = 180$

$m\angle 2 + m\angle 3 = 180$

$m\angle 3 + m\angle 4 = 180$

$m\angle 4 + m\angle 1 = 180$

$m\angle 5 + m\angle 6 = 180$

$m\angle 6 + m\angle 7 = 180$

$m\angle 7 + m\angle 8 = 180$

$m\angle 8 + m\angle 5 = 180$

## Vertical Angles are Equal

m∠1 = m∠3    m∠2 = m∠4    m∠5 = m∠7    m∠6 = m∠8

### Angle Locations

AB ∥ CD

## Corresponding Angles are Equal

m∠1 = m∠5    m∠2 = m∠6    m∠3 = m∠7    m∠4 = m∠8

## Alternate Interior Angles are Equal

m∠3 = m∠5    m∠4 = m∠6

## Alternate Exterior Angles are Equal

m∠2 = m∠8    m∠1 = m∠7

## Interior Angles on the Same Side of the Transversal are Supplementary

m∠3 + m∠6 = 180    m∠4 + m∠5 = 180

## Exterior Angles on the Same Side of the Transversal are Supplementary

m∠1 + m∠8 = 180    m∠2 + m∠7 = 180

**Examples:**

1. If $AB \parallel CD$ and $m\angle 1 = 35$,

    A) find $m\angle 5$
    B) find $m\angle 8$
    C) find $m\angle 6$

**Solutions:**

1. A) $\angle 1$ and $\angle 5$ are corresponding angles.

$$m\angle 1 = m\angle 5 = 35$$

    B) $\angle 8 + \angle 1$ are exterior angles on the same side of the transversal. They are supplementary.

$$m\angle 8 + m\angle 1 = 180$$
$$m\angle 8 + 35 = 180$$
$$m\angle 8 = 145$$

    Note: $\angle 8$ is also supplementary to $\angle 5$.

    C) $\angle 6$ is a vertical angle with $\angle 8$.

$$m\angle 6 = m\angle 8 = 145$$

2. If $AB$ is parallel to $CD$ and $m\angle 1 = 3x + 1$ and $m\angle 2 = 2x - 6$, find the value of $x$.

2. $\angle 1$ and $\angle 2$ are interior angles on the same side of the transversal.

$$m\angle 1 + m\angle 2 = 180$$
$$(3x + 1) + (2x - 6) = 180$$
$$5x - 5 = 180$$
$$5x = 185$$
$$x = 37$$

3. The measures of two supplementary angles are represented by $7x - 27$ and $2x$. Find the measure of the smaller angle.

3.

$$(7x - 27) + (2x) = 180$$
$$9x - 27 = 180$$
$$9x = 207$$
$$x = 23$$
$$2x = 2(23) = 46$$
$$7x - 27 = 7(23) - 27 = 134$$

4. If $AB \parallel CD$, $m\angle A = 35$ and $m\angle C = 45$, find the $m\angle AEC$.

4.

Construct a line parallel to $AB$ and $CD$ through $E$. The $m\angle x = m\angle A = 35$, alternate interior angles. The $m\angle y = m\angle C = 45$, alternate interior angles.

$$m\angle AEC = m\angle x + m\angle y = 35 + 45 = 80$$

5. If $\angle a$ and $\angle b$ are in the ratio of $2:3$, find the measure of angle $a$.

5. $m\angle a + m\angle b = 180$   Let $m\angle a = 2x$

$\qquad\quad 2x + 3x = 180 \qquad\quad m\angle b = 3x$

$\qquad\qquad\quad 5x = 180$

$\qquad\qquad\quad\; x = 36$

$\quad m\angle a = 2x = 2(36) = 72$

6. Find the value of $x$.

6. $2x = 3x - 12 \qquad$ Vertical angles

$\quad x = 12$

# Practice Problems 10.1

1. If $AB \parallel CD$, find $m\angle x$ when $m\angle y = 20$.

   A) 20  B) 25  C) 70
   D) 110  E) 160

2. Two complementary angles have measures, in degrees, represented by $x + 20$ and $2x + 1$. Find $x$.

   A) 21  B) 23  C) 37
   D) 53  E) 113

3. If parallel lines $l$ and $m$ are cut by a transversal and angles $a$ and $b$ are in the ratio of $4:5$, find the measure of $\angle b$.

   A) 10  B) 20  C) 50
   D) 80  E) 100

4. In the diagram, $AB \perp BC$. If $XBY$ is a straight line and $m\angle XBC = 37$, find $m\angle ABY$.

   A) 37  B) 53  C) 63  D) 127
   E) 143

Solutions on page 128

## 10.2 Triangles

A triangle is a closed three sided figure. Triangles are classified by sides and/or angles.

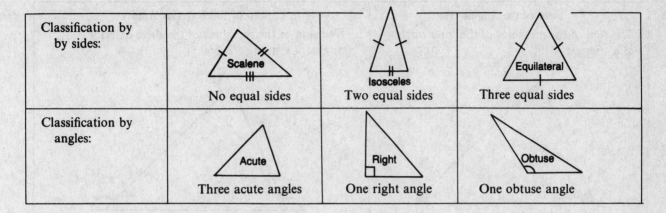

| Classification by by sides: | Scalene No equal sides | Isosceles Two equal sides | Equilateral Three equal sides |
| Classification by angles: | Acute Three acute angles | Right One right angle | Obtuse One obtuse angle |

# Parts of a Triangle

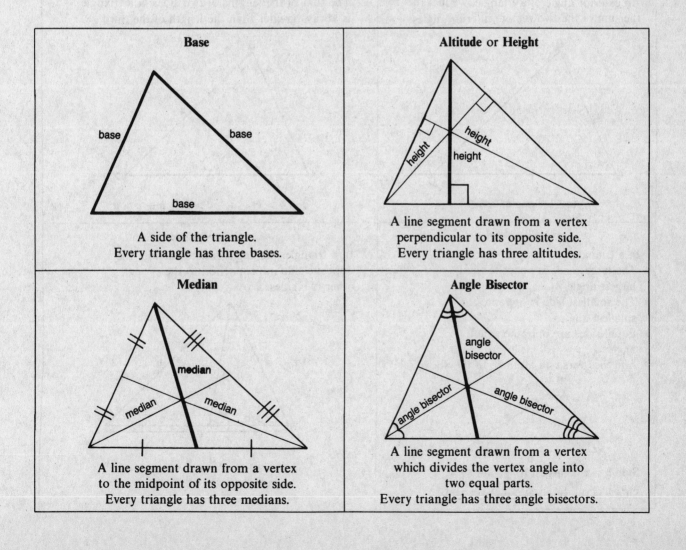

**Base**

A side of the triangle.
Every triangle has three bases.

**Altitude or Height**

A line segment drawn from a vertex
perpendicular to its opposite side.
Every triangle has three altitudes.

**Median**

A line segment drawn from a vertex
to the midpoint of its opposite side.
Every triangle has three medians.

**Angle Bisector**

A line segment drawn from a vertex
which divides the vertex angle into
two equal parts.
Every triangle has three angle bisectors.

# General Properties

## Sum of the Angles

The sum of the measures of the three angles of a triangle is 180.

$$m\angle N + m\angle J + m\angle L = 180$$

## Sum of the Exterior Angles

The sum of the measures of the three exterior angles of a triangle is 360.

$$m\angle r + m\angle s + m\angle t = 360$$

## Exterior Angle

An exterior angle of a triangle is equal to the sum of the two remote interior angles.

$$m\angle l = m\angle a + m\angle b$$

## Lengths of Sides

The sum of the lengths of two sides of a triangle is always greater than the length of the third side.

$$a + b > c, \quad a + c > b, \quad b + c > a$$

## Sides vs Angles

In a triangle:
- The largest side is opposite the largest angle.
- The smallest side is opposite the smallest angle.
- Equal sides are opposite equal angles.
- And vice versa on all three of the above.

Side *b* is the largest (opposite 87°)

Side *c* is the smallest (opposite 33°)

## Angle Bisector

In a triangle, the angle bisector divides the opposite side in proportion to the other two sides.

$$\frac{d}{e} = \frac{a}{b}$$

## Equilateral Triangle:

Has **three equal sides** and **three equal angles**.
Each angle is 60°.

 An easy way to calculate the area of an equilateral triangle is:

$$A = \frac{(\text{side})^2}{4}\sqrt{3}$$

## Right Triangle:

**Pythagorean Theorem**

$$(\text{leg 1})^2 + (\text{leg 2})^2 = (\text{hypotenuse})^2$$

This is the **Pythagorean theorem** and is a property of every right triangle.

A Pythagorean triple is a set of **integer values** which represent the sides of a right triangle, e.g., $7^2 + 24^2 = 25^2$.

Any multiple or division of a triple which results in integral values is also a triple, e.g., multiplying the 3, 4, 5 triple by 2 $\Rightarrow$ 6, 8, 10:

$$6^2 + 8^2 = 10^2$$

| Common Pythagorean Triples | | |
|---|---|---|
| leg | leg | hypotenuse |
| 3 | 4 | 5 |
| 5 | 12 | 13 |
| 7 | 24 | 25 |
| 8 | 15 | 17 |
| 9 | 40 | 41 |
| 11 | 60 | 61 |

examples: $8^2 + 15^2 = 17^2$
$11^2 + 60^2 = 61^2$

# Special Right Triangles

| **45° − 45° − 90°** | **30° − 60° − 90°** |
|---|---|
| **General form:**  Each leg is equal. The hypotenuse is leg times $\sqrt{2}$. | **General form:**  The leg opposite the 30° angle is half the hypotenuse. The leg opposite the 60° angle is half the hypotenuse times $\sqrt{3}$. |
| **Numerical master:**  | **Numerical master:**  |

# Similar Triangles

Two triangles are similar ($\sim$) if two angles of one triangle are equal to two angles of a second triangle.

☞ **Similar**    ($\sim$) means the same shape, not necessarily the same size
$$\triangle \sim \triangle \quad \square \sim \square \quad \odot \sim \odot$$

**Equal**    ($=$) means the same size (i.e., area), not necessarily the same shape

$A = 8 \quad A = 8$

**Congruent**    ($\cong$) means the same size and shape (identical)

---

**If two figures are similar, their corresponding parts are proportional.**

---

**Line Ratio** $= \dfrac{a}{x} = \dfrac{b}{y} = \dfrac{c}{z} = \dfrac{\text{any line in } \triangle I}{\text{any corresponding line in } \triangle II}$

$\qquad\qquad = \dfrac{\text{perimeter } \triangle I}{\text{perimeter } \triangle II}$

☞ Perimeter is made of lines and follows the line ratio.

$$\text{Area Ratio} = (\text{line ratio})^2 = \left(\frac{a}{x}\right)^2 = \left(\frac{b}{y}\right)^2 = \left(\frac{c}{z}\right)^2 \cdots$$

☞ Area ratio is the square of the line ratio, area is *not* made of lines.

 All parts of the diagram are named after the large right triangle. All three right triangles are similar.

| Rule 1: | $(\text{altitude})^2 = (\text{piece 1})(\text{piece 2})$ |
|---|---|
| Rule 2: | $(\text{leg})^2 = (\text{whole hypotenuse})(\text{piece next to leg})$<br>$(\text{leg 1})^2 = (\text{hypotenuse})(\text{piece 1})$<br>$(\text{leg 2})^2 = (\text{hypotenuse})(\text{piece 2})$ |
| Rule 3: | **Pythagorean Theorem:**<br>$(\text{leg 1})^2 + (\text{leg 2})^2 = (\text{hypotenuse})^2$<br>$(\text{piece 1})^2 + (\text{altitude})^2 = (\text{leg 1})^2$<br>$(\text{piece 2})^2 + (\text{altitude})^2 = (\text{leg 2})^2$ |

**Examples:**

1. The legs of a right triangle are 5 and 8. Find the length of the hypotenuse.

**Solutions:**

1. By the Pythagorean theorem

$$(\text{leg})^2 + (\text{leg})^2 = (\text{hyp.})^2$$
$$5^2 + 8^2 = x^2$$
$$25 + 64 = x^2$$
$$89 = x^2$$
$$x = \sqrt{89}$$

2. If a rectangle of width 10 has a diagonal of length 26, find the length of the rectangle.

2. $(\text{leg})^2 + (\text{leg})^2 = (\text{hyp.})^2$
$$10^2 + x^2 = 26^2$$
$$100 + x^2 = 676$$
$$x^2 = 576$$
$$x = \sqrt{576} = 24$$

3. Given the two similar triangles as shown below,
   a) find the perimeter of $\triangle I$, if the perimeter of $\triangle II = 12$.
   b) find the area of $\triangle I$ if the area $\triangle II = 12$.

3.

a) The line ratio = $\dfrac{\text{side } \triangle I}{\text{Corresponding side } \triangle II} = \dfrac{1}{2}$

$$\frac{1}{2} = \frac{\text{perim. } \triangle I}{\text{perim. } \triangle II} = \frac{x}{12}$$

$$2x = 12$$

$$x = 6$$

b) The area ratio = (line ratio)$^2$ = $\left(\dfrac{1}{2}\right)^2$

$$\frac{\text{area } \triangle I}{\text{area } \triangle II} = \left(\frac{1}{2}\right)^2 = \frac{1}{4}$$

$$\frac{x}{12} = \frac{1}{4}$$

$$4x = 12$$

$$x = 3$$

4. In $\triangle ABC$, $NL \parallel AC$, find the length of $AC$.

4. $\triangle NBL \sim \triangle ABC$

$$\frac{2}{5} = \frac{4}{x}$$

$$2x = 20$$

$$x = 10$$

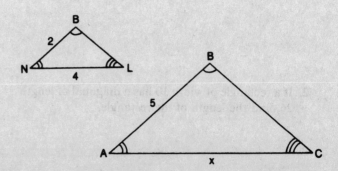

5. The measure of the base angles of an isosceles triangle are $3x - 1$ and $2x + 16$. Find the measure of the vertex angle.

5.

The measure of the base angles of an isosceles triangle are equal.

$$3x - 1 = 2x + 16$$
$$x = 17$$

The measure of the base angles are:

$$3x - 1 = 3(17) - 1 = 50$$
$$2x + 16 = 2(17) + 16 = 50$$

The vertex angle is the non-equal angle. Since the measures of the three angles must add up to 180 and the two base angles total 100, the measure of the vertex angle $= 180 - 100 = 80$.

6. Three angles of a triangle are represented by $3x - 1$, $10x + 11$ and $x + 2$. Find the measure of the smallest angle.

6.

The sum of the measures of the angles is 180.

$$(3x - 1) + (10x + 11) + (x + 2) = 180$$
$$14x + 12 = 180$$
$$14x = 168$$
$$x = 12$$

The measures of the three angles are:

$$3x - 1 = 3(12) - 1 = 35$$
$$10x + 11 = 10(12) + 11 = 131$$
$$x + 2 = 12 + 2 = 14$$

31

**Ans.: 14**

7. An isosceles triangle has sides 13, 13, and 10. Find the area.

7. The altitude is also the median. By the 5-12-13 triple $x = 12$.

Area $= \frac{1}{2}bh = \frac{1}{2}(12)(10) = 60$

Alternative Solution:

Semi-perimeter $= \dfrac{13 + 13 + 10}{2} = 18$

$$A = \sqrt{s(s - a)(s - b)(s - c)}$$
$$= \sqrt{18(18 - 13)(18 - 13)(18 - 10)}$$
$$= \sqrt{18(5)(5)(8)}$$
$$= 5\sqrt{(18)(8)} = 5\sqrt{9 \cdot 2 \cdot 2 \cdot 4} = 5 \cdot 3 \cdot 2 \cdot 2$$
$$= 60$$

8. In $\triangle ABC$, $BD \perp AC$. If $AC = 13$ and $AD = 4$, find the length of $BD$.

8.

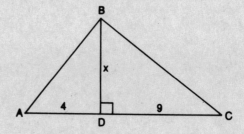

$(\text{altitude})^2 = (\text{piece 1})(\text{piece 2})$
$$x^2 = (4)(9) = 36$$
$$x = 6$$

9. Find the area of an equilateral triangle whose side is 10.

9. $A = \dfrac{s^2}{4}\sqrt{3} = \dfrac{10^2}{4}\sqrt{3}$

$= 25\sqrt{3}$

10. In the given △*NJL*, find the lengths of *j* and *l*.

10. Using the numerical master of the 45°-45°-90°

$$\text{line ratio} = \frac{\text{master}}{\text{problem}} = \frac{1}{17} = \frac{1}{l}$$

$$l = 17$$

$$\frac{\text{master}}{\text{problem}} = \frac{1}{17} = \frac{\sqrt{2}}{j}$$

$$j = 17\sqrt{2}$$

11. The angles of a triangle are in the ratio of 2:3:4. What is the classification of the triangle (by angles and by sides)?

11. $2x + 3x + 4x = 180$

$$9x = 180$$
$$x = 20$$
$$2x = 2(20) = 40$$
$$3x = 3(20) = 60$$
$$4(x) = 4(20) = 80$$

The triangle: is acute (all three angles are acute)

is scalene (no equal angles implies no equal sides)

12. In △*ABC*, shown, find the length of side *a*.

12. Using the numerical master:

$$\text{line ratio} = \frac{\text{master}}{\text{problem}} = \frac{\sqrt{3}}{15} = \frac{1}{a}$$

$$a\sqrt{3} = 15$$

$$a = \frac{15}{\sqrt{3}}\left(\frac{\sqrt{3}}{\sqrt{3}}\right) = \frac{15\sqrt{3}}{3} = 5\sqrt{3}$$

13. In the figure, $XZ$ is the angle bisector of $< X$. If $WX$ is 12, $XY$ is 10, and $WZ = 6$, find $ZY$.

13. The angle bisector divides the opposite sides in proportion to the legs.

$$\frac{6}{a} = \frac{12}{10}$$

$$12a = 60$$

$$a = 5$$

14. Find $x$, the length of $AD$.

14. In $\triangle ABC$

$$(\text{leg})^2 + (\text{leg})^2 = (\text{hyp.})^2$$
$$1^2 + 1^2 = (AC)^2$$
$$2 = (AC)^2$$
$$AC = \sqrt{2}$$

In $\triangle ACD$

$$(\text{leg})^2 + (\text{leg})^2 = (\text{hyp.})^2$$
$$(\sqrt{2})^2 + 1^2 = x^2$$
$$2 + 1 = x^2$$
$$3 = x^2$$
$$x = \sqrt{3}$$

# Practice Problems 10.2

1. The measures of the angles of a triangle are in the ratio $1:5:6$. This triangle is

   A) acute   B) right   C) obtuse
   D) equilateral   E) isosceles

2. The measure of the vertex angle of an isosceles triangle is 50. Find the number of degrees in each base angle.

   A) 50   B) 60   C) 65
   D) 55   E) 130

3. In △*ABC*, the measure of angle *A* is 6 more than 3 times the measure of angle *B*. The measure of the exterior angle at *C* is 106. Find the degree measure of angle *A*.

A) 25  B) 30  C) 74
D) 81  E) 106

4. In △*ABC*, *AB* = *BC*. If $m\angle A = 4x - 30$ and $m\angle C = 2x + 10$, find the number of degrees in ∠*B*.

A) 20  B) 40  C) 50
D) 80  E) 100

5. Find the altitude of an equilateral triangle whose side is 20.

A) 10  B) $20\sqrt{3}$  C) $10\sqrt{3}$
D) $20\sqrt{2}$  E) $10\sqrt{2}$

6. In △*RST*, *UV* ∥ *RS*, find *x*.

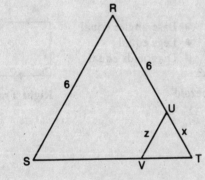

A) 2  B) 3  C) 4  D) 8  E) 9

7. In the figure, what is the value of *x*?

A) 8  B) 10  C) 12  D) 15  E) 20

8. Which of the following sets of numbers can represent the sides of a triangle?

A) {7, 8, 9}  B) {3, 5, 8}  C) {3, 3, 7}
D) {3, 10, 6}  E) {4, 5, 10}

Solutions on page 129

## 10.3 Quadrilaterals

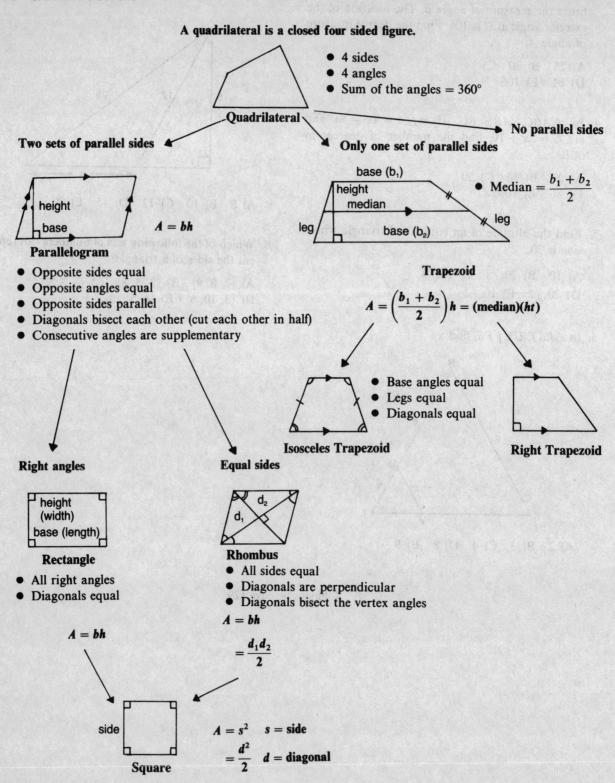

**A quadrilateral is a closed four sided figure.**

- 4 sides
- 4 angles
- Sum of the angles = 360°

Quadrilateral

**Two sets of parallel sides**

**Only one set of parallel sides**

**No parallel sides**

$A = bh$

**Parallelogram**
- Opposite sides equal
- Opposite angles equal
- Opposite sides parallel
- Diagonals bisect each other (cut each other in half)
- Consecutive angles are supplementary

base ($b_1$)
height
median
leg
base ($b_2$)
leg

- Median $= \dfrac{b_1 + b_2}{2}$

**Trapezoid**

$A = \left(\dfrac{b_1 + b_2}{2}\right) h = \text{(median)}(ht)$

- Base angles equal
- Legs equal
- Diagonals equal

**Isosceles Trapezoid**

**Right Trapezoid**

**Right angles**

**Equal sides**

height (width)
base (length)

**Rectangle**
- All right angles
- Diagonals equal

$A = bh$

$d_2$
$d_1$

**Rhombus**
- All sides equal
- Diagonals are perpendicular
- Diagonals bisect the vertex angles

$A = bh$

$= \dfrac{d_1 d_2}{2}$

side

$A = s^2 \quad s = \text{side}$

$= \dfrac{d^2}{2} \quad d = \text{diagonal}$

**Square**

Each figure in this tree has all the properties of the figures above it. For example, a rhombus has diagonals which bisect each other because its a parallelogram, and has 360° since it is a quadrilateral

▱ or Ⓟ is the symbol for a parallelogram.
▱ABCD means label the vertices of the parallelogram *consecutively* A, B, C, D

**Examples:**

1. In ▱ABCD, if $m\angle A = 4x - 10$ and $m\angle C = 100 - 6x$, find x.

**Solutions:**

1. Opposite angles are equal.

$$4x - 10 = 100 - 6x$$
$$10x = 110$$
$$x = 11$$

2. In quadrilateral ABCD the angles are in the ratio of 1:3:5:6. Find the measure of each angle.

2. The sum of the angles of a quadrilateral is 360°.

$$x + 3x + 5x + 6x = 360$$
$$15x = 360$$
$$x = 24$$

$$m\angle A = x = 24$$
$$m\angle B = 3x = 72$$
$$m\angle C = 5x = 120$$
$$m\angle D = 6x = 144$$

3. The diagonals of a rhombus are 6 and 10. Find a side of the rhombus.

3. ● Diagonals bisect each other (because it is a parallelogram).
   ● Diagonals are perpendicular.

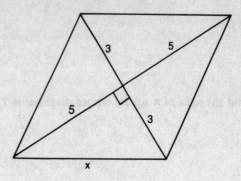

$$3^2 + 5^2 = x^2$$
$$9 + 25 = x^2$$
$$x^2 = 34$$
$$x = \sqrt{34}, \text{ all four sides are equal}$$

**4.** In rectangle $WXYZ$, $WY = 3x - 1$ and $XZ = 5x - 13$. Find the length of the diagonals.

**4.** The diagonals of a rectangle are equal.

$$3x - 1 = 5x - 13$$
$$12 = 2x$$
$$6 = x$$

Each diag $= 5(6) - 13 = 17$
$$= 3(6) - 1 = 17$$

**5.**

Find the area of the trapezoid $ABCD$ if $BC \parallel AD$

**5.**

$$\text{line ratio} = \frac{\text{master}}{\text{problem}} = \frac{2}{6} = \frac{1}{h}$$
$$2h = 6$$
$$h = 3$$
$$A = \left(\frac{b_1 + b_2}{2}\right) h$$
$$= \left(\frac{8 + 16}{2}\right)(3) = 12(3) = 36$$

**6.** Find the area of a square whose diagonal is 7.

**6.**

$$A = \frac{d^2}{2} = \frac{7^2}{2} = \frac{49}{2}$$

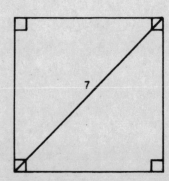

7. Find the area of a rhombus whose diagonals are 10 and 12.

7. $A_{rhombus} = \dfrac{d_1 d_2}{2} = \dfrac{10(12)}{2} = 60$

8. If two consecutive sides of a square are $7x + 4$ and $13x - 18$, find the area of the square.

8.

Sides of a square are equal.

$$7x + 4 = 13x - 8$$
$$12 = 6x$$
$$2 = x$$

∴ Each side $= 7(2) + 4 = 18$

or $13(2) - 8 = 18$

Area $= s^2 = 18^2 = 324$

# Practice Problems 10.3

1.

In □$RSTU$, $ST = 10$, $RS = 4$ and $m\angle R = 60$. Find the area of the parallelogram.

A) 20  B) $20\sqrt{3}$  C) 40
D) $40\sqrt{3}$  E) 60

2. The diagonals of a rhombus are 2 and 4. Find the perimeter of the rhombus.

A) 4  B) $\sqrt{5}$  C) $4\sqrt{5}$
D) 8  E) 16

3. If the consecutive angles of a parallelogram are represented by $4t + 6$ and $5t - 15$, then the parallelogram could be a

I) rectangle  II) rhombus  III) square
IV) trapezoid

A) I only  B) II only  C) I, II, III only
D) II and III only
E) I and III only

4. In □$ABCD$, $m\angle 1 = 2x + 12$ and the $m\angle 2 = 7x - 8$. Find $x$.

A) 4  B) 8  C) 9
D) 20  E) 28

Solutions on page 130

## 10.4 Polygons

A **convex polygon** is a closed figure with straight sides and

- Each side intersects exactly two other sides, one at each end point.
- None of the vertices lies in the interior of the polygon.

| No. of Sides | 3 | 4 | 5 | 6 |
|---|---|---|---|---|
| Graphic | | | | |
| Name | **Triangle** | **Quadrilateral** | **Pentagon** | **Hexagon** |

| 7 | 8 | 10 | 12 | No. of Sides |
|---|---|---|---|---|
| | | | | Graphic |
| **Heptagon** | **Octagon** | **Decagon** | **Dodecagon** | Name |

---

**Equiangular:** All equal angles
**Equilateral:** All equal sides
**Regular:** All equal sides and all equal angles

---

| Rectangle | Rhombus | Square |
|---|---|---|
| Equiangular | Equilateral | Regular |
| (not equilateral) | (not equiangular) | |

| | | |
|---|---|---|
| **Any Convex Polygon** | $n$ = number of sides | |
| | Sum of the measures of the exterior angles | = 360 |
| | **an interior angle + adjacent exterior angle = 180°** | |
| | Sum of the measures of the interior angles | = $(n-2)(180)$ |
| **Regular Polygons Only** | The measure of each exterior angle | = $\dfrac{360}{n}$ |
| | The measure of each interior angle | = $\dfrac{(n-2)(180)}{n}$ |

exterior angle    interior angle

Bottom view of a polygon

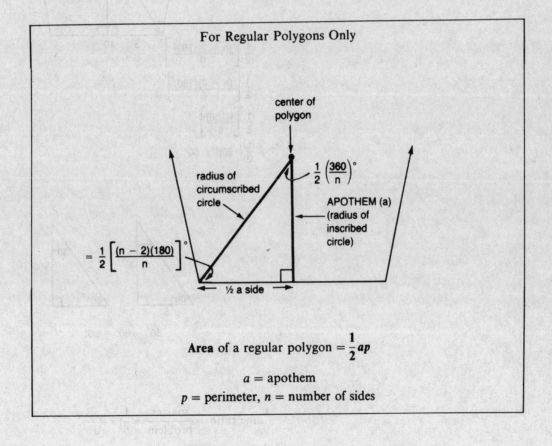

For Regular Polygons Only

center of polygon

radius of circumscribed circle

$\dfrac{1}{2}\left(\dfrac{360}{n}\right)^{\circ}$

APOTHEM (a) (radius of inscribed circle)

$=\dfrac{1}{2}\left[\dfrac{(n-2)(180)}{n}\right]^{\circ}$

½ a side

**Area** of a regular polygon $=\dfrac{1}{2}ap$

$a$ = apothem

$p$ = perimeter, $n$ = number of sides

**Examples:**

1. Given a polygon with 22 sides, find the sum of the measures of the exterior angles.

**Solutions:**

1. 360. The sum of the measures of the exterior angles of any polygon is 360.

2. Given a polygon with 22 sides, find the sum of the measures of the interior angles.

2. $(n - 2)(180) = (22 - 2)(180) = 3600$

3. Find the measure of an interior angle of a regular pentagon.

3. The measure of each interior

$$= \frac{(n - 2)(180)}{n} = \frac{(5 - 2)(180)}{5}$$

$$= \frac{3(180)}{5} = 108$$

4. Find the area of a regular hexagon whose side is 10.

4. $A = \frac{1}{2}ap$

$$\frac{1}{2}\left(\frac{360}{n}\right)^\circ = \frac{1}{2}\left(\frac{360}{6}\right)^\circ = \frac{1}{2}\left(60\right)^\circ = 30^\circ$$

$$\frac{1}{2}\left[\frac{(n-2)(180)}{n}\right]^\circ$$

$$= \frac{1}{2}\left[\frac{(n-2)(180)}{6}\right]^\circ$$

$$= \frac{1}{2}\left[\frac{4(180)}{6}\right]$$

$$= \frac{1}{2}(120)^\circ = 60^\circ$$

$$30^\circ - 60^\circ - 90^\circ$$

$$\text{line ratio} = \frac{\text{master}}{\text{problem}} = \frac{1}{5} = \frac{\sqrt{3}}{a}$$

$$a = 5\sqrt{3}$$

perimeter of hexagon (six sides) $= 6 \cdot 10 = 60$

Area $= \frac{1}{2}ap = \frac{1}{2}(5\sqrt{3})(60) = 150\sqrt{3}$

# Practice Problems 10.4

1. Find the measure of each exterior angle of a regular decagon.

   A) 36  B) 144  C) 180
   D) 360  E) 720

2. Find the sum of the measures of the interior angles of a convex polygon of 12 sides.

   A) 30  B) 36  C) 360
   D) 1800  E) 2160

3. If a regular polygon has an exterior angle of 72°, how many sides does it have?

   A) 4  B) 5  C) 6  D) 7  E) 8

4. If the sum of the measures of the interior angles of a regular polygon is equal to the sum of the measures of the exterior angles, how many sides does the polygon have?

   A) 4  B) 6  C) 8  D) 10  E) 12

Solutions on page 131

## 10.5 Circles

A circle is the set of points in a plane equidistant from a given point called the center.

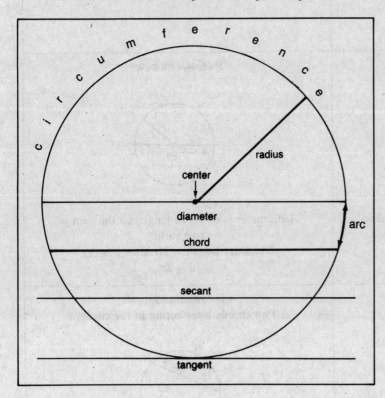

| | |
|---|---|
| **circumference** | The boundary of the circle. |
| **radius:** | A line segment from the center to the circumference. |
| **diameter:** | A line segment from the circumference to the circumference through the center. |
| **chord:** | A line segment from a point on the circumference to any other point on the circumference |
| **secant:** | The line determined by a chord. A line that intersects the circle at two points. |
| **tangent:** | A line outside the circle which touches the circle at one point. |
| **arc:** | A piece of the circumference. |

**Notation**

A circle is named for its center.
⊙ *T* is shown on the left.

Chord **AB** is represented as **AB**
Arc **AB** is represented as **ÂB**

# Basic Circle Rules

The number of degrees of arc in a circle is **360**.

| **Central Angle** | **Radii** |
|---|---|
|  |  |
| The measure of a central angle is equal to the measure of its arc. $$m \angle J = m\widehat{NL}$$ $$x = a$$ | $r$ = radius **All radii of a circle are equal in length.** |
| **Inscribed Angle** | **Radius/Diameter** |
|  |  |
| The measure of an inscribed angle is equal to one-half the measure of its intercepted arc. $$m \angle J = \tfrac{1}{2}m\widehat{NL}$$ $$x = \tfrac{1}{2}a \quad \text{or} \quad a = 2x$$ | A diameter is equal in length to the sum of two radii. **Diameter length = 2(radius length)** $$d = 2r$$ |
| **Angle formed by:** **Two chords intersecting in the circle** | **Line relationship of:** **Two chords intersecting in the circle** |
|  |  |
| $m \angle KML$ is one-half the sum of the measures of the intercepted arcs. $$m \angle KML = \tfrac{1}{2}(m\widehat{NJ} + m\widehat{KL})$$ $$x = \tfrac{1}{2}(a + b)$$ | The product of the segments of one chord equals the product of the segments of the other chord. $$(JM)(MK) = (NM)(ML)$$ $$p \cdot q = r \cdot s$$ |

| | |
|---|---|
| **Angle formed by:**<br>● **Two tangents intersecting outside the circle** | **Line relationship of:**<br>● **Two tangents intersecting outside the circle** |
| 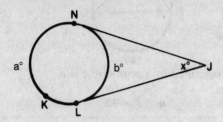<br><br>$m \angle J = \frac{1}{2}(m\widehat{NKL} - m\widehat{NL})$<br>$x = \frac{1}{2}(a - b)$ | 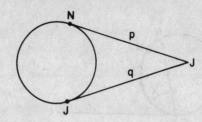<br><br>$\therefore \overline{NJ} = \overline{LJ}$<br>$p = q$ |
| ● **A tangent and secant intersecting outside the circle** | ● **A tangent and secant intersecting outside the circle** |
| <br><br>$m \angle J = \frac{1}{2}(m\widehat{NL} - m\widehat{NK})$<br>$x = \frac{1}{2}(a - b)$ | <br><br>$(NJ)(NJ) = (LJ)(KJ)$<br>$p \cdot p = (r + q)(q)$ |
| ● **Two secants intersecting outside the circle** | ● **Two secants intersecting outside the circle** |
| <br><br>$m \angle J = \frac{1}{2}(m\widehat{NL} - m\widehat{MK})$<br>$x = \frac{1}{2}(a - b)$ | <br><br>$(NJ)(MJ) = (LJ)(KJ)$<br>$(p + s)s = (r + q)q$ |
| The measure of the external angle at $J$ equals one-half the difference of its intercepted arcs. | The whole secant or tangent times the external segment equals the whole other secant or tangent times its external segment. |

## Angle formed by:
## A tangent and a chord drawn to the point of contact

$m \angle NJL$ is one-half the measure of its intercepted arc.

$$m \angle NJL = \tfrac{1}{2}m\widehat{NMJ}$$
$$x = \tfrac{1}{2}a$$

## Angle inscribed in a semi-circle

An angle inscribed in a semi-circle is a right angle.

$$m \angle J = 90$$

## Radius drawn to a tangent at the point of contact

A radius drawn to a tangent at the point of contact forms a right angle with the tangent.

$$m \angle NJL = 90$$

## Radius and Chord

- A radius drawn ⊥ to a chord bisects the chord and its arc.
- A radius which bisects a chord is ⊥ to the chord.
- A line ⊥ to a chord and bisecting it goes through the center of the circle.

## Chords and Arcs

- Chords of equal length intercept arcs of equal measure.
- Arcs of equal measure determine chords of equal length.

## Quadrilateral Inscribed in a Circle

If a quadrilateral is inscribed in a circle, the opposite angles are supplementary.

- $m \angle M + m \angle K = 180$
  $$w + y = 180$$
- $m \angle J + m \angle L = 180$
  $$x + z = 180$$

## Parallel Chords

Parallel chords intercept arcs of equal measure.

$$m\widehat{NL} = m\widehat{MK}$$
$$x = a$$

| Circumference length | Area |
|---|---|
| $C = 2\pi r$ <br> $\quad = \pi d$ <br> 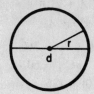 <br><br> The length of the circumference of a circle <br><br> $\quad = 2\pi r$ or $\pi d$. | $A = \pi r^2$ <br> $\quad = \dfrac{\pi d^2}{4}$ <br>  <br><br> The area of a circle $= \pi r^2$ or $\dfrac{\pi d^2}{4}$. |

| Arc Length (degrees) | Area of a Sector (degrees) |
|---|---|
| Arc length $= \dfrac{x}{360} \cdot 2\pi r$   <br><br> The arc length is $\dfrac{\text{central angle}}{360}$ of the whole circumference. <br><br>  The curvature of the arc is measured in degrees. It is equal to the measure of its central angle. The length of the arc is measured in linear units such as inches, centimeters, etc., and is a piece of the circumference length. | Area of sector $= \dfrac{x}{360} \cdot \pi r^2$   <br><br> A sector contains $\dfrac{\text{central angle}}{360}$ of the whole circle area. |

| Arc Length (radians) | Area of Sector (radians) |
|---|---|
| $r$ is the radius <br> $\theta$ is the central angle in radians <br> $S$ is the arc length <br> $S = r\theta$   | $r$ is the radius <br> $\theta$ is the central angle in radians <br><br> $\underset{\text{sector}}{A} = \dfrac{1}{2} r^2 \theta$ |

**Examples:**

1. In $\odot N$, $m\angle ANB = 40$. Find the $m\angle C$.

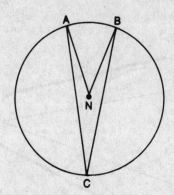

**Solutions:**

1. $\angle ANB$ is a central angle.
   $m\widehat{AB} = m\angle ANB = 40$
   $\angle C$ is an inscribed angle.
   $m\angle C = \frac{1}{2} m\widehat{AB} = \frac{1}{2}(40) = 20$

2. Find the value of x.

2. $m\angle BED = \frac{1}{2}(m\overset{\frown}{CA} + m\overset{\frown}{BD})$

   $x = \frac{1}{2}(m\overset{\frown}{CA} + 210)$

   However, we need $m\overset{\frown}{CA}$.

   $m\overset{\frown}{CA} + 20 + 210 + 70 = 360$

   $m\overset{\frown}{CA} = 60$

   $\therefore x = \frac{1}{2}(60 + 210) = 135$

3. The $m\overset{\frown}{AB} = 200$, $m\angle C = 50$. Find the $m\overset{\frown}{AD}$.

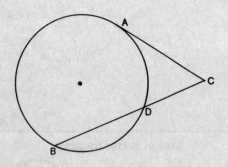

3. $m\angle C = \frac{1}{2}(m\overset{\frown}{AB} - m\overset{\frown}{AD})$

   $50 = \frac{1}{2}(200 - x)$

   $100 = 200 - x$

   $x = 100$

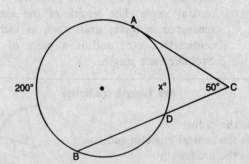

4. $ED = 5$, $DC = 4$.
   If $AC = 12$, find the length of $BC$.

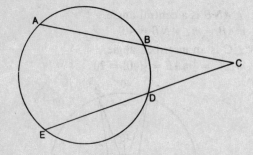

4. The whole secant times the segment outside equals the whole other secant times the segment outside.

   $(EC)(DC) = (AC)(BC)$

   $(9)(4) = (12)x$

   $36 = 12x$

   $x = 3$

5. A chord of length 12 intercepts an arc of 60°. Find the radius of the circle.

5. First construct the radii as shown.

∠ B is a central angle
$$m \angle B = m\widehat{AC} = 60$$

Triangle *ABC* is isosceles because *AB* = *BC* (both are radii)
$$\therefore m \angle A = m \angle C$$
$$m \angle A + m \angle B + m \angle C = 180$$
$$x + 60 + x = 180$$
$$2x = 120$$
$$x = 60$$

Triangle *ABC* is equilateral
$$\therefore AB = BC = AC = 12$$

6. Find the area of a circle whose circumference is 18π.

6. $C = 2\pi r = 18\pi$
$$2r = 18$$
$$r = 9$$
$$A = \pi r^2 = \pi(9)^2 = 81\pi$$

7. Find the area of a sector of a circle whose central angle is 72° and whose radius is 5.

7. $A = \frac{72}{360} \cdot \pi(5)^2$
$$= \frac{1}{5} \cdot 25\pi = 5\pi$$

8. If a circle of radius 3 has a central angle of $\frac{\pi}{6}$ radians, find the arc length associated with the central angle.

8. $S = r\theta$

$$S = \frac{\pi}{6}(3) = \frac{3\pi}{6} = \frac{\pi}{2}$$

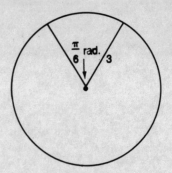

9. Find $x$, the length of $AE$.

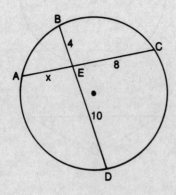

9. The product of the lengths of the segments of one chord equals the product of the lengths of the segments of the other chord.

$$(AE)(EC) = (BE)(ED)$$
$$x(8) = 4(10)$$
$$x = 5$$

10. In the Figure, $m\angle B = 6x - 20$ and $m\angle D = 4x + 10$
Find the value of $x$.

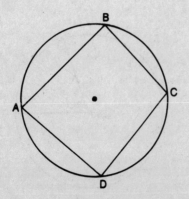

10. If a quadrilateral is inscribed in a circle, the opposite angles are supplementary.

$$m\angle B + m\angle D = 180$$
$$(6x - 20) + (4x + 10) = 180$$
$$10x - 10 = 180$$
$$10x = 190$$
$$x = 19$$

# Practice Problems 10.5

1. In the figure, what is the value of x.

A) 40   B) 60   C) 70
D) 80   E) 140

2. In the figure, what is the value of x.

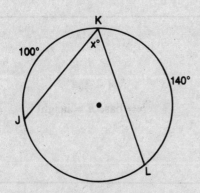

A) 30   B) 60   C) 90
D) 120   E) 150

3. $m\widehat{AB} : m\widehat{BC} : m\widehat{CNA} = 2:3:7$.
   Find $m \angle P$.

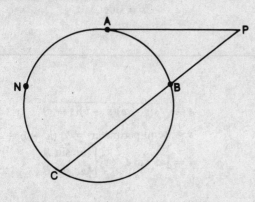

A) $22\frac{1}{2}$   B) 30   C) 60   D) 75   E) 150

4. The area of a circle is $12\pi$. Find the length of the circumference.

A) $6\pi$   B) $12\pi$   C) $36\pi$   D) $6\pi\sqrt{2}$   E) $4\pi\sqrt{3}$

5. A square is inscribed in a circle of radius 4. Find the shaded area.

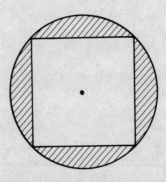

A) $16\pi$   B) $8\pi - 32$   C) $8\pi - 64$
D) $16\pi - 32$   E) $16\pi - 64$

6. $AC$ is 12 and $B$ is the midpoint of $AC$.
   $\widehat{ANB}$ and $\widehat{BMC}$ are semi-circles. Find the length of the curved path $ANBMC$.

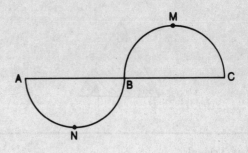

A) $6\pi$   B) $10\pi$   C) $12\pi$   D) $18\pi$   E) $24\pi$

7. A chord of length 4 is 6 units from the center of a circle. Find the radius of the circle.

A) $10\sqrt{2}$   B) $2\sqrt{10}$   C) 8   D) $2\sqrt{13}$   E) 6

8. $AC$, $CE$ and $AE$ are tangents. If $CD = 4$, $DE = 5$ and $AE = 9$, find the length of $AC$.

A) 6   B) 7   C) 8   D) 9   E) 10

Solutions on page 132

## 10.6 Area

| | |
|---|---|
| **Triangle:** | $A = \frac{1}{2}bh$ <br> $b$ = base $h$ = height |
| **Equilateral Triangle:** | $A = \frac{s^2}{4}\sqrt{3}$ <br><br> $s$ = side |
| **Any Triangle:** | $A = \sqrt{s(s-a)(s-b)(s-c)}$ <br> $s$ = semi-perimeter = $\frac{1}{2}(a + b + c)$ <br> $A = \frac{1}{2}ab \sin C$ or $\begin{cases} \frac{1}{2}bc \sin A \\ \frac{1}{2}ac \sin B \end{cases}$ |

**Any Triangle:**

$$A = \tfrac{1}{2}|(x_1y_2 + x_2y_3 + x_3y_1) - (y_1x_2 + y_2x_3 + y_3x_1)|$$

**Parallelogram:**

$A = bh$

$b = $ base    $h = $ height

$A = ab \sin C$

**Rectangle:**

$A = bh$

$b = $ base    $h = $ height

**Rhombus:**

$$A = \frac{d_1 d_2}{2}$$

$d_1 = $ diagonal 1    $d_2 = $ diagonal 2

**Square:**

$$A = s^2$$
$$s = \text{side}$$

---

$$A = \frac{d^2}{2}$$

$$d = \text{diagonal}$$

**Trapezoid:**

$$A = \left(\frac{b_1 + b_2}{2}\right) h$$

$b_1 = \text{base 1}$   $b_2 = \text{base 2}$   $h = \text{height}$

**Any Quadrilateral:**

$$A = \tfrac{1}{2}(d_1)(d_2) \sin \theta$$
$d_1 = \text{diagonal 1}$   $d_2 = \text{diagonal 2}$
$\theta = \text{angle between the diagonals}$

**Any Regular Polygon:**

$$A = \tfrac{1}{2}ap$$
$a = \text{apothem}$   $p = \text{perimeter}$

---

$$A = \tfrac{1}{4}ns \cot \left(\frac{180}{n}\right)^\circ$$

$n = \text{number of sides}$
$s = \text{length of a side}$

**Circle:**

$$A = \pi r^2 \quad r = \text{radius}$$

$$A = \frac{\pi d^2}{4} \quad d = \text{diameter}$$

# Practice Problems 10.6

1.

A circle of radius 6 is inscribed in a square. Find the area of the shaded region.

A) $36\pi - 144$   B) $36\pi - 36$   C) $144 - 12\pi$
D) $144\pi - 36$   E) $144 - 36\pi$

2. Find the area of triangle *NJL*.

A) 455   B) 24   C) 12   D) $\sqrt{455}$

E) $\dfrac{\sqrt{455}}{2}$

3.

In isosceles trapezoid *ABCD*, $m\angle D = 60$. If $AB = 10$ and $AD = 8$, find the area of the trapezoid.

A) 40   B) $56\sqrt{2}$   C) $56\sqrt{3}$   D) 80   C) 140

4.

Find the area of triangle *NJL*.

A) $\dfrac{\sqrt{53}\sqrt{61}}{2}$   B) 26   C) 20   D) 13   E) $17\frac{1}{2}$

Solutions on page 133

# 11. Lateral Area, Total Area, Volume

## 11.1 Fundamental Facts

**Lateral Area (L.A.)**    the sum of the areas of the lateral faces

**Total Area**   (T.A.)       the sum of the areas of all the faces (lateral and bases)
**Surface Area (S.A.)**

**Volume**  (V)    the cubic content of the solid

<table>
<tr><td colspan="2">

**Right Prism**

$L.A. = ph$
$T.A. = L.A. + 2B$
$\quad V = Bh$

$B$ = area of base
$p$ = perimeter of base
$h$ = height

</td><td colspan="2">

**Right Cylinder**

$L.A. = 2\pi rh$
$T.A. = L.A. + 2B$
$\quad V = \pi r^2 h$

$B$ = area of base = $\pi r^2$
$r$ = radius of base
$h$ = height

</td></tr>
<tr><td colspan="2">

**Oblique Prism**

$V = Bh$

$B$ = area of base
$p$ = perimeter of base
$h$ = height

</td><td colspan="2">

**Oblique Cylinder**

$V = \pi r^2 h$

$B$ = area of base
$r$ = radius of base
$h$ = height

</td></tr>
<tr><td colspan="2">

**Regular Pyramid**

$L.A. = \frac{1}{2}pl$
$T.A. = L.A. + B$
$\quad V = \frac{1}{3}Bh$

$l$ = slant height
$B$ = area of base
$p$ = perimeter of base
$h$ = height

</td><td colspan="2">

**Right Cone**

$L.A. = \pi rl$
$T.A. = L.A. + B$
$\quad V = \frac{1}{3}\pi r^2 h$

$l$ = slant height
$B$ = area of base = $\pi r^2$
$r$ = radius of base
$h$ = height

</td></tr>
</table>

| Sphere | Cube |
|---|---|
| $T.A. = 4\pi r^2$<br>$V = \frac{4}{3}\pi r^3$<br><br>$r$ = radius of sphere  | $L.A. = 4e^2$<br>$T.A. = 6e^2$<br>$V = e^3$<br><br>$e$ = edge of cube  |

**Examples:**

1. If the total area of a cube is 36, find its volume.

2. The bases of a right prism, whose height is 8, are equilateral triangles with side 10. Find the lateral area, total area and volume of the prism.

3. A regular square pyramid has a base edge of 8 and a height of 4. Find the lateral area, total area and volume of the pyramid.

**Solutions:**

1. $T.A. = 6e^2 = 36 \qquad V = e^3$
   $e^2 = 6 \qquad\qquad V = (\sqrt{6})^3 = 6\sqrt{6}$
   $e = \sqrt{6}$

2. $L.A. = ph$
   perimeter of base $= 30$
   $\qquad\qquad$ height $= 8$
   $L.A. = 30(8) = 240$
   $T.A. = L.A. + 2B$

   $B$ = area of base $= \dfrac{S^2}{4}\sqrt{3} = \dfrac{10^2}{4}\sqrt{3}$

   $\qquad\qquad = 25\sqrt{3}$
   $T.A. = 240 + 2(25\sqrt{3})$
   $\qquad = 240 + 50\sqrt{3}$

   ---
   $V = Bh = (25\sqrt{3})8 = 200\sqrt{3}$

3.

   $l^2 = 4^2 + 4^2$
   $l = 4\sqrt{2}$
   $p = 4(8) = 32$
   $B = 8^2 = 64$

   $L.A. = \frac{1}{2}pl = \frac{1}{2}(32)(4\sqrt{2}) = 64\sqrt{2}$
   $T.A. = L.A. + B = 64\sqrt{2} + 64$
   $V = \frac{1}{3}Bh = \frac{1}{3}(64)(4) = \frac{256}{3}$

# Practice Problems 11.1

1. Find the T.A. of a cylinder whose height is 10 and whose radius is 3.

   A) $69\pi$   B) $78\pi$   C) $60\pi$   D) $18\pi$   E) $9\pi$

2. If the height and radius of a cylinder are of equal length and the volume is $125\pi$, find the radius of the base.

   A) 5   B) 10   C) 25   D) $5\pi$   E) $62\frac{1}{2}$

3. Find the volume of the given sphere.

   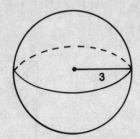

   A) $36\pi$   B) $9\pi$   C) $18\pi$   D) $27\pi$   E) $64\pi$

4. If the radius of a right cone is 8 and the height (altitude) is 15, find the lateral area.

   A) $8\pi$   B) $15\pi$   C) $120\pi$   D) $136\pi$   E) $17\pi$

Solutions on page 134

# 12. Locus

## 12.1 Basics and Beyond

Locus: The region formed by the set of all points satisfying a given condition.
Loci: The plural of locus

| | | |
|---|---|---|
| The locus of points (in a plane) **a distance _r_ from a given point _P_**  **A circle** with center at _P_ and radius _r_ | The locus of points (in a plane) **a distance _d_ from a given line _l₁_**  **Two lines** each parallel to the given line at a distance _d_ | The locus of points (in a plane) **equidistant from two parallel lines**  **A line** parallel to the two given lines and mid-way between them |
| The locus of points (in a plane) **equidistant from two fixed points _A_ and _B_**  The **perpendicular bisector** of the line segment connecting points _A_ and _B_. | The locus of points (in a plane) **equidistant from the ends of a line segment**  The **perpendicular bisector** of the given line segment | The locus of points (in a plane) **equidistant from the rays of a given angle _ABC_**  The **angle bisector** of angle _ABC_ |
| The locus of points (in a plane) **equidistant from two intersecting lines _l₁_ and _l₂_**  **Two lines** _l₃_ and _l₄_, the angle bisectors of the angles formed between _l₁_ and _l₂_ | The locus of points (in a plane) **equidistant from the vertices of a triangle**  **A point.** The intersection of the perpendicular bisectors of the sides of the triangle | The locus of points (in a plane) **equidistant from the sides of a triangle**  **A point.** The intersection of the angle bisectors of the angles of the triangle. |

The locus of points (in a plane) such that the distance from a fixed point *F* is equal to the distance to a fixed line *l*

LOCUS

**A parabola**

The locus of points (in a plane) such that the sum of the distances to each of two fixed points $F_1$ and $F_2$ is a positive constant

**An ellipse**

The locus of points in a plane such that the absolute value of the difference of the distances from two fixed points $F_1$ and $F_2$ is constant

**A hyperbola**

☞ **Compound locus:** Do each piece of the problem separately. The intersection of the individual solutions is the solution set or locus for the compound example.

**Examples:**

1. Find the locus of points 3 units from a given line and 4 inches from a given point on the line.

**Solutions:**

1. Do each piece of the problem separately.

3 units from a given line

4 inches from a given point on the line

Composite

Four points (shown by Xs)

2. Given two points *N* and *L* 8 units apart, what is the locus of points 6 units from *N* and 3 units from *L*?

2. Do each piece of the problem separately.

6 units from point *N*

3 units from point *L*

Composite

Two points (designated by the Xs)

## Practice Problems 12.1

1. In a plane, parallel lines *l* and *m* are 6 units apart and *P* is a point on line *l*. The total number of points in the plane that are equidistant from *l* and *m* and also 3 units from *P* is

A) 0   B) 1   C) 2   D) 3   E) 4

2. The equation of the locus of points in a plane equidistant from the points $(7, 3)$ and $(-3, 3)$ is

A) $x^2 + y^2 = 100$   B) $x = 5$   C) $x = 2$
D) $y = 5$   E) $y = 2$

3. In a plane the locus of points 4 units from the point $N(-2, 3)$ is

A) $x = -2$   B) $y = 3$   C) $x^2 + y^2 = 16$
D) $(x - 2)^2 + (y + 3)^2 = 16$
E) $(x + 2)^2 + (y - 3)^2 = 16$

4. In a plane, how many points are equidistant from two parallel lines and also equidistant from two points on one of these lines?

A) 0   B) 1   C) 2   D) 3   E) 4

Solutions on page 135

# Cumulative Review
# Chapters 1-12

1. If $\log a = p$, express $\log 10a^2$ in terms of $p$.

   A) $2p$  B) $p^2$  C) $1 + 2p$  D) $1^{2p}$
   E) $1 + p^2$

2. The square of $(2 - 2i)$ is

   A) $0$  B) $-8i$  C) $4 - 4i$  D) $4$
   E) $8 - 8i$

3. In the accompanying diagram, secant $ADB$ and tangent $AC$ are drawn to circle $O$ from external point $A$. If $AD = 3$ and $DB = 9$, what is the length of $AC$?

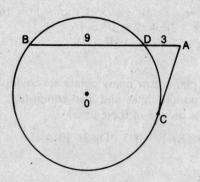

   A) $6$  B) $\sqrt{6}$  C) $27$  D) $\sqrt{27}$  E) $\sqrt{12}$

4. In the accompanying diagram $AB \parallel CD$, $EF$ is a transversal and $m\angle 1 = 110$. What is the $m\angle 2$?

   A) $110$  B) $70$  C) $50$  D) $80$  E) $20$

5. Given point $P$ on a line. In a plane containing the line, the total number of points which are at a distance of 4 units from $P$ and also a distance of 3 units from the given line is

   A) $0$  B) $1$  C) $2$  D) $3$  E) $4$

6. $i^{219} = 7$.

   A) $0$  B) $1$  C) $i$  D) $-1$  E) $-i$

7. A cylindrical pail has a radius of 3 units and a height of 6 units. Approximately how many gallons will it hold if there are 27 cubic units to a gallon?

   A) $1$  B) $3$  C) $\pi$  D) $2\pi$  E) $\pi^2$

8. In the accompanying diagram, right triangle $JSL$ has altitude $h$ drawn to the hypotenuse, dividing it into segments of 8 and 10. What is the value of $h$?

   A) $4\sqrt{5}$  B) $80$  C) $3\sqrt{2}$  D) $\sqrt{2}$  E) $2\sqrt{10}$

9. The range of the function shown in the given graph is

(0, −2)

A) all real numbers   B) imaginary   C) $x > 0$
D) $y \geqslant -2$   E) $y \leqslant -2$

10. Find the value of $(a + 1)^0 + 4(a)^{-1/2}$ when $a = 1$.

A) 6   B) 5   C) −2   D) −1   E) $1\frac{1}{2}$

11. What is the measure of the supplement of an angle whose complement is 70°

A) 20   B) 110   C) 160   D) 140   E) 130

12. Express $\log \dfrac{a}{b\sqrt{c}}$ in terms of the sums and differences of logarithms.

A) $\dfrac{\log a}{\log b + \frac{1}{2} \log c}$   B) $\log a - \log b + \frac{1}{2} \log c$

C) $\log a - \dfrac{\log b}{\log \sqrt{c}}$

D) $\log a - (\log b + \frac{1}{2} \log c)$

E) $\log a + \log b - \frac{1}{2} \log c$

13. In a given plane, the total number of points which are equidistant from two intersecting lines and

also 5 units from their point of intersection is

A) 0   B) 1   C) 2   D) 3   E) 4

14. What is the smallest integral value of $x$ for which $\sqrt{2 - x}$ is imaginary?

A) −1   B) 0   C) 1   D) 2   E) 3

15. The base of a rectangular tank is 6 units by 10 units and its height is 12 units. Find the number of cubic units in the tank when it is $\frac{1}{4}$ empty.

A) 270   B) 180   C) 150   D) 540   E) 720

16. Find the value of $\sqrt{19}$ to the nearest tenth.

A) 4.2   B) 4.3   C) 4.4   D) 4.42   E) 4.5

17. Simplify $\left(\dfrac{49m^{-2}}{16r^{-2}}\right)^{-1/2}$

A) $\dfrac{49m}{16r}$   B) $\dfrac{4m}{7r}$   C) $\dfrac{4r}{7m}$

D) $\dfrac{7m}{4r}$   E) $\dfrac{7m}{2r}$

18. In triangle $ABC$, $m \angle C = 25$ and $BC = 13$. If the measure of an exterior angle at $B = 50$, find $AB$.

A) 13   B) 26   C) $7\frac{1}{2}$   D) $6\frac{1}{2}$   E) 130

19. The sum of $-3 - 2i$ and $2 - 3i$ is a complex number which, when graphically represented, lies in quadrant

A) I   B) II   C) III   D) IV
E) on the $x$-axis

20. Which set of numbers could represent the lengths of the sides of a right triangle?

A) {1, 7, 8}   B) {7, 8, 12}   C) {7, 9, 11}
D) {8, 15, 17}   E) {6, 8, 9}

Solutions on page 136

# Diagnostic Chart for Cumulative Review Test

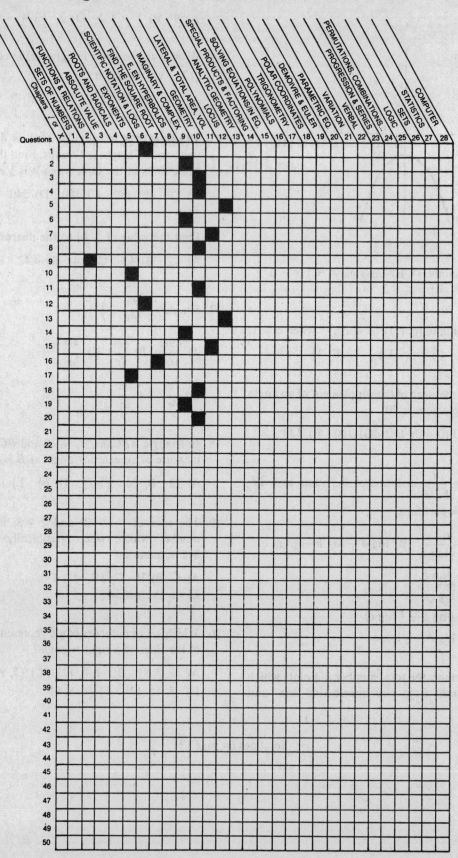

# Solutions to Problems in Unit Three

## Practice Problems 9.1

1. **(A)**

$$\frac{(25)}{(4+3i)} \frac{(4-3i)}{(4-3i)} = \frac{25(4-3i)}{16-9i^2} = \frac{25(4-3i)}{25}$$

$$= 4 - 3i$$

2. **(C)** $i^{25} = i^1 = i$

   *Note:* $\frac{25}{4} = 6r1$

3. **(D)**

   $(3+2i) + (3+6i) + (-4-4i)$

   $= (3+3-4) + (2+6-4)i$

   $= 2 + 4i$

4. **(B)**

   $\sqrt{-3} = \sqrt{3}i$

   $3\sqrt{-27} = 3(3\sqrt{3}i)$

   $\sqrt{-12} = 2\sqrt{3}i$

   $\sqrt{-3} + 3\sqrt{-27} - \sqrt{-12}$

   $= \sqrt{3}i + 9\sqrt{3}i - 2\sqrt{3}i$

   $= 8\sqrt{3}i$

   $= 0 + 8\sqrt{3}i$

5. **(A)**

$$\frac{(3+2i)}{i} \cdot \frac{i}{i} = \frac{3i + 2i^2}{i^2} = \frac{3i + 2(-1)}{-1}$$

$$= \frac{3i - 2}{-1} = 2 - 3i$$

6. **(E)**

   $(3 + \sqrt{2}i)(3 - \sqrt{2}i)$

   $= 9 - 3\sqrt{2}i + 3\sqrt{2}i - 2i^2$

   $= 9 - 2i^2 = 9 + 2$

   $= 11$

7. **(B)**     reals = reals

   imaginary = imaginary

   $a + 3i = 7 - bi$

   $\therefore a = 7$

   $b = -3$

8. **(A)**

   $(3 + 7i)(2 - i) = 6 - 3i + 14i - 7i^2$

   $= 6 + 11i - 7(-1)$

   $= 13 + 11i$

127

# Practice Problems 9.2

1.

From the graph, the solution is $1 + i$.

2. $(6 + 2i) - (-2 + 3i)$ becomes
$(6 + 2i) + (2 - 3i)$

From the graph, the solution is $8 - i$.

# Practice Problems 10.1

1. **(A)** $m \angle x = 20$.
   Alternate interior angles

$4x + 5x = 180$
$9x = 180$
$x = 20$
$m \angle b = 5x = 5(20) = 100$

4. **(D)**
   $\angle XBC$ and $\angle XBA$ are complementary.
   $37 + m \angle XBA = 90$
   $m \angle XBA = 53$
   $\angle XBA$ and $\angle ABY$ are supplementary.
   $53 + m \angle ABY = 180$
   $m \angle ABY = 127$

2. **(B)** Complementary angles add to 90°.
   $(x + 20) + (2x + 1) = 90$
   $3x + 21 = 90$
   $3x = 69$
   $x = 23$

3. **(E)** Angles $a$ and $b$ are supplementary.
   $m \angle a = 4x$
   $m \angle b = 5x$

# Practice Problems 10.2

**1. (B)**  $1x + 5x + 6x = 180$
$$12x = 180$$
$$x = 15$$

$1x = 15$
$5x = 5(15) = 75$
$6x = 6(15) = 90$

The triangle is a right triangle.

**2. (C)** The base angles are equal.
$$x + x + 50 = 180$$
$$2x + 50 = 180$$
$$2x = 130$$
$$x = 65$$

**3. (D)**

$$x + 3x + 6 = 106$$
$$4x + 6 = 106$$
$$4x = 100$$
$$x = 25$$

$$m\angle A = 3x + 6 = 3(25) + 6 = 81$$

**4. (D)**  $4x - 30 = 2x + 10$
$$2x = 40$$
$$x = 20$$

$m\angle A = 4(20) - 30 = 50$
$m\angle C = 2(20) + 10 = 50$
$m\angle B = 180 - (50 + 50) = 80$

**5. (C)**

☞ An equilateral △ is isosceles. The altitude is also the median and angle bisector.

$$\text{line ratio} = \frac{\text{master}}{\text{problem}} = \frac{\sqrt{3}}{x} = \frac{2}{20}$$
$$2x = 20\sqrt{3}$$
$$x = 10\sqrt{3}$$

**6. (B)**

$$\triangle RST \sim \triangle UVT$$

$$\text{line ratio} = \frac{6}{2} = \frac{6 + x}{x}$$
$$6x = 12 + 2x$$
$$4x = 12$$
$$x = 3$$

**7. (C)** This is a 5 multiple of 3, 4, 5 ⇒ 15, 20, 25.

This is an 8, 15, 17
Pythagorean triple.

$x = 20 - 8 = 12$

**8. (A)** The sum of the lengths of two sides of a triangle must be greater than the length of the third side.

A) $7 + 8 > 9$   $7 + 9 > 8$   $8 + 9 > 7$
B) $3 + 5$ is not greater than 8
C) $3 + 3$ is not greater than 7
D) $3 + 6$ is not greater than 10
E) $4 + 5$ is not greater than 10

# Practice Problems 10.3

**1. (B)**

$$\text{line ratio} = \frac{\text{master}}{\text{problem}} = \frac{2}{4} = \frac{\sqrt{3}}{h}$$

$$2h = 4\sqrt{3}$$

$$h = 2\sqrt{3}$$

$$A = (\text{base})(ht)$$

$$= 10(2\sqrt{3}) = 20\sqrt{3}$$

2. **(C)**

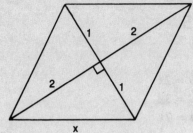

$$1^2 + 2^2 = x^2$$
$$5 = x^2$$
$$x = \sqrt{5} = \text{each side}$$
$$\text{perimeter} = 4 \times \sqrt{5} = 4\sqrt{5}$$

3. **(E)**

Consecutive angles of a parallelogram are supplementary.

$$(4t + 6) + (5t - 15) = 180$$
$$9t - 9 = 180$$
$$9t = 189$$
$$t = 21$$
$$4t + 6 = 4(21) + 6 = 90$$
$$5t - 15 = 5(21) - 15 = 90$$

A parallelogram with 90° angles is either a rectangle or a square.

4. **(A)**

Since $AD \parallel BC$
$\angle 1$ and $\angle 2$ are alternate interior angles.

$$2x + 12 = 7x - 8$$
$$20 = 5x$$
$$4 = x$$

# Practice Problems 10.4

1. **(A)** The measure of each exterior angle $= \dfrac{360}{n}$.

decagon = 10 sides $\therefore n = 10$

$$\frac{360}{10} = 36$$

2. **(D)** The sum of the measures of the interior angles $= (n - 2)180$

$$n = 12$$
$$(12 - 2)(180) = 1800$$

3. **(B)** The measure of each exterior $= \dfrac{360}{n}$.

$$72 = \frac{360}{n}$$
$$n = 5$$

4. **(A)**

$$\frac{\text{The sum of the measures of the interior angles}}{} = \frac{\text{The sum of the measures of the exterior angles}}{}$$

$$(n - 2)(180) = 360$$
$$180n - 360 = 360$$
$$180n = 720$$
$$n = 4$$

# Practice Problems 10.5

**1. (C)**

∠S is a central angle. $m\angle S = m\widehat{RT} = 40$
△RST is isosceles  ∴$m\angle R = m\angle T = x$
$40 + x + x = 180$
$2x = 140$
$x = 70$

$$m\widehat{AB} = 2x$$
$$m\widehat{BC} = 3x$$
$$m\widehat{CNA} = 7x$$

$$2x + 3x + 7x = 360$$
$$12x = 360$$
$$x = 30$$

$$m\angle P = \frac{m\widehat{ANC} - m\widehat{AB}}{2} = \frac{7(30) - 2(30)}{2}$$

$$= \frac{210 - 60}{2}$$

$$= \frac{150}{2} = 75$$

**2. (B)**

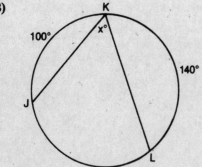

∠JKL is an inscribed angle
$m\angle JKL = x = \frac{1}{2}m\widehat{JL}$
$100 + 140 + m\widehat{JL} = 360$
$m\widehat{JL} = 120$
$x = \frac{120}{2} = 60$

**4. (E)** $A = \pi r^2 = 12\pi$
$r^2 = 12$
$r = \sqrt{12}$

$C = 2\pi r = 2(\sqrt{12})\pi = 2\sqrt{12}\pi$
$= 2(2\sqrt{3})\pi$
$= 4\pi\sqrt{3}$

*Note:* $\sqrt{12} = \sqrt{4 \cdot 3}$
$= \sqrt{4} \cdot \sqrt{3}$
$= 2\sqrt{3}$

**5. (D)** Shaded area = Area ⊙ − Area ☐

Area circle $= \pi r^2 = \pi(4)^2 = 16\pi$

diagonal of square = 8

Area square $= \frac{d^2}{2} = \frac{8^2}{2} = 32$

Shaded area $= 16\pi - 32$

**3. (D)**

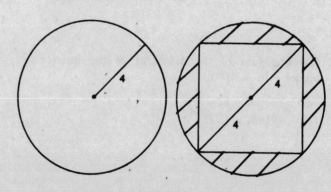

**6. (A)** The radius of each semi-circle is 3.

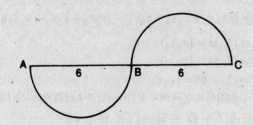

length semicircle path $= \frac{1}{2}$ circumference

length $\overset{\frown}{AB} = \frac{1}{2} \cdot 2\pi r$
$= \frac{1}{2}(2\pi(3))$
$= 3\pi$

length $BC = \frac{1}{2} \cdot 2\pi r$
$= \frac{1}{2}(2\pi(3))$
$= 3\pi$

$3\pi + 3\pi = 6\pi$

**7. (B)** A line from the center of a circle, perpendicular to a chord, bisects the chord. Draw the radius to complete a right triangle.

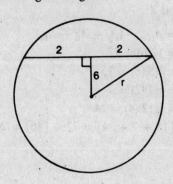

By the Pythagorean theorem

$2^2 + 6^2 = r^2$
$4 + 36 = r^2$
$r^2 = 40$
$r = \sqrt{40} = \sqrt{4 \cdot 10} = \sqrt{4}\sqrt{10} = 2\sqrt{10}$

**8. (C)**

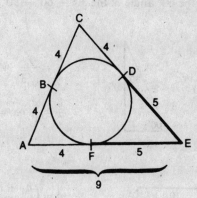

$DE = FE = 5$
Two tangents drawn from an outside point are equal in length.

$AF = 9 - FE$  $\therefore AF = 4$
$AB = AF = 4$
Two tangents drawn from an outside point are equal in length.
$BC = CD = 4$
Two tangents drawn from an outside point are equal in length.
$AC = AB + BC = 4 + 4 = 8$

## Practice Problems 10.6

**1. (E)** Shaded area = area $\square$ minus area $\bigcirc$
Circle
$A_\square = S^2 = (12)^2 = 144$
$A_\bigcirc = \pi r^2 = \pi(6)^2 = 36\pi$
$A = 144 - 36\pi$

**2. (D)**

$A = \sqrt{s(s-a)(s-b)(s-c)}$
$s = \frac{1}{2}(6 + 8 + 12) = 13$
$A = \sqrt{13(13-6)(13-8)(13-12)}$
$= \sqrt{13(7)(5)(1)}$
$= \sqrt{455}$

3. **(C)**

Using the 60° angle of the 30° − 60° − 90° triangle

$$\frac{8}{2} = \frac{y}{\sqrt{3}} \quad y = 4\sqrt{3}$$

$$\frac{8}{2} = \frac{x}{1} \quad x = 4$$

$$A = \left(\frac{b_1 + b_2}{2}\right) h$$

$$= \left(\frac{10 + 18}{2}\right) 4\sqrt{3}$$

$$= 14(4\sqrt{3})$$

$$= 56\sqrt{3}$$

4. **(B)** Method 1:

$$A = \tfrac{1}{2}|(x_1 y_2 + x_2 y_3 + x_3 y_1) - (y_1 x_2 + y_2 x_3 + y_3 x_1)|$$

$$N = (1, 1) = (x_1 y_1)$$
$$J = (7, 6) = (x_2 y_2)$$
$$L = (3, -6) = (x_3 y_3)$$
$$= \tfrac{1}{2}|(1(6) + 7(-6) + 3(1)) - (1(7) + 6(3) + (-6)(1)|$$
$$= \tfrac{1}{2}|[6 - 42 + 3] - [7 + 18 - 6]|$$
$$= \tfrac{1}{2}|-33 - 19| = \tfrac{1}{2}|-52| = \tfrac{52}{2} = 26$$

Method 2:

$$A_\triangle = A_{\text{rect}} - (A_I + A_{II} + A_{III})$$
$$A_{\text{rect}} = 12(6) = 72$$
$$A_I = \tfrac{1}{2}(5)(6) = 15$$
$$A_{II} = \tfrac{1}{2}(7)(2) = 7$$
$$A_{III} = \tfrac{1}{2}(12)(4) = 24$$
$$72 - [15 + 7 + 24] = 72 - [46] = 26$$

# Practice Problems 11.1

1. **(B)**

$$\text{T.A.} = \text{L.A.} + 2B$$
$$= 2\pi rh + 2(\pi r^2)$$
$$= 2\pi(3)(10) + 2(\pi(3^2))$$
$$= 60\pi + 18\pi = 78\pi$$

2. **(A)** Let $x$ = height and radius.

$$V = \pi r^2 h$$
$$125\pi = \pi(x^2)(x)$$
$$\pi x^3 = 125\pi$$
$$x^3 = 125$$
$$x = 5$$

3. **(A)**

$V = \frac{4}{3}\pi r^3$

$\quad = \frac{4}{3}\pi(3)^3$

$\quad = 36\pi$

4. **(D)**

L.A. $= \pi r l$

L.A. $= \pi(8)(17)$

$\quad = 136\pi$

*Note:* $l^2 = 15^2 + 8^2$

$\qquad l^2 = 225 + 64$

$\qquad l^2 = 289$

$\qquad l = 17$

# Practice Problems 12.1

1. **(B)** Do each part separately:

Equidistant from *l* and *m*

3 units from point *P* on *l*

Composite

One point (indicated by the **X**)

2. **(C)**

$x = 2.$

The locus in the perpendicular bisector of the line segment joining the two points.

3. **(E)**

A circle with center $(-2, 3)$ and radius $= 4$

$(x - h)^2 + (y - k)^2 = r^2$

$(x + 2)^2 + (y - 3)^2 = 16$

4. **(B)** Do each part separately:

Equidistant from two parallel lines

Equidistant from two points on one of the lines

Composite

One point (indicated by the **X**)

# Cumulative Review 1–12

1. **(C)**

$$\log 10a^2 = \underbrace{\log 10}_{1} + 2\underbrace{\log a}_{p}$$

$$= 1 + 2p$$

2. **(B)**

$$(2 - 2i)(2 - 2i) = 4 - 4i - 4i + 4i^2$$

$$= 4 - 8i + 4(-1)$$

$$= -8i$$

3. **(A)**

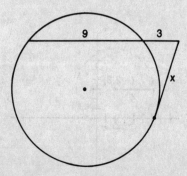

(whole)(segment outside)

$$= (\text{whole})(\text{segment outside})$$

$$(9 + 3)(3) = x(x)$$

$$x^2 = 12 \cdot 3 = 36$$

$$x = 6$$

4. **(B)** Angles 1 and 2 are exterior angles on the same side of the transversal.

$$\underbrace{m\angle 1 + m\angle 2}_{110} = 180$$

$$m\angle 2 = 70$$

5. **(E)**

Four points (indicated by Xs)

6. **(E)**

$$\begin{array}{r} 54\,r3 \\ 4\overline{)219} \\ \underline{20} \\ 19 \\ \underline{16} \\ 3 \end{array} \qquad i^{219} = i^3 = -i$$

7. **(D)**

$$V = \pi r^2 h = \pi(3)^2(6) = 54\pi$$

$$\frac{54\pi \ \text{units}^3}{27 \dfrac{\text{units}^3}{\text{gal}}} = 2\pi \ \text{gal}$$

8. **(A)**

$$h^2 = 8 \cdot 10 = 80$$

$$h = \sqrt{80} = \sqrt{16}\sqrt{5} = 4\sqrt{5}$$

9. **(E)** Range is all values of $y$ for which the function or relation exists. From the graph $y$ only has values at $-2$ or less.

$$\therefore \ \text{range:} \ y \leqslant -2$$

10. **(B)** $(a + 1)^0 + 4(a)^{-1/2}$

$\qquad (1 + 1)^0 + 4(1)^{-1/2} \qquad 1^{\text{any power}} = 1$

$\qquad\qquad 2^0 + 4(1) \qquad\qquad\qquad 2^0 = 1$

$\qquad\qquad\quad 1 + 4 = 5$

11. **(C)** The angle whose complement is $70°$, is $20°$. The supplement of $20°$ is $160°$.

12. **(D)**

$$\log \frac{a}{b\sqrt{c}} = \log a - \log b\sqrt{c}$$

$$\log a - (\log b + \tfrac{1}{2}\log c)$$

13. **(E)**

Four points (indicated by **X**s)

14. **(E)** The $\sqrt{2-x}$ is imaginary when

$$2 - x < 0 \quad \therefore x > 2.$$

The smallest integer value greater than 2, is 3.

15. **(D)** $\frac{1}{4}$ empty is $\frac{3}{4}$ full.

The water rises to $\frac{3}{4}(12) = 9$ units.

Volume = (base area)(height)

$$= (6 \cdot 10)(9)$$
$$= 540 \text{ cubic units}$$

16. **(B)**

$$
\begin{array}{r}
4. \underline{3}\ \underline{5} \\
\sqrt{19.0000}
\end{array}
$$

$$
\begin{array}{r}
16 \\
\hline
83 \mid 300 \\
\hline
249 \\
865 \mid 5100
\end{array}
$$

*Note:*  
$$
\begin{array}{r}
865 \\
\times 5 \\
\hline
4325
\end{array}
$$

$\therefore$ to the nearest tenth = 4.4

17. **(B)**

$$\left(\frac{49m^{-2}}{16r^{-2}}\right)^{-1/2} = \frac{49^{-1/2}m^1}{16^{-1/2}r^1} = \frac{\frac{1}{7}m}{\frac{1}{4}r}$$

$$= \frac{4m}{7r}$$

18. **(A)**

$$m\angle ABC = 180 - 50 = 130$$
$$\therefore m\angle A = 180 - (25 + 130)$$
$$= 25$$

$\therefore \triangle ABC$ is isosceles and $AB = BC = 13$

19. **(C)**
$$(-3 - 2i) + (2 - 3i) = -1 - 5i$$

20. **(D)** To be a right triangle

$$(\text{leg})^2 + (\text{leg})^2 = (\text{hyp.})^2$$
$$8^2 + 15^2 = 17^2$$
$$64 + 225 = 289$$
$$289 = 289 \ \checkmark$$

# UNIT FOUR

# 13. Analytic Geometry

## 13.1 Basic Coordinate Geometry Equations

$$\text{BASIC EQUATIONS}$$

ordered pair
$(x, y)$

absissa    ordinate

midpoint

$$\text{slope} = m = \frac{\Delta y}{\Delta x} = \frac{y_2 - y_1}{x_2 - x_1} = \tan\theta$$

Slope is the tilt of the line from the horizontal.

$$\text{distance} = d = \sqrt{(x_2 - x_1)^2 + (y_2 - y_1)^2}$$

Distance between two points is also the *length* of the line segment connecting them.

$$\text{midpoint} = M = \left(\frac{x_1 + x_2}{2}, \frac{y_1 + y_2}{2}\right)$$

Midpoint is the *middle* of the line segment.

### Types of slope

Slope is the slanting or tilting of a line from the horizontal.

$m = 0$    This is a **horizontal line**

$m = $ "$+$"    This line has a **positive slope**.

$m = $ "$-$"    This line has a **negative slope**.

$m = $ **no slope**
$= $ **undefined** } **Vertical line**
$= \infty$

### Parallel lines

Parallel lines have equal slope.

$m_1 = m_2$

### Perpendicular lines

Perpendicular lines have slopes that are negative reciprocals. (Multiply to $-1$)

$m_1 \cdot m_2 = -1$

Example: $m_1 = \frac{2}{3}, m_2 = \frac{-3}{2}$

☞ Colinear:    Three points $A$, $B$ and $C$ are colinear (along the same line) if

$$m_{AB} = m_{BC} = m_{AC}$$

## Miscellaneous Equations

| **Angle between two lines** | **The distance from a point to a line** |
|---|---|
| If $\theta$ is the angle between two lines, $\theta$ is measured counter-clockwise. | |

$$tan\ \theta = \frac{m_2 - m_1}{1 + m_1 m_2}$$

$$d = \left| \frac{Ax_1 + By_1 + C}{\sqrt{A^2 + B^2}} \right|$$

☞ $\dfrac{0}{\text{Number}} = 0, \quad \dfrac{\text{Number}}{0} = \text{Undefined}$

**Examples:**

1. Given the points $A(1, 3)$ and $B(7, 1)$, find the midpoint, the distance between them and the slope of the line that joins them.

**Solutions:**

1.

$$\text{midpoint} = M_{AB} = \left( \frac{x_1 + x_2}{2}, \frac{y_1 + y_2}{2} \right)$$

$$= \left( \frac{1 + 7}{2}, \frac{3 + 1}{2} \right) = (4, 2)$$

$$d = \sqrt{(x_2 - x_1)^2 + (y_2 - y_1)^2}$$

$$= \sqrt{(7 - 1)^2 + (1 - 3)^2}$$

$$= \sqrt{(6)^2 + (-2)^2} = \sqrt{36 + 4}$$

$$= \sqrt{40} = 2\sqrt{10}$$

$$m = \frac{\Delta y}{\Delta x} = \frac{y_2 - y_1}{x_2 - x_1} = \frac{1 - 3}{7 - 1} = \frac{-2}{6}$$

$$= -\frac{1}{3}$$

2. Given the points $C(-1, -5)$ and $D(1, -4)$, find the midpoint, the length of the line that connects them and the slope of the line.

2.

$$M_{AB} = \left(\frac{x_1 + x_2}{2}, \frac{y_1 + y_2}{2}\right)$$

$$= \left(\frac{-1 + 1}{2}, \frac{-5 + (-4)}{2}\right)$$

$$= \left(\frac{0}{2}, \frac{-9}{2}\right) = \left(0, -\frac{9}{2}\right)$$

$$d = \sqrt{(x_2 - x_1)^2 + (y_2 - y_1)^2}$$

$$= \sqrt{(-1 - 1)^2 + (-5 - (-4))^2}$$

$$= \sqrt{(-2)^2 + (-1)^2} = \sqrt{4 + 1} = \sqrt{5}$$

$$m = \frac{y_2 - y_1}{x_2 - x_1} = \frac{-4 - (-5)}{1 - (-1)}$$

$$= \frac{-4 + 5}{1 + 1} = \frac{1}{2}$$

3. Given the coordinates below, are the line segments $AB$ and $CD$ perpendicular?

$A(-3, -1)$, $B(2, 2)$, $C(-3, 4)$, $D(3, -6)$

3.

$$m_{AB} = \frac{y_2 - y_1}{x_2 - x_1} = \frac{2 - (-1)}{2 - (-3)} = \frac{2 + 1}{2 + 3} = \frac{3}{5}$$

$$m_{CD} = \frac{y_2 - y_1}{x_2 - x_1} = \frac{-6 - 4}{3 - (-3)} = \frac{-10}{6} = \frac{-5}{3}$$

$m_{AB}$ and $m_{CD}$ are negative reciprocals. The lines are perpendicular.

4. Find the distance between the line $3x - 2y - 2 = 0$ and the point $(5, -1)$.

4. $d = \left| \dfrac{Ax_1 + By_1 + C}{\sqrt{A^2 + B^2}} \right| = \left| \dfrac{3(5) - 2(-1) - 2}{\sqrt{3^2 + (-2)^2}} \right|$

$\left| \dfrac{15 + 2 - 2}{\sqrt{9 + 4}} \right| = \left| \dfrac{15}{\sqrt{13}} \right|$

$= \dfrac{15}{\sqrt{13}} \left( \dfrac{\sqrt{13}}{\sqrt{13}} \right) = \dfrac{15\sqrt{13}}{13}$

# Practice Problems 13.1

1. Find the slope of the line joining the points $N(-3, -10)$ and $L(4, -4)$.

   A) $\dfrac{6}{7}$  B) $\dfrac{-6}{7}$  C) $\dfrac{-1}{14}$  D) $\dfrac{-14}{1}$  E) $\dfrac{7}{6}$

2. Find the length of the line joining the points $N(-7, 0)$ and $L(-4, -2)$

   A) 13  B) $\sqrt{13}$  C) $\sqrt{41}$  D) $\sqrt{97}$  E) $\sqrt{125}$

3. Find the slope of the line joining $R(5, 2)$ and $S(-5, -2)$.

   A) 1  B) $-1$  C) $\dfrac{-3}{7}$  D) $\dfrac{-7}{3}$  E) $\dfrac{2}{5}$

4. Find the midpoint of the line joining $J(-6, -2)$ and $K(-8, 4)$.

   A) $(1, -7)$  B) $(-5, -1)$  C) $(-7, 1)$
   D) $(-1, -3)$  E) $(3, -1)$

5. If the midpoint $M$ of a line segment $AB$ is $M(3, 2)$ and endpoint $A$ is $(8, 3)$, find the coordinates of point $B$, the other endpoint.

   A) $(\frac{9}{2}, 3)$  B) $(-2, 1)$  C) $(\frac{5}{2}, \frac{1}{2})$
   D) $(-8, -3)$  E) $(1, -2)$

6. Are the points $N(-2, -11)$, $J(0, -7)$ and $L(4, 1)$ colinear?

   A) Yes  B) No

7. If the center of a circle has coordinates $C(2, 4)$ and a point $S(6, 1)$ is on the circumference, find the length of the radius.

   A) $\sqrt{5}$  B) $\sqrt{13}$  C) 7  D) 5  E) $\sqrt{89}$

8. If a line has a slope of $\frac{-5}{2}$, the slope of a perpendicular line is

   A) $\frac{-5}{2}$  B) $\frac{-2}{5}$  C) $\frac{5}{2}$  D) $\frac{2}{5}$  E) $-1$

Solutions on page 253

# 13.2 Lines

The equation of a line is a first degree equation.

 This means no $x^2$, $y^2$ or $xy$ terms.

| Format | Name | Comment |
|---|---|---|
| $Ax + By + C = 0$ | General form | $AB \neq 0$ |
| | | $m = -\dfrac{A}{B}$ If $A = 0$, line is horizontal. If $B = 0$, line is vertical. |
| | | $b = -\dfrac{C}{B}$ If $C = 0$, line passes through origin. |
| | | $m = \text{slope}, \ b = y - \text{intercept}$ |
| $y = mx + b$ | Slope-intercept | $m = \text{slope}, \quad b = y\text{-intercept}$ |
| $(y - y_1) = m(x - x_1)$ | Point-slope | $m = \text{slope}, \quad (x_1, y_1)$ is a point on the line |
| $\dfrac{x}{a} + \dfrac{y}{b} = 1$ | Intercept form | $a = x\text{-intercept}, \quad b = y\text{-intercept}$ |
| $x = c$ | Vertical line | A vertical line at the value $x = c$ |
| $y = c$ | Horizontal line | A horizontal line at the value $y = c$ |

 **y-intercept: where the line crosses the y-axis.** At this point $x$ has the value 0.

**x-intercept: where the line crosses the x-axis.** At this point $y$ has the value 0.

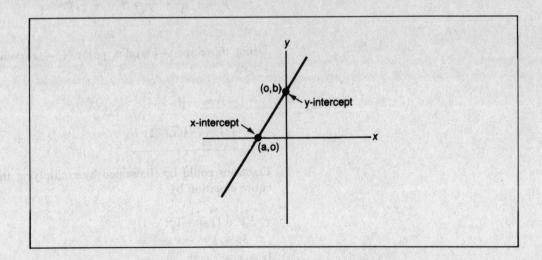

**Examples:**

1. Find the equation of the line passing through the points $N(-2, -1)$ and $L(1, 3)$.

**Solutions:**

1. Given two points, first find the slope:

$$m = \frac{y_2 - y_1}{x_2 - x_1} = \frac{3 - (-1)}{1 - (-2)} = \frac{4}{3}$$

Then use the point-slope form.

Using $m = \frac{4}{3}$ and *either* point, e.g., $L(1,3)$

$$y - y_1 = m(x - x_1)$$

$$\boxed{y - 3 = \tfrac{4}{3}(x - 1)}$$

2. Given the equation $3y = 6x - 9$, find the slope and $y$-intercept.

2. Rearrange the equation into the slope-intercept form $y = mx + b$:

$$3y = 6x - 9$$

Divide by 3:

$$y = \frac{6x}{3} - \frac{9}{3} = 2x - 3$$

$$m = 2 \quad b = -3$$

3. Write the equation of the line parallel to $2x + 6y = 11$ and passing through the point $V(1, -1)$.

3. To be parallel to $2x + 6y = 11$, the line we are looking for must have the same slope. First find the slope of the given line:

$$2x + 6y = 11$$
$$6y = -2x + 11$$
$$y = -\frac{1x}{3} + \frac{11}{3} \Rightarrow m = -\frac{1}{3}, b = \frac{11}{3}$$

Using the slope $-\frac{1}{3}$ and a point $(1, -1)$ (point-slope):

$$y - (-1) = -\tfrac{1}{3}(x - 1)$$

$$y + 1 = -\tfrac{1}{3}(x - 1)$$

Fractions could be eliminated by multiplying the entire equation by 3:

$$3(y + 1) = -1(x - 1)$$
$$3y + 3 = -x + 1$$
$$3y + x + 2 = 0$$

4. Find the *x*- and *y*-intercepts of $y = 3x - 7$.

4. The easiest way to find *x*-intercepts is to let $y = 0$.
The easiest way to find *y*-intercepts is let $x = 0$.

$$y = 3x - 7$$

Let $y = 0$.
$$0 = 3x - 7$$
$$x = \tfrac{7}{3}$$

The *x*-intercept is $\tfrac{7}{3}$.

$(\tfrac{7}{3}, 0)$

Let $x = 0$.
$$y = 3(0) - 7$$
$$y = -7$$

The *y*-intercept is $-7$.

$(0, -7)$

# Graphing A Line

The three most common techniques are:

1) $y = mx + b$    Determine $m =$ slope   $b = $ *y*-intercept.
2) **intercepts**    Determine the *x*- and *y*-intercepts.
3) **table**    Create a table of coordinate values.

Graph $-3x + 2y = 6$ **using** $y = mx + b$

$$-3x + 2y = 6$$
$$2y = 3x + 6$$
$$y = \tfrac{3}{2}x + 3$$
$$m = \tfrac{3}{2} \quad b = 3$$

- Rearrange the equation into $y = mx + b$ form.
- Determine the slope (*m*) and *y*-intercept (*b*).
- On the graph, mark the *y*-intercept.
- The slope (*m*) should be written in fraction form.

$$\left(\frac{3}{2}, \frac{-3}{2}, \frac{2}{1}, \frac{-1}{5}, \frac{-6}{1}\right)$$

☞   $m = \dfrac{\Delta y}{\Delta x} = \dfrac{\text{change in } y}{\text{change in } x} = \begin{cases} \text{a plus number means go up } (\uparrow) \\ \text{a minus number means go down } (\downarrow) \\ \text{a plus number means go right } (\rightarrow) \\ \text{a minus number means go left } (\leftarrow) \end{cases}$

$\dfrac{-3}{2}$   go 3 down ($\downarrow$) and then 2 to the right ($\rightarrow$)

$\dfrac{2}{5}$   go 2 up ($\uparrow$) then 5 to the right ($\rightarrow$)

- Beginning at the y-intercept use the slope to determine the next point. Repeat this step, progressing from one point to the next.
- Connect the points, and label the line.

Since $m = \frac{3}{2}$
go 3 up and 2 to the right

---

> Graph $-3x + 2y = 6$ **using the x- and y-intercepts**

- Let $x = 0$ and solve for $y$.
- Plot the point (0, y-intercept).

$-3x + 2y = 6$
$-3(0) + 2y = 6$
$y = 3$
plot (0,3)

- Let $y = 0$ and solve for $x$.
- Plot the point (x-intercept, 0).
- Connect the points and label the line.

$-3x + 2y = 6$
$-3x + 2(0) = 6$
$x = -2$
plot $(-2,0)$

 If the line goes through the origin, that is, if both the x- and y-intercepts are (0, 0), then determine a third point; **usually** either let $x = 1$ and solve for $y$ or let $y = 1$ and solve for $x$.

---
Graph $-3x + 2y = 6$ **using a table (chart)**
---

- Prepare a chart with 3 to 5 values of $x$; these values are arbitrary. (Try to include 0.)
- Substitute each $x$ into the equation and solve for the respective $y$ value.
- Plot the points, join them, label the graph.

$$-3x + 2y = 6$$
$$-3(-1) + 2y = 6$$
$$2y = 3$$
$$y = \tfrac{3}{2}$$

$$-3(-2) + 2y = 6$$
$$2y = 0$$
$$y = 0$$

| $x$ | $y$ |
|---|---|
| $-1$ | $\frac{3}{2}$ |
| $-2$ | $0$ |
| $0$ | $3$ |
| $1$ | $\frac{9}{2}$ |

$$-3(0) + 2y = 6$$
$$y = 3$$

$$-3(1) + 2y = 6$$
$$2y = 9$$
$$y = \tfrac{9}{2}$$

# Special Line Graphs

**Horizontal Lines**

$$y = c$$

$y = 3$

$y = -1$

$y = 0$

This is a horizontal line at $y = 3$.

Note: this is a graph of the $x$-axis.

**Vertical Lines**

$$x = c$$

x = 1

x = -3

x = 0

This is a vertical
line at $x = 1$.

Note: this is the
graph of the
$y$-axis.

# Practice Problems 13.2

1. Write the equation of the line perpendicular to $y = \frac{1}{2}x - 7$ and passing through the point $(-1, 5)$.

   A) $y = -2x + 7$   B) $y + 1 = -2(x - 5)$   C) $y - 5 = -2(x - 1)$
   D) $y - 5 = -2(x + 1)$   E) $y + 1 = 2(x + 5)$

2. Write the equation of the line joining the points $N(5, 0)$ and $L(0, 5)$.

   A) $5x - 5y = 0$   B) $5y + 5x = 1$   C) $\frac{x}{5} + \frac{y}{5} = 1$

   D) $5x - 5y = 1$   E) $y = 5x + 5$

3. Find the slope of the line $7 + 3y = -4x$.

   A) $-4$   B) $7$   C) $\frac{-4}{3}$   D) $\frac{-7}{3}$   E) $\frac{-3}{4}$

4. The coordinates of the $x$-intercept of the line passing through $N(0, 3)$ and $L(1, 4)$ are

   A) $(-3, 0)$   B) $(0, 0)$   C) $(1, 0)$   D) $(3, 0)$   E) $(4, 0)$

5. Write the equation of the line parallel to $\frac{x}{3} - \frac{y}{2} = 1$ and passing through the point $(1, 3)$.

   A) $(x - 3) = \frac{2}{3}(y - 1)$   B) $(y - 3) = \frac{2}{3}(x - 1)$   C) $(y - 1) = \frac{2}{3}(x - 3)$
   D) $(x - 1) = \frac{2}{3}(y - 3)$   E) $(y + 3) = \frac{2}{3}(x + 1)$

6. Each of the following points is on the line $2y = 4 - x$ except

   A) $(2, 1)$   B) $(-4, 4)$   C) $(-2, -3)$   D) $(0, 2)$   E) $(4, 0)$

7. If the points $N(1, 0)$, $J(3, 6)$ and $L(-4, k)$ are colinear, the value of $k$ is

   A) 4   B) 2   C) $-8$   D) $-10$   E) $-15$

8. If $y = 3x + b$ and the point $(-3, 1)$ is on the line, find the value of $b$.

   A) $-6$   B) $-8$   C) $-10$   D) 10   E) 6

Solutions on page 254

## 13.3  Conic Sections—A Pictorial View

Conic sections are formed by the intersection of a plane with a right circular cone with two nappes.

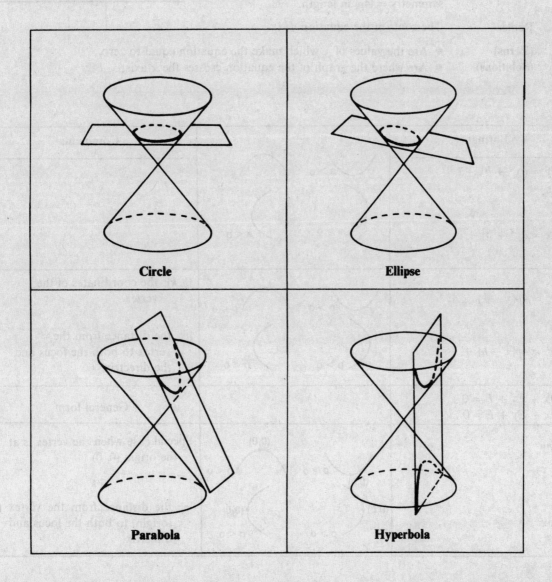

| Circle | Ellipse |
| --- | --- |
| **Parabola** | **Hyperbola** |

# 13.4 Parabola/Quadratic Equation

The equation of a parabola is called a quadratic equation. It is a second degree equation in which only one variable is squared and the other is linear.

**Vertex:** The turning point (minimum/maximum point). Has the coordinates $(h, k)$.

**Axis of Symmetry:** A line through the middle of the parabola. The line containing the focus and perpendicular to the directrix.

**Focus:**
**Directrix:**
- The focus is a point (on the axis of symmetry) inside the parabola.
- The directrix is a line outside the parabola.
- Both the focus and directrix are a distance "$p$" from the vertex.

**Latus Rectum:** A chord of the parabola through the focus and perpendicular to the axis of symmetry $= |4p|$ in length.

**Roots:** The roots of the equation $f(x)$:

**(Zeros)**
**(Solutions)**
- Are the values of $x$ which make the equation equal to zero.
- Are where the graph of the equation crosses the $x$-axis.

| Format | | Comments |
|---|---|---|
| $f(x) = y = ax^2 + bx + c$ <br><br> $f(y) = x = ay^2 + by + c$ | | — |
| $(x - h)^2 = 4p(y - k)$ <br><br><br> $(y - k)^2 = 4p(x - h)$ | | $(h, k)$: the coordinates of the vertex <br><br> $|p|$: the distance from the vertex to both the focus and the directrix |
| $x^2 + Dx + Ey + F = 0$ <br> $y^2 + Dx + Ey + F = 0$ | — | General form |
| $x^2 = 4py$ <br><br><br> $y^2 = 4px$ | | Special case when the vertex is at the origin $(0, 0)$ <br><br> $|p|$: the distance from the vertex (origin) to both the focus and directrix |

# Equations/Properties

| Variable | x is the independent variable | | y is the independent variable | |
|---|---|---|---|---|
| **Equation** | $f(x)=y$ $=ax^2+bx+c$ | $(x-h)^2=4p(y-k)$ | $f(y)=x$ $=ay^2+by+c$ | $(y-k)^2=4p(x-h)$ |
| **Graph** | $a>0$ or $p>0$ | $a<0$ or $p<0$ | $a>0$ or $p>0$ | $a<0$ or $p<0$ |
| **Axis of Symmetry** | $x=\dfrac{-b}{2a}$ | $x=h$ | $y=\dfrac{-b}{2a}$ | $y=k$ |
| **Vertex** (Turning Point) (Min/Max Point) | $\left(\dfrac{-b}{2a},\ f\left(\dfrac{-b}{2a}\right)\right)$ | $(h,k)$ | $\left(f\left(\dfrac{-b}{2a}\right),\ \dfrac{-b}{2a}\right)$ | $(h,k)$ |
| **Focus** | — | $p>0$  $p<0$ $F\colon(h,\ k+p)$ | — | $p>0$  $p<0$ $F\colon(h+p,\ k)$ |
| **Directrix** | — | $y=k-p$ | — | $x=h-p$ |
| **Length of Latus Rectum** | — | $\lvert 4p\rvert$ | — | $\lvert 4p\rvert$ |
| **Eccentricity (e)** | Definition: $\dfrac{\text{the distance focus to vertex}}{\text{the distance vertex to directrix}}$ $\Rightarrow$ the eccentricity of a parabola $=\dfrac{p}{p}=1$ | | | |
| **Roots, Solutions, Zeros** | • The values of $x$ which make $f(x)=0$ *or* • The $x$-intercepts | — | • The values of $y$ which make $f(y)=0$ *or* • The $y$-intercepts | — |

# Roots/Solutions/Zeros

The words **roots, solutions** and **zeros** are **used interchangeably.** There are *three* basic techniques to find the roots of a quadratic equation:

- Factoring
- Completing the square
- Quadratic formula

Without actually finding the roots, several **characteristics** of the roots can be determined:

- **Sum** of the roots
- **Product** of the roots
- **Nature** of the roots

> **Characteristics of the roots**
> $(r_1, r_2)$

Write the equation as:

$$f(x) = y = ax^2 + bx + c \ \text{ or}$$

$$f(y) = x = ay^2 + by + c$$

*a* is the **coefficient of the $x^2$ or $y^2$** (including the + or −)
*b* is the **coefficient of the $x$ or $y$** (including the + or −)
*c* is the **numerical term without a variable** (include the + or −)

| Sum of the roots $(r_1 + r_2)$ | What the roots add to | $-\dfrac{b}{a}$ |
|---|---|---|
| **Product of the roots** $(r_1 \cdot r_2)$ | What the roots multiply to | $\dfrac{c}{a}$ |
| **Nature of the roots** (discriminant) | Are the roots <br> • imaginary or real <br> • The same or different (equal or unequal) <br> • Rational or irrational | $b^2 - 4ac$ |

Example: $y = x^2 - 3x - 1$

$a = 1$    the coefficient of $x^2$
$b = -3$ the coefficient of $x$
$c = -1$ the numerical term

$$\text{Sum} = \frac{-b}{a} = \frac{-(-3)}{1} = 3$$

$$\text{Product} = \frac{c}{a} = \frac{-1}{1} = -1$$

$$\text{Discriminant} = b^2 - 4ac$$
$$= (-3)^2 - 4(1)(-1)$$
$$= 9 + 4$$
$$= 13$$

# Analysis of the Discriminant: $(b^2 - 4ac)$

| The roots are: | $b^2 - 4ac < 0$ (a negative value) | $b^2 - 4ac = 0$ | $b^2 - 4ac > 0$ (a positive value) |
|---|---|---|---|
| • Real or imaginary | Imaginary | Real | Real |
| • The same (equal) or different (unequal) | Different (Unequal) | Same (Equal) | Different (Unequal) |
| • Rational or irrational (has a $\sqrt{\ }$) | | Rational | Rational—If the value is a perfect square, i.e., 1, 4, 9, 16, 25, 36 ... Irrational—All other values |
| Typical graphic representation (Roots are where the graph of the equation crosses the x-axis for $f(x)$) | Discriminant < 0 No real roots | Discriminant = 0 Two equal roots | Discriminant > 0 $r_1$ $r_2$ |
| Comments on graphic interpretation | Note: The curve does NOT cross the x-axis; the roots are imaginary. | Note: The curve is TANGENT to the x-axis. Since there are two equal (the same) roots, the curve can only touch the one value given by this root. | Note: The curve crosses the x-axis at two different values. The two crossings can be both positive, both negative, one of each or zero. |

# Solving a Quadratic Equation

**Factoring:** The easiest and quickest, *if* the equation can be factored

**Completing the Square:** Useful to obtain $(x - h)^2$ or $(y - k)^2$ terms

**Quadratic formula:** Always works, may take a while

## Factoring

Find the roots of the equation $f(x) = x^2 - 5x + 6$ **by factoring.**

1. To find the roots set $f(x) = 0$.
2. Factor.
3. Set each factor equal to 0.
4. Solve for $x$.

$$x^2 - 5x + 6 = 0$$
$$(x - 2) \quad (x - 3) = 0$$
$$x - 2 = 0 \quad | \quad x - 3 = 0$$
$$x = 2 \quad | \quad x = 3$$

The roots are $x = 2$, $x = 3$.

## Completing the Square

Find the roots of $f(x) = x^2 - 5x + 6$ **by completing the square.**

1. To find the roots set $f(x) = 0$:
$$x^2 - 5x + 6 = 0$$

2. Algebraically move the numerical term to the other side of the equal sign:
$$x^2 - 5x \quad = -6$$

3. Divide the coefficient of the $x$ by 2 and add this term squared to both sides of the equation:
$$x^2 - 5x + \left(\frac{-5}{2}\right)^2 = -6 + \left(\frac{-5}{2}\right)^2$$
$$\left(x + \frac{-5}{2}\right)^2 = \frac{1}{4}$$

☞ Don't forget to carry the sign of the $x$ terms coefficient.

4. a) Express the left side of the equation as $(x + a)^2$, where $a$ is the value from step 3 found by dividing the coefficient by 2.
   b) Express the right side as a single numerical value.

5. Take the square root of each side of the equation.
$$\sqrt{(x - \tfrac{5}{2})^2} = \pm\sqrt{\tfrac{1}{4}}$$
$$x - \tfrac{5}{2} = \pm\tfrac{1}{2}$$

☞ Don't forget: if $x^2 = b$, then $x = \pm\sqrt{b}$

6. Separate the equation into two parts and solve for $x$:
$$x - \tfrac{5}{2} = +\tfrac{1}{2} \quad | \quad x - \tfrac{5}{2} = -\tfrac{1}{2}$$
$$x = \tfrac{6}{2} = 3 \quad | \quad x = \tfrac{4}{2} = 2$$

The roots are $x = 2$, $x = 3$.

**Quadratic Formula**

Find the roots of $f(x) = y = x^2 - 5x + 6$ **using the quadratic formula.**

---

**QUADRATIC FORMULA**

$$f(x) = ax^2 + bx + c = 0 \qquad\qquad f(y) = ay^2 + by + c = 0$$

$$x = \frac{-b \pm \sqrt{b^2 - 4ac}}{2a} \qquad\qquad y = \frac{-b \pm \sqrt{b^2 - 4ac}}{2a}$$

---

To find the roots set $f(x) = 0$:

$$x^2 - 5x + 6 = 0$$

Determine the values of $a$, $b$ and $c$.

$a$ is the coefficient of the squared term (including sign):

$b$ is the coefficient of the linear term (including sign):

$c$ is the numerical term (including sign):

$$a = 1$$
$$b = -5$$
$$c = 6$$

Substitute into the quadratic formula:

$$x = \frac{-b \pm \sqrt{b^2 - 4ac}}{2a}$$

Solve for both values of $x$:

$$= \frac{-(-5) \pm \sqrt{(-5)^2 - 4(1)(6)}}{2(1)}$$

$$= \frac{5 \pm \sqrt{25 - 24}}{2} = \frac{5 \pm \sqrt{1}}{2}$$

$$= \frac{5 \pm 1}{2}$$

$$x = \frac{5 + 1}{2} = \frac{6}{2} \quad\Big|\quad x = \frac{5 - 1}{2} = \frac{4}{2}$$

$$= 3 \qquad\qquad\quad = 2$$

# Graphing a Quadratic Equation

The basic technique for graphing a parabola is to **develop a table with seven ordered pairs.** The turning point should be the center value and three points before and after should be used.

Graph the equation $f(x) = y = 2x^2 - 8x + 3$.

Arrange the equation into $y = ax^2 + bx + c$ form:

$$y = 2x^2 - 8x + 3$$

Determine the axis of symmetry:

$$x = \frac{-b}{2a} = \frac{-(-8)}{2(2)} = 2$$

$$x = -\frac{b}{2a}$$

(This is also the $x$ value of the vertex)

Set up a chart with the value from step 2 at the center.

| x | $2x^2 - 8x + 3$ | y |
|---|---|---|
| | | |
| | | |
| | | |
| **2** | | |
| | | |
| | | |
| | | |

4. Choose three values before and after the starting value. Use increments of 1.

5. Determine all seven values of y. Graph the seven ordered pairs; label the ctrve. Graph and label the axis of symmetry (as a dotted line). Label the vertex.

☞ Don't forget to label the axes x and y respectively.

| x | $2x^2 - 8x + 3$ | y |
|---|---|---|
| −1 | $2(-1)^2 - 8(-1) + 3 =$ | 13 |
| 0 | $2(0)^2 - 8(0) + 3 =$ | 3 |
| 1 | $2(1)^2 - 8(1) + 3 =$ | −3 |
| 2 | $2(2)^2 - 8(2) + 3 =$ | −5 |
| 3 | $2(3)^2 - 8(3) + 3 =$ | −3 |
| 4 | $2(4)^2 - 8(4) + 3 =$ | 3 |
| 5 | $2(5)^2 - 8(5) + 3 =$ | 13 |

**Examples:**

1. If $y = x^2 - 2x + 4$, find the sum and product of the roots, the nature of the roots and the solution set.

**Solutions:**

1. $y = x^2 - 2x + 4 \Rightarrow a = 1, b = -2, c = 4$

$$\text{Sum} = \frac{-b}{a} = \frac{-(-2)}{1} = 2$$

$$\text{Product} = \frac{c}{a} = \frac{4}{1} = 4$$

$$\text{Discriminant} = b^2 - 4ac = (-2)^2 - 4(1)(4)$$
$$= 4 - 16 = -12$$

A negative value indicates the roots are imaginary.

$$x = \frac{-b \pm \sqrt{b^2 - 4ac}}{2a}$$

$$= \frac{-(-2) \pm \sqrt{(-2)^2 - 4(1)(4)}}{2(1)}$$

$$= \frac{2 \pm \sqrt{-12}}{2}$$

$$= \frac{2 \pm 2i\sqrt{3}}{2} = 1 \pm i\sqrt{3} \quad \text{or} \quad 1 \pm \sqrt{3}i$$

*Note:* $\sqrt{12} = \sqrt{4}\sqrt{3} = 2\sqrt{3}$

2. If $y = x^2 - 2x - 4$, find the coordinates of the vertex and the equation of the axis of symmetry.

2. $y = x^2 - 2x - 4 \Rightarrow a = 1, b = -2, c = -4$

Axis of symmetry:

$$x = \frac{-b}{2a} = \frac{-(-2)}{2(1)} = 1$$

$$y = f(x) = x^2 - 2x - 4$$
$$y = f(1) = 1^2 - 2(1) - 4 = -5$$

Vertex: $(1, -5)$

Or change the form:

$$y = x^2 - 2x - 4$$
$$y + 4 = x^2 - 2x$$

Complete the square:

$$y + 4 + (-1)^2 = x^2 - 2x + (-1)^2$$
$$y + 5 = (x - 1)^2$$
$$(x - 1)^2 = y + 5$$
$$(x - h)^2 = 4p(y - k)$$

Axis of symmetry is $x = h$ $\therefore x = 1$
Vertex is $(h, k)$ $\therefore (1, -5)$

3. If $2y + 49 = 16x - y^2$, find the coordinates of the vertex and focus, the equations of the axis of symmetry and the directrix, and also the length of the latus rectum.

3. Rearrange the equation:

$$2y + 49 = 16x - y^2$$
$$y^2 + 2y = 16x - 49$$

Complete the square:

$$y^2 + 2y + (1)^2 = 16x - 49 + (1)^2$$
$$(y + 1)^2 = 16x - 48$$
$$(y + 1)^2 = 16(x - 3)$$
$$(y - k)^2 = 4p(x - h)$$

$\therefore k = -1$, $h = 3$ and $p = 4$

Vertex: $(h, k) \Rightarrow (3, -1)$
Focus: $(h + p, k) \Rightarrow (7, -1)$
Axis of symmetry: $y = k \Rightarrow y = -1$
Directrix: $x = h - p \Rightarrow x = -1$
Latus Rectum: $|4p| \Rightarrow |16| = 16$

4. If the roots of an equation are 3 and $-1$, find the equation.

4. Method 1:

If the roots are $x = 3$ and $x = -1$, the factors were $(x - 3)(x + 1) = 0$. Therefore the equation was

$$x^2 - 3x + x - 3 = 0$$
$$x^2 - 2x - 3 = 0$$

Method 2:

$r_1 = 3$   $r_2 = -1$
$x^2 + bx + c = 0$   (Let $a = 1$)

$$\text{Sum} = \frac{-b}{1} = -b$$

$$\text{Sum} = r_1 + r_2 = 3 + (-1) = 2$$
$$-b = 2 \quad b = -2$$

$$\text{Product} = \frac{c}{1} = c$$

$$\text{Product} = r_1 \cdot r_2 = 3(-1) = -3$$

$$c = -3$$
$$x^2 - 2x - 3 = 0$$

5. If 6 is a root of the equation $y = 2x^2 - 7x + k$, find the value $k$.

5. A root is the value of $x$ which makes the equation equal to zero:

$$y = 2(6)^2 - 7(6) + k = 0$$
$$72 - 42 + k = 0$$
$$30 + k = 0$$
$$k = -30$$

# Practice Problems 13.4

1. The quadratic equation whose roots are 1 and 2 is

A) $x^2 + 2x + 3$   B) $x^2 - 2x + 3$
C) $x^2 + 3x + 2$   D) $x^2 - 3x + 2$
E) $x^2 - 2x - 3$

2. For which value of $c$ will the roots of the equation $x^2 + 4x + c = 0$ be real and equal?

A) 1   B) 2   C) 3   D) 4   E) 5

3. Find the focus of the equation $y = x^2 - 4x$.

A) $(2, -3)$   B) $(2, -4)$   C) $(2, -5)$
D) $(1, -3)$   E) $(1, -4)$

4. The directrix of $x^2 = 12(y - 4)$ is

A) $y = 0$   B) $y = 1$   C) $y = 3$
D) $x = 1$   E) $x = 3$

5. The solution set of $f(x) = x^2 - 4x - 5$ is

A) $\{2, -3\}$   B) $\{1, 5\}$   C) $\{-1, 5\}$
D) $\{-5, 1\}$   E) $\{-1, -5\}$

6. If $x = 2$ is a root of $y = x^2 + kx - 2$, find the value of $k$.

A) $-1$   B) 0   C) 1   D) 2   E) 3

7. For which values of $k$ is the graph $f(x) = kx^2 - 4x + k$ tangent to the $x$-axis?

A) 1 and 4   B) 2 and $-2$   C) 2 only
D) 0 and 5   E) $-2$ only

8. Find the roots of $2x^2 - 4 = 5x$.

A) $\left\{\dfrac{-5 \pm \sqrt{57}}{4}\right\}$   B) $\left\{\dfrac{5 \pm \sqrt{57}}{4}\right\}$

C) $\left\{\dfrac{5 \pm \sqrt{-7}}{4}\right\}$   D) $\left\{\dfrac{-5 \pm \sqrt{-7}}{4}\right\}$   E) $\{\pm 3\}$

Solutions on page 255

## 13.5 Circles

A circle is represented by a second degree equation in both $x$ and $y$. When the $x^2$ and $y^2$ terms are on the same side of the equation, they have the **same sign** and the **same coefficients**.

| Format | | |
|---|---|---|
| $(x - h)^2 + (y - k)^2 = r^2$ $\dfrac{(x-h)^2}{r^2} + \dfrac{(y-k)^2}{r^2} = 1$ | Standard form | center: $(h, k)$ $r$ = radius center: $(h, k)$ $r$ = radius |
| $x^2 + y^2 + Dx + Ey + F = 0$ | General form | center: $\left(\dfrac{-D}{2}, \dfrac{-E}{2}\right)$ $r = \frac{1}{2}\sqrt{D^2 + E^2 - 4F}$ |
| $x^2 + y^2 = r^2$ | — | center: $(0, 0)$ $r$ = radius |

## Graphing a Circle

The basic technique is to determine the center and radius. From the center mark points to the right, left, up and down of center at a distance equal to the radius.

Graph the equation $x^2 + 2x + y^2 - 4y = 4$.

1. Arrange the equation into $(x - h)^2 + (y - k)^2 = r^2$ form.

$$x^2 + 2x \quad + y^2 - 4y \quad = 4$$

Complete the square
on $x$ and $y$.

$$x^2 + 2x + (1)^2 + y^2 - 4y + (-2)^2 = 4 + (1)^2 + (-2)^2$$

$$(x + 1)^2 + (y - 2)^2 = 9$$

2. Determine the center and radius.

Center: $(-1, 2)$ $r = 3$

3. Graph the center; locate the four convenient points at $r = 3$.
4. Connect the graphed points.

**Examples:**

1. Write the equation of the circle with center: (0,4) and radius $= \sqrt{5}$.

**Solutions:**

1. $(x - h)^2 + (y - k)^2 = r^2$

$$(x - 0)^2 + (y - 4)^2 = (\sqrt{5})^2$$

$$x^2 + (y - 4)^2 = 5$$

2. Write the equation of the circle with center: (1, −2) and passing through (3, 5).

2. Method 1

radius = distance (1, −2) to (3, 5)

$$r = d = \sqrt{(5 - (-2))^2 + (3 - 1)^2}$$
$$= \sqrt{7^2 + 2^2}$$
$$= \sqrt{53}$$

$$(x - 1)^2 + (y + 2)^2 = 53$$

Method 2:

$$(x - 1)^2 + (y + 2)^2 = r^2$$

Substitute the values of $x$ and $y$ into the equation:

$$(3 - 1)^2 + (5 + 2)^2 = r^2$$
$$4 + 49 = r^2$$
$$r^2 = 53$$
$$\therefore (x - 1)^2 + (y + 2)^2 = 53$$

3. Find the coordinates of the center and the length of the radius of $x^2 + y^2 - 2x - 2y + 1 = 0$.

3. Rearrange and complete the square:

$$x^2 - 2x + (-1)^2 + y^2 - 2y + (-1)^2$$
$$= -1 + (-1)^2 + (-1)^2$$
$$(x - 1)^2 + (y - 1)^2 = 1$$
$$Center: (1, 1) \quad r = 1$$

# Practice Problems 13.5

1. Find the equation of a circle with *Center*: (−1, 1) and $r = 3$.

   A) $(x - 1)^2 + (y + 1)^2 = 3$
   B) $(x + 1)^2 + (y - 1)^2 = 3$
   C) $(x - 1)^2 + (y - 1)^2 = 9$
   D) $(x + 1)^2 + (y + 1)^2 = 9$
   E) $(x + 1)^2 + (y - 1)^2 = 9$

2. The radius of the circle $x^2 + y^2 = 8$ is

   A) 2   B) $2\sqrt{2}$   C) $4\sqrt{2}$   D) 4   E) 8

3. Find the center of $x^2 + y^2 - 6x + 4y + 10 = 0$.

   A) (3, −2)   B) (2, −3)   C) (2, 3)
   D) (−2, 3)   E) (−3, 2)

4. Write the equation of the circle with center: (−1, −1) and passing through the point (4, −5).

   A) $x^2 + y^2 = 41$
   B) $(x - 1)^2 + (y - 1)^2 = \sqrt{41}$
   C) $(x + 1)^2 + (y + 1)^2 = \sqrt{41}$
   D) $(x - 1)^2 + (y - 1)^2 = 41$
   E) $(x + 1)^2 + (y + 1)^2 = 41$

Solutions on page 255

## 13.6 Ellipses

An ellipse is represented by a second degree equation in both $x^2$ and $y^2$. When the $x^2$ and $y^2$ terms are put on the same side of the equation, they have the **same sign** and **different coefficients**.

**Center:** Has coordinates $(h, k)$. Is at the midpoint of both the major vertex and minor vertex.

**Major Axis:** The longer line segment between vertices and passing through the center of the ellipse.

**Minor Axis:** The shorter line segment between vertices and passing through the center. The perpendicular bisector of the major axis.

**Major Vertices:** The endpoints of the major axis. A distance **"a"** from the center.

**Minor Vertices:** The endpoints of the minor axis. A distance **"b"** from the center.

**Focus:** A distance **"c"** from the center.

$$b^2 + c^2 = a^2$$

**Latus Rectum:** A chord of the ellipse through the focus and perpendicular to the major axis.

$$\text{Length} = \left| \frac{2b^2}{a} \right|$$

**Directrix:** A line outside the ellipse perpendicular to the major axis at a distance $\frac{a^2}{c}$ or $\frac{a}{e}$ from the center. Their equations are:

$$y = k \pm \frac{a}{e}$$

$$x = h \pm \frac{a}{e}$$

**Eccentricity (*e*):** A measurement defined as $\frac{c}{a}$.

☞ The focus, directrix and latus rectum are associated with the major axis.

☞ The eccentricity (*e*) for an ellipse is less than 1: $e < 1$.

| | center: $(h, k)$ |
|---|---|
| $\dfrac{(x - h)^2}{a^2} + \dfrac{(y - k)^2}{b^2} = 1$ ⎫ <br><br><br> $\dfrac{(y - k)^2}{a^2} + \dfrac{(x - h)^2}{b^2} = 1$ ⎭ Standard form | "*a*" distance from center to major vertex <br> "*b*" distance from center to minor vertex <br><br> $a^2 > b^2$ |
| $Ax^2 + Cy^2 + Dx + Ey + F = 0$ <br> (*A* and *C* have the same sign) | General form |
| $\dfrac{x^2}{a^2} + \dfrac{y^2}{b^2} = 1$ <br><br> $\dfrac{y^2}{a^2} + \dfrac{x^2}{b^2} = 1$ | Special case: Center is at origin $(0, 0)$ |
| ☞ $a^2$ always represents the larger of the two denominator values and is always associated with the major axis. | |

# Graphing an Ellipse

The usual approach to sketching an ellipse is to determine its center, and the coordinates of its major and minor vertices.

Sketch a graph of the ellipse given by the equation $4x^2 + 9y^2 + 16x - 18y - 11 = 0$.

1. Rearrange the equation into a more useful form.

$$4x^2 + 9y^2 + 16x - 18y = 11$$

Associate the *x* and *y* terms separately:

$$4x^2 + 16x \qquad + 9y^2 - 18y \qquad = 11$$

Complete the square.

☞ When completing the square the $x^2$ or $y^2$ terms should have a coefficient of 1.

$$4(x^2 + 4x + (+2)^2) + 9(y^2 + 2y + (+1)^2)$$
$$= 11 + 4(+2)^2 + 9(+1)^2$$

☞ The factored out numbers, in this case a 4 and 9, multiply the $(+2)^2$ and $(+1)^2$, respectively, adding $4(+2)^2$ and $9(+1)^2$ to the other side of the equation.

$$4(x + 2)^2 + 9(y + 1)^2 = 11 + 16 + 9$$
$$= 36$$

$$\frac{4(x + 2)^2}{36} + \frac{9(y + 1)^2}{36} = \frac{36}{36}$$

$$\frac{(x + 2)^2}{9} + \frac{(y + 1)^2}{4} = 1$$

2. Determine the center and vertices.

Center: $(-2, -1)$

Major vertices $\pm 3$ from center
Minor vertices $\pm 2$ from center

The major axis is parallel to the $x$ axis.

3. Graph the information from step 2 and sketch the ellipse.

**Examples:**

1. Find the equation of an ellipse having foci at $(3, 8)$ and $(3, 2)$, and whose eccentricity is $\frac{3}{5}$.

**Solutions:**

1. Draw a sketch to assist.

Since the foci are on the major axis the form is

$$\frac{(y-k)^2}{a^2} + \frac{(x-h)^2}{b^2} = 1$$

Since the foci are equidistant from the center, the center is the midpoint between the foci.

$$\text{center} = MP = \left(\frac{3+3}{2}, \frac{8+2}{2}\right) = (3, 5)$$

The distance "$c$" center to focus is 3.

$$c = 3$$

The eccentricity $e = \dfrac{3}{5} = \dfrac{c}{a}$

Substituting $c = 3 \Rightarrow a = 5$

$$b^2 + c^2 = a^2$$
$$b^2 \quad = 5^2 - 3^2 = 16$$

*Center*: $(3, 5)$   $a = 5$,   $b = 4$

$$\frac{(y-5)^2}{25} + \frac{(x-3)^2}{16} = 1$$

2. Find the coordinates of the center, foci, major vertices, minor vertices and the length of the latus rectum for the ellipse $2x^2 + 3y^2 - 4x - 18y + 23 = 0$.

2. Rearrange the equation.
First separate the variables and complete the square.

 Be sure the $x^2$ and $y^2$ terms have a coefficient of 1 when completing the square

$$2x^2 - 4x + 3y^2 - 18x = -23$$
$$2(x^2 - 2x + (-1)^2) + 3(y^2 - 6x + (-3)^2)$$
$$= -23 + 2(-1)^2 + 3(-3)^2$$
$$2(x-1)^2 + 3(y-3)^2 = -23 + 2 + 27$$
$$= 6$$
$$\frac{2(x-1)^2}{6} + \frac{3(y-3)^2}{6} = \frac{6}{6}$$
$$\frac{(x-1)^2}{3} + \frac{(y-3)^2}{2} = 1$$

● Center: $(1, 3)$

$$a = \sqrt{3}$$
$$b = \sqrt{2}$$

● Major axis is parallel to the $x$ axis
$$b^2 + c^2 = a^2$$
$$(\sqrt{2})^2 + c^2 = (\sqrt{3})^2$$
$$2 + c^2 = 3$$
$$c = 1$$

- Foci: (2, 3)
    (0, 3)

- Major vertices: $(1 + \sqrt{3}, 3)$
    $(1 - \sqrt{3}, 3)$

- Minor vertices: $(1, 3 + \sqrt{2})$
    $(1, 3 - \sqrt{2})$

- Length of latus rectum:

$$\left|\frac{2b^2}{a}\right| = \frac{2(2)}{\sqrt{3}} = \frac{4\sqrt{3}}{3}$$

# Practice Problems 13.6

1. Find the equation of the ellipse with $V(0, \pm 6)$ and $F(0, \pm 2)$.

   A) $\dfrac{x^2}{36} + \dfrac{y^2}{32} = 1$   B) $\dfrac{y^2}{36} + \dfrac{x^2}{32} = 1$

   C) $\dfrac{y^2}{36} + \dfrac{x^2}{4} = 1$   D) $\dfrac{x^2}{36} + \dfrac{y^2}{4} = 1$

   E) $\dfrac{y^2}{32} + \dfrac{x^2}{4} = 1$

2. Find the distance from the center to the major vertex for the ellipse with equation

$$\frac{(x - 1)^2}{4} + \frac{(y - 2)^2}{3} = 3$$

   A) $\sqrt{3}$   B) 2   C) $\dfrac{2\sqrt{3}}{3}$   D) $\dfrac{4}{3}$   E) $\sqrt{12}$

3. If $4x^2 + 9y^2 = 36$, find the length of the latus rectum.

   A) $\dfrac{4}{3}$   B) $\dfrac{10}{3}$   C) $\dfrac{8}{3}$   D) $\dfrac{81}{2}$   E) 9

4. Find the distance from the center to a focus for the ellipse

$$\frac{25x^2}{9} + 4y^2 = 1$$

   A) $\dfrac{\sqrt{11}}{10}$   B) $\dfrac{\sqrt{11}}{6}$   C) $\dfrac{5}{2}$   D) $\dfrac{4}{3}$   E) $\dfrac{7}{6}$

Solutions on page 256

# 13.7 Hyperbola

A hyperbola is represented by a second degree equation in both $x^2$ and $y^2$. When the $x^2$ and $y^2$ terms are on the same side of the equation, the two terms have **different signs.**

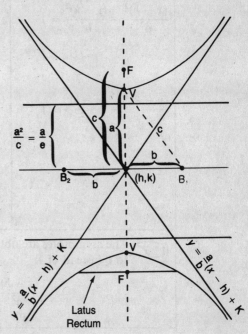

| | |
|---|---|
| **Center:** | Has the coordinates $(h, k)$. Is at the midpoint of the line segment connecting the two vertices. |
| **Vertices:** | The turning points. A distance **'a'** from the center. |
| **Foci:** | A distance **'c'** from the center. |
| **Transverse Axis:** | The line through the foci, vertices and center. |
| **Conjugate Axis:** | The line through the center perpendicular to the transverse axis. It extends between two points $(B_1, B_2)$ each a distance **'b'** from the center. |

☞ The hyperbola does not cross the conjugate axis but points $B_1$ and $B_2$ are useful in its construction.

| | |
|---|---|
| **Latus Rectum:** | A chord of the hyperbola perpendicular to the transverse axis, passing through the focus. Its length is $\left\|\dfrac{2b^2}{a}\right\|$. |
| **Directrices:** | Two lines, each a distance $\dfrac{a^2}{c}$ or $\dfrac{a}{e}$ from the center. |
| **Asymptotes:** | Two lines which the hyperbola branches approach but do not touch. |

Their equation are $\begin{cases} y = \pm b/a(x - h) + k & \text{for} \quad \text{⤭} \quad \text{hyperbolas} \\ y = \pm a/b(x - h) + k & \text{for} \quad \text{⤭} \quad \text{hyperbolas} \end{cases}$

$$\boxed{b^2 + a^2 = c^2}$$

**Eccentricity:** A measurement defined as $\dfrac{c}{a}$.
**(e)**

 The eccentricity ($e$) of a hyperbola is greater than $e > 1$.

| | |
|---|---|
| $\dfrac{(x-h)^2}{a^2} - \dfrac{(y-k)^2}{b^2} = 1$ <br><br><br> $\dfrac{(y-k)^2}{a^2} - \dfrac{(x-h)^2}{b^2} = 1$ | **Standard form** <br> center: $(h,k)$ <br><br> 'a' distance from center to vertex <br><br> 'b' distance from center to a point on the conjugate axis <br>   |

 The asymptotes are obtained from setting the equation of the hyperbola equal to 0 instead of 1:

$$\frac{(x-h)^2}{a^2} - \frac{(y-k)^2}{b^2} = 0, \quad \frac{(y-k)^2}{a^2} - \frac{(x-h)^2}{b^2} = 0$$

| | |
|---|---|
| $Ax^2 + Cy^2 + Dx + Ey + F = 0$ <br> ($A$ and $C$ have different signs) | **General form** |
| $\dfrac{x^2}{a^2} - \dfrac{y^2}{b^2} = 1$ | Special case: center: $(0,0)$ <br><br> Asymptotes are: $y = \pm\dfrac{a}{b}x$ |
| $\dfrac{y^2}{a^2} - \dfrac{x^2}{b^2} = 1$ | Asymptotes are: $y = \pm\dfrac{b}{a}x$ |

 $a^2$ is always associated with the positive term.

| | |
|---|---|
| $xy = a$ | The hyperbola whose asymptotes are the $x$ and $y$ axis <br><br>  <br> $a > 0$   the curve is in the I and III quadrants <br> $a < 0$   The curve is in the II and IV quadrants |

# Graphing a Hyperbola

The usual approach to sketching a hyperbola is to determine its center, vertices and the points $B_1$ and $B_2$ on the conjugate axis. Then the asymptotes are constructed by completing a rectangle containing the vertices, $B_1$ and $B_2$. The diagonals of the rectangle form the asymptotes.

Sketch a graph of the hyperbola given by the equation $4x^2 - 9y^2 + 16x - 18y - 29 = 0$

1. Rearrange the equation into a more useful form.

$$4x^2 - 9y^2 + 16x - 18y = 29$$

Associate the $x$ and $y$ terms separately and complete the square.

$$4x^2 + 16x - 9y^2 - 18y = 29$$
$$4(x^2 + 4x + (2)^2) - 9(y^2 + 2y + (1)^2)$$
$$= 29 + 4(2)^2 - 9(1)^2$$
$$4(x + 2)^2 - 9(y + 1)^2 = 29 + 16 - 9$$
$$= 36$$
$$4(x + 2)^2 - 9(y + 1)^2 = 36$$

divide by 36 to make the equation equal to 1

$$\frac{(x + 2)^2}{9} - \frac{(y + 1)^2}{4} = 1$$

2. Determine the center, vertices, '$a$' and '$b$'.

*Center*: $(-2, -1)$

Vertices are a distance 3 from the center.

$$a = 3$$

The points on the conjugate axis are a distance 2 from the center.

$$b = 2$$

$$B_1(-2, 1) \qquad B_2(-2, -3)$$

3. Graph the center, vertices, $B_1$ and $B_2$, complete the rectangle and draw the asymptotes.

4. Sketch the hyperbola.

**Examples:**

1. Find the center, foci, vertices and eccentricity of the hyperbola $4x^2 - 9y^2 - 8x + 36y = 68$.

**Solutions:**

1. Put in standard form by completing the square.

$$(4x^2 - 8x) \qquad -(9y^2 + 36y) \qquad = 68$$
$$4(x^2 - 2x + (-1)^2) - 9(y^2 - 4y + (-2)^2)$$
$$= 68 + 4(-1)^2 - 9(-2)^2$$

$$4(x-1)^2 - 9(y-2)^2 = 36$$

divide each term by 36 to make the equation equal to 1.

$$\frac{(x-1)^2}{9} - \frac{(y-2)^2}{4} = 1$$

center: $(1, 2)$ $\qquad a = 3 \qquad b = 2$

$$c^2 = a^2 + b^2$$
$$= 9 + 4$$
$$= 13$$
$$c = \sqrt{13}$$

center: $(1, 2)$

Vertices: $(-2, 2), (4, 2)$

Foci: $(1 - \sqrt{13}, 2), (1 + \sqrt{13}, 2)$

Eccentricity: $e = \dfrac{c}{a} = \dfrac{\sqrt{13}}{3}$

2. Write the equation of the hyperbola with $C: (0, 0)$, one vertex at $(0, 6)$ and focus at $(0, 10)$.

2.

The format is $\dfrac{y^2}{a^2} - \dfrac{x^2}{b^2} = 1$.

$$a = 6 \qquad c = 10$$

$$\therefore b^2 + a^2 = c^2$$
$$b^2 = 10^2 - 6^2$$
$$= 64$$
$$b = 8$$

$$\dfrac{y^2}{36} - \dfrac{x^2}{64} = 1$$

3. Find the equation of the asymptotes for the hyperbola $4(y - 4)^2 - 9(x - 2)^2 = 36$.

3. $4(y-4)^2 - 9(x-2)^2 = 36$
divide each term by 36 to make the equation equal to 1

$$\dfrac{(y - 4)^2}{9} - \dfrac{(x - 2)^2}{4} = 1$$

center: $(2, 4)$   $a = 3, b = 2$

$$y = \pm\tfrac{3}{2}(x - 2) + 4$$

# Practice Problems 13.7

1. Find the length of the latus rectum for the hyperbola

$$\dfrac{x^2}{25} - \dfrac{y^2}{9} = 1$$

A) $\tfrac{50}{9}$   B) $\tfrac{18}{5}$   C) 3   D) $\tfrac{50}{3}$   E) $\sqrt{34}$

2. Find the coordinates of one of the foci of

$$\dfrac{(x - 1)^2}{2} - \dfrac{(y + 11)^2}{2} = 1$$

A) $(1, -11)$   B) $(-1, -11)$   C) $(2, -11)$
D) $(1, 13)$   E) $(1, 9)$

3. Find the equation of one of the asymptotes for the hyperbola

$$16x^2 + 96x - 9y^2 + 36y = 36$$

A) $3y - 4x - 18 = 0$   B) $4y - 3x - 17 = 0$
C) $96x + 36y = 36$   D) $9y - 16x - 62 = 0$
E) $9y + 16x + 30 = 0$

4. The graph of $xy = -3$ is in which quadrants?

A) I and II only   B) I and III only
C) II and IV only   D) I and IV only
E) II only

Solutions on page 257

## 13.8 Summary of Characteristics

| Type of Relation | Characteristics of Equations | Examples | Usual Format | Graphic Example | Misc. |
|---|---|---|---|---|---|
| **Line** | All linear terms | $x=4,\; y=7$ <br> $3x+2y=9$ <br> $y=5x+3$ | $y=mx+b$ | | |
| **Parabola** | Only one variable is squared, the other is linear | $y^2=8,\; y^2+x=7y$ <br> $y=3x^2-9x+2$ | $(y-k)=4p(x-h)^2$ <br> and <br> $(x-h)=4p(y-k)^2$ <br> or <br> $y=ax^2+bx+c$ <br> and <br> $x=ay^2+by+c$ | | $e=1$ |

**With $x^2$ and $y^2$ on the same side of the equation**

| Type | Characteristics | Examples | Usual Format | Graphic | Misc. |
|---|---|---|---|---|---|
| **Circle** | Both variables squared, same signs, same coefficients | $x^2+y^2=9$ <br> $3x^2+3y^2=9$ <br> $(x-2)^2+(y+1)^2=9$ | $(x-h)^2+(y-k)^2=r^2$ | | |
| **Ellipse** | Both variables squared, same signs, different coefficients | $x^2+2y^2=9$ <br> $3x^2+4y^2=9$ <br> $\dfrac{x^2}{2}+\dfrac{y^2}{4}=9$ <br> $\dfrac{(x-1)^2}{2}+\dfrac{(y+2)^2}{3}=1$ | $\dfrac{(x-h)^2}{a^2}+\dfrac{(y-k)^2}{b^2}=1$ <br> $\dfrac{(y-k)^2}{a^2}+\dfrac{(x-h)^2}{b^2}=1$ | | $e<1$ |
| **Hyperbola** | Both variables squared, different signs (coefficients can be the same or different) <br><br> $xy=a$ | $x^2-2y^2=9$ <br> $\dfrac{y^2}{3}-\dfrac{x^2}{4}=1$ <br> $\dfrac{(x-1)^2}{4}-\dfrac{(y+2)^2}{3}=1$ <br><br> $xy=6,\quad x=\dfrac{2}{y}$ | $\dfrac{(x-h)^2}{a^2}-\dfrac{(y-k)^2}{b^2}=1$ <br> $\dfrac{(y-k)^2}{a^2}-\dfrac{(x-h)^2}{b^2}=1$ <br><br> $xy=a$ | | $e>1$ |

# Practice Problems 13.8

1. The equation

$$2x^2 + 4y^2 + 2x + 4y + 2 = 0$$

is a(n)

A) circle   B) parabola   C) ellipse
D) hyperbola   E) line

2. The equation

$$2x^2 + 2y^2 + 2x + 2y + 2 = 0$$

is a(n)

A) circle   B) parabola   C) ellipse
D) hyperbola   E) line

3. The equation

$$2x^2 - 4y^2 + 2x + 4y + 2 = 0$$

is a(n)

A) circle   B) parabola   C) ellipse
D) hyperbola   E) line

4. The equation

$$2x^2 - 4y + 2x + 2 = 0$$

is a(n)

A) circle   B) parabola   C) ellipse
D) hyperbola   E) line

Solutions on page 257

# 14. Special Products, Factoring and Applications of Factoring

## 14.1 Special Products

These are frequently used algebraic multiplications. They serve as a model for more complex products of the same form. Because of their frequency of occurrence math students may find it useful to commit these to memory. These products are also used in reverse to convert from a polynomial into factors.

$$(x + y)^2 = (x + y)(x + y) = x^2 + 2xy + y^2$$
$$(x - y)^2 = (x - y)(x - y) = x^2 - 2xy + y^2$$
$$(x + y)(x - y) = x^2 - y^2$$

$$(x + y)(x^2 - xy + y^2) = x^3 + y^3$$
$$(x - y)(x^2 + xy + y^2) = x^3 - y^3$$

 $(a+b)^2 \neq a^2 + b^2$ unless $ab = 0$
$(a-b)^2 \neq a^2 + b^2$ unless $ab = 0$

**Examples:**

1. Multiply $(3a + 2b)^2$.

2. Multiply $(7n - 8d)(7n + 8d)$.

**Solutions:**

1. Using the model
$(x+y)^2 = (x+y)(x+y) = x^2 + 2xy + y^2$
where $x = 3a$ and $y = 2b$
$(3a + 2b)^2 = (3a)^2 + 2(3a)(2b) + (2b)^2$
$\qquad\qquad = 9a^2 + 12ab + 4b^2$

2. Using the model
$(x - y)(x + y) = x^2 - y^2$
where $x = 7n$ and $y = 8d$
$(7n - 8d)(7n + 8d) = (7n)^2 - (8d)^2$
$\qquad\qquad\qquad = 49n^2 - 64d^2$

3. Multiply $(7l + 3)(49l^2 - 21l + 9)$.

3. Using the model

$$(x+y)(x^2 - xy + y)^2 = x^3 + y^3$$

where $x = 7l$ and $y = 3$

$$(7l + 3)(49l^2 - 21l + 9) = (7l)^3 + (3)^3$$
$$= 343l^3 + 27$$

## Practice Problems 14.1

1. Multiply $(3f - 6g^2)^2$.

   A) $9f^2 - 36fg^2 + 36g^4$  B) $9f^2 - 18fg^2 + 36g^4$
   C) $9f - 18fg + 36g^2$
   D) $9f^2 + 36g^4$  E) $9f^2 - 36g^4$

2. Multiply $(5m^3 - 3)(5m^3 + 3)$

   A) $25m^6 - 9$  B) $25m^9 - 9$
   C) $25m^6 - 6$  D) $25m^9 + 6$
   E) $10m^6 - 9$

3. Multiply $(5 - 3x)(25 + 15x + 9x^2)$.

   A) $27x^3 - 125$  B) $27x^2 + 45x - 30$
   C) $27x^3 - 12x + 125$  D) $125 - 27x^3$
   E) $9x^2 + 12x + 30$

4. Multiply $(7\sqrt{a} + 3\sqrt{b})(7\sqrt{a} - 3\sqrt{b})$.

   A) $58$  B) $49a - 9b$  C) $49a^2 - 9b^2$
   D) $49a^2 + 9b^2$
   E) $49a + 9b$

Solutions on page 258

## 14.2 Factoring/Factoring by Greatest Common Factor

Factors are numbers, terms or a series of terms which divide without remainder into the original expression. Expressions which can be factored may be stated as a product.

There are three basic types of factoring:

- **Greatest Common Factor (GCF)**
- **Special Products** $\Rightarrow x^2 + 2xy + y^2$
  $$x^2 - 2xy + y^2$$
  $$x^2 - y^2$$
  $$x^3 + y^3$$
  $$x^3 - y^3$$
- **Trinomial** $(ax^2 + bx + c)$

The basic starting point is to attempt factoring in the order provided above: GCF first, Special Products second and Trinomial last.

*Factoring by* **Greatest Common** *Factor:*

- Find the greatest common monomial factor of each term.
- Divide the GCF into each term and obtain a quotient.
- Express the original expression as the product of the GCF and quotient obtained (Original Expression) = (GCF)(Quotient).

**Examples:**

1. Find the GCF of $18x^3y^2$ and $27x^5y$.

2. Factor $18x^3y^2 + 27x^5y$.

3. Factor $3a^2b^3 - 6a^3b^2 + 9a^4$.

**Solutions:**

1. 9 is the largest common numerical factor.
   $x^3$ is the highest common power of $x$.
   $y$ is the highest common power of $y$.
   The GCF $= 9x^3y$.

2. 1. The GCF $= 9x^3y$.

   2. $\dfrac{18x^3y^2 + 27x^5y}{9x^3y} = \dfrac{18x^3y^2}{9x^3y} + \dfrac{27x^5y}{9x^3y}$

      $\qquad\qquad\qquad = 2y + 3x^2$

   3. $18x^3y^2 + 27x^5y = 9x^3y(2y + 3x^2)$

3. 3 is the greatest common numerical factor.
   $a^2$ is the highest common power of $x$.
   There is no common factor of $b$.

   1. The GCF $= 3a^2$.

   2. $\dfrac{3a^2b^3 - 6a^3b^2 + 9a^4}{3a^2} = b^3 - 2ab^2 + 3a^2$

   3. $3a^2 - 6a^3b^2 + 9a^4 = 3a^2(b^3 - 2ab^2 + 3a^2)$

# Practice Problems 14.2

1. Factor $9x^2 - 9x^2y$.

   A) $9(x^2 - y)$   B) $9x^2(1 - y)$
   C) $9y(1 - x^2)$   D) $9x^2(y)$
   E) $9x^2(y - 1)$

2. Factor $3m^2n^2 + 3m^3n^3$.

   A) 3   B) $3m^2n^2$   C) $3m^2n^2(1 + mn)$
   D) $3m^2n^2(mn)$   E) $3mn(1 + mn)$

3. Factor $4a^2 + 20a + 5a^3b$.

   A) $a(4a + 20 + 5a^2b)$
   B) $4a(a + 5 - a^2b)$   C) $5a(a + 4 - a^2b)$
   D) $a(4a^2 + 20 + ab)$   E) $ab(4a + 1)$

4. Factor $3n^2l^2 + 6nl + 9n^3l^3$.

   A) $nl(nl + 3 + 3n^2l^2)$
   B) $n^2l^2(3 + 2nl + 3n^2l^2)$
   C) $3(7nl + 3n^3l^3)$
   D) $3nl(nl + 2 + 3n^2l^2)$
   E) $3n^3l^3(3nl + 2n^2l^2 + 1)$

Solutions on page 258

## 14.3 Factoring by Special Products

- Match the pattern of the polynomial to be factored to the formats of the special products.

| Special Product | Factors | Name of Special Product |
|---|---|---|
| $x^2 + 2xy + y^2$ | $= (x+y)(x+y) = (x+y)^2$ | Perfect square |
| $x^2 - 2xy + y^2$ | $= (x-y)(x-y) = (x-y)^2$ | trinomials |
| $x^2 - y^2$ | $= (x+y)(x-y)$ | Difference of two squares |
| $x^3 + y^3$ | $= (x+y)(x^2 - xy + y^2)$ | Sum of two cubes |
| $x^3 - y^3$ | $= (x-y)(x^2 + xy + y^2)$ | Difference of two cubes |

● If there is a match, assign appropriate values to the $x$ and $y$ and use the chart above.

 Students will find memorizing the special products and their factors useful.

**Examples:**

1. Factor $4a^2 - 1$.

2. Factor $8a^3 - 1$.

3. Factor $4a^2 + 4ab + b^2$.

**Solutions:**

1. Using the format

$$x^2 - y^2$$

where $x = 2a$ and $y = 1$

$$x^2 - y^2 = (x + y)(x - y)$$
$$4a^2 - 1 = (2a + 1)(2a - 1)$$

2. Using the format

$$x^3 - y^3$$

where $x = 2a$ and $y = 1$

$$x^3 - y^3 = (x - y)(x^2 + xy + y^2)$$
$$8a^3 - 1 = (2a - 1)(4a^2 + 2a + 1)$$

3. Using the format

$$x^2 + 2xy + y^2$$

where $x = 2a$ and $y = b$

$$y = b$$

$$x^2 + 2xy + y^2 = (x + y)(x + y)$$
$$4a^2 + 4ab + b^2 = (2a + b)(2a + b)$$

# Practice Problems 14.3

1. Factor $9a^2 - 16$.

   A) $(9a)(4)$  B) $(9a - 16)(9a + 16)$
   C) $(3a + 4)(3a + 4)$  D) $(3a - 4)(3a - 4)$
   E) $(3a + 4)(3a - 4)$

2. Factor $16 - 9a^2$.

   A) $(3a + 4)(3a - 4)$  B) $(4 + 3a)(4 - 3a)$
   C) $(3a - 4)(3a - 4)$  D) $(4 - 3a)(4 - 3a)$
   E) $(4 + 3a)(4 + 3a)$

3. Factor $a^2 + 10a + 25$.

    A) $(a + 1)(a + 25)$   B) $(a + 4)(a + 5)$
    C) $(a + 5)(a - 5)$   D) $(a + 5)(a + 5)$
    E) $(a - 5)(a - 5)$

4. Factor $27a^3 - 8$.

    A) $(3a + 2)(3a - 2)$   B) $(3a + 2)(9a^2 - 6a + 4)$
    C) $(3a - 2)(9a^2 - 6a + 4)$
    D) $(3a - 2)(9a^2 + 6a + 4)$
    E) $(3a - 2)(9a^2 + 6a + 2)$

5. Factor $27a^3 + 8$.

    A) $(3a + 2)(3a - 2)$
    B) $(3a + 2)(9a^2 - 6a + 4)$
    C) $(3a - 2)(9a^2 - 6a + 4)$
    D) $(3a - 2)(9a^2 + 6a + 4)$
    E) $(3a - 2)(9a^2 + 6a + 12)$

6. Factor $9a^2 - 12a + 4$.

    A) $(3a + 2)(3a + 2)$   B) $(3a + 2)(3a - 2)$
    C) $(3a - 2)(3a - 2)$   D) $(3a - 2)(3a - 10)$
    E) $(9a + 1)(a + 4)$

Solutions on page 258

## 14.4 Factoring Trinomials and Factoring Completely

Factoring **trinomials** of the form $ax^2 + bx + c$ is a trial and error process. The approach varies slightly depending on the value of $a$.

 Always see if a greatest common monomial factor, different from 1, exists. If so, factor out the GCF before attempting trinomial factoring.

> There are four types to be considered:
>
> - $ax^2 + bx + c$  when $a = 1$  $\Rightarrow$  $x^2 + bx + c$
> - $ax^2 + bx + c$  when $a$ is a  $\Rightarrow$  $2x^2 + bx + c$
>         **prime number**        $3x^2 + bx + c$
>         **greater than 1**     $5x^2 + bx + c$
>                            $7x^2 + bx + c$
>                            $11x^2 + bx + c$
> - $ax^2 + bx + c$  when $a$ is a  $\Rightarrow$  $4x^2 + bx + c$
>         **non-prime**         $6x^2 + bx + c$
>         **number**           $8x^2 + bx + c$
>                              $9x^2 + bx + c$
> - $ax^2 + bx + c$  **non-factorable**

**Factoring $ax^2 + bx + c$ when $a = 1$**

1. Write the factors as $(x\quad)(x\quad)$.
2. Determine all the corresponding pairs of factors of $c$. Note: If $c > 0$ use pairs of positive factors and pairs of negative factors.
   If $c < 0$ use pairs of factors with different signs.
3a. Insert each pair of factors into the parentheses on a trial and error basis. Do one pair at a time.
3b. Insert the first pair in order. Put the first number of the pair into the first set of parentheses and the

Factor $x^2 + 6x + 5$.

1. $(x\quad)(x\quad)$
2. $+5, +1$
   $-5, -1$

      $(+5)(x) = +5x$
3. $(x + 5)(x + 1)$
      $x(+1) = +1x$

    $+5x + 1x = 6x\ \checkmark$

second number of the pair into the second set of parentheses.

3c.  Test for the middle term by multiplying the inner and outer terms (including signs). Add the results.

3d.  Does the sum equal the middle term of the trinomial? If yes—these are the factors.
If no—continue the process with the next pair of factors.

The factors are

$$(x + 5)(x + 1)$$

**Examples:**

1.  Factor $x^2 + 2x - 8$.

**Solutions:**

1.  $c = -8$, middle term $= +2x$

1.  $(x \quad )(x \quad )$   the trinomial has $a = 1$.

2.  $-8, +1$   Identify the pairs of factors of $c = -8$.
$+8, -1$
$-4, +2$
$+4, -2$

3.  Insert each pair and test.

$\overset{-8x}{(x - 8)(x + 1)} \quad -8x + 1x = -7x \neq 2x$
$\underset{+1x}{}$

$\overset{8x}{(x + 8)(x - 1)} \quad 8x + (-1x) = 7x \neq 2x$
$\underset{-1x}{}$

$\overset{-4x}{(x - 4)(x + 2)} \quad -4x + 2x = -2x \neq 2x$
$\underset{+2x}{}$

$\overset{+4x}{(x + 4)(x - 2)} \quad +4x + (-2x) = 2x = 2x \checkmark$
$\underset{-2x}{}$

Factors are $(x + 4)(x - 2)$.

2.  Factor $x^2 + 9x + 20$.

2.  $c = 20$, middle term $= +9x$

1.  $(x \quad )(x \quad )$   The trinomial has $a = 1$.

2.  $+4, +5$   Identify the pairs of factors of $c = 20$.
$-4, -5$
$+10, +2$
$-10, -2$
$+20, +1$
$-20, -1$

3.  Insert each pair and test.

$\overset{+4x}{(x + 4)(x + 5)} \quad 4x + 5x = 9x = 9x \checkmark$
$\underset{+5x}{}$

Factors are $(x + 4)(x + 5)$.

## Factoring $ax^2 + b + c$ when $a$ is a prime number greater than one

1. Write the factors as $(ax\quad)(x\quad)$.
2. Determine all the corresponding pairs of factors of $c$. Note: If $c > 0$ use pairs of positive factors and pairs of negative factors.
   If $c < 0$ use pairs of factors with different signs.
3a. Insert each pair of factors into the parentheses on a trial and error basis. Do one pair at a time.
3b. Insert the first pair in order. Put the first number of the pair into the first set of parentheses and the second number of the pair into the second set of parentheses.
3c. Test for the middle term by multiplying the inner and outer terms (including signs). Add the results.
3d. Does the sum equal the middle term of the trinomial? If yes—these are the factors.
   If no—reverse the pair of numbers. Put the first number of the pair in the second set of parentheses and the second number of pair in the first set of parentheses.
3e. Test for the middle term by multiplying the inner and outer terms (including signs). Add the results.
3f. Does the sum equal the middle term of the trinomial? If yes—these are the factors.
   If no—continue the process with the next pair of factors.

Factor $3x^2 - 5x - 12$.

1. $(3x\quad)(x\quad)$
2. $-12, +1$
   $+12, -1$
   $-4, +3$
   $+4, -3$
   $-6, +2$
   $+6, -2$

3. $(3x \underset{+3x}{\overset{-12x}{- 12)(x + 1}})$

$(-12x) + (+3x) = -9x \neq -5x$

$(3x \underset{-36x}{\overset{+1x}{+ 1)(x - 12}})$

$(+1x) + (-36x) = -35x \neq -5x$

$(3x \underset{-3x}{\overset{12x}{+ 12)(x - 1}})$

$12x + (-3x) = 9x \neq -5x$

$(3x \underset{36x}{\overset{-x}{- 1)(x + 12}})$

$-x + 36x = 35x \neq -5x$

$(3x \underset{+9x}{\overset{-4x}{- 4)(x + 3}})$

$-4x + 9x = 5x \neq -5x$

$(3x \underset{-12x}{\overset{3x}{+ 3)(x - 4}})$

$3x(-12x) = -9x \neq -5x$

$(3x \underset{-9x}{\overset{4x}{+ 4)(x - 3}})$

$4x + (-9x) = -5x = -5x \checkmark$

Factors are $(3x + 4)(x - 3)$.

**Examples:**

3. Factor $5x^2 - 24x - 5$.

**Solutions:**

3. $c = -5$, middle term $= -24x$

   1. $(5x \quad )(x \quad )$    5 is a prime number.

   2. $-5, +1$ Identify the pairs of factors of $c = 5$.

   3. Insert each pair and test. Also, be sure to reverse each pair.

$$(5x \underset{+5x}{\overset{-5x}{- 5)(x + 1)}}$$

$$-5x + 5x = 0 \neq -24x$$

$$(5x \underset{-25x}{\overset{x}{+ 1)(x - 5)}}$$

$$x + (-25x) = -24x = -24x$$

The factors are $(5x + 1)(x - 5)$.

4. Factor $11x^2 - 35x + 6$.

4. $c = 6$, middle term $= -35x$

   1. $(11x \quad )(x \quad )$    11 is a prime number.

   2. $+6, +1$   Identify the pairs of factors of $c = +6$.
      $-6, -1$
      $+3, +2$
      $-3, -2$

   3. Insert each pair of factors and test. Be sure to reverse each pair.

$$(11x \underset{11x}{\overset{6x}{+ 6)(x + 1)}}$$

$$6x + 11x = 17x \neq -35x$$

$$(11x \underset{66x}{\overset{x}{+ 1)(x + 6)}}$$

$$x + 66x = 67x \neq -35x$$

$$(11x \underset{-11x}{\overset{-6x}{- 6)(x - 1)}}$$

$$-6x + (-11x) = -17x \neq -35x$$

$$(11x \underset{-66x}{\overset{-x}{- 1)(x - 6)}}$$

$$-x + (-66x) = -67x \neq -35x$$

$$(11x \underset{22x}{\overset{3x}{+ 3)(x + 2)}}$$

$$3x + 22x = 25x \neq -35x$$

$$(11x + \underset{33x}{\overset{2x}{\underline{2}}})(x + 3)$$

$$2x + 33x = 35x \neq -35x$$

---

$$(11x \underset{-22x}{\overset{-3x}{\underline{-3}}})(x - 2)$$

$$-3x + (-22x) = -25x \neq -35x$$

---

$$(11x \underset{-33x}{\overset{-2x}{\underline{-2}}})(x - 3)$$

$$-2x + (-33x) = -35x = -35x \checkmark$$

Factors are $(11x - 2)(x - 3)$.

---

### Factoring $ax^2 + bx + c$ when $a$ is a non-prime number

1. List all the positive sets of factors for $a$. If for some reason $a$ is negative, factor out a GCF of $-1$ before starting.
2. Write the factors as $(a'x \quad )(a''x \quad )$, where $a'$ and $a''$ represent the first set of positive factors of $a$.
3. Determine all the corresponding pairs of factors of $c$. Note: If $c > 0$ use pairs of positive factors and pairs of negative factors.
   If $c < 0$ use pairs of factors with different signs.
4a. Insert each pair of factors of $c$ into the parentheses on a trial and error basis. Do one pair at a time.
4b. Insert the first pair in order. Put the first number of the pair into the first set of parentheses and the second number of the pair into the second set of parentheses.
4c. Test for the middle term by multiplying the inner and outer terms (including signs). Add the results.
4d. Does the sum equal the middle term of the trinomial? If yes—these are the factors.
   If no—reverse the pair of numbers. Put the first number of the pair in the second set of parentheses and the second number of the pair in the first set of parentheses
4e. Test for the middle term by multiplying the inner and outer terms (including signs). Add the results.
4f. Does the sum equal the middle term of the trinomial? If yes—these are the factors.
   If no— continue the process with the next pair of factors of $c$.
5. If none of the factors or their reverses work, change the original $a'$ and $a''$ to the next possible set of positive factors of $a$ and repeat step 4a.

Factor $4x^2 + x - 3$.

1. 4, 1
   2, 2

2. $(4x \quad )(1x \quad )$

3. $-3, +1$  middle term $= +1x$
   $+3, -1$

4. $(4x \underset{4x}{\overset{-3x}{\underline{-3}}})(1x + 1)$

$$-3x + 4x = x = x \checkmark$$

The factors are $(4x - 3)(x + 1)$.

**Examples:**

5. Factor $6x^2 + 7x - 5$.

**Solutions:**

5. $a = 6$, $c = -5$, middle term $= 7x$

1. List all the positive pairs of factors of $a$
   6, 1
   3, 2

2. Use the first positive pair.
   $(6x \quad )(1x \quad )$

3. Pairs of factors of $c$
   $-5, 1$
   $5, -1$

4. $(6x \underset{+6x}{\overset{-5x}{- 5)(1x + 1}})$   Insert and test.

   $-5x + 6x = x \neq 7x$   Be sure to reverse each pair.

---

   $(6x \underset{-30x}{\overset{x}{+ 1)(1x - 5}})$

   $x + (-30x) = -29x \neq 7x$

---

   $(6x \underset{-6x}{\overset{5x}{+ 5)(1x - 1}})$

   $5x + (-6x) = -x \neq 7x$

---

   $(6x \underset{+30x}{\overset{-x}{- 1)(1x + 5}})$

   $-x + (30x) = 29x \neq 7x$

---

5. Use the next set of factors of $a$ and repeat step 4a.

   $(3x \underset{+3x}{\overset{-10x}{- 5)(2x + 1}})$

   $-10x + 3x = -7x \neq 7x$

---

   $(3x \underset{-15x}{\overset{2x}{+ 1)(2x - 5}})$

   $2x + (-15x) = -13x \neq 7x$

---

   $(3x \underset{-3x}{\overset{10x}{+ 5)(2x - 1}})$

   $10x + (-3x) = 7x = 7x \checkmark$

   Factors are: $(3x + 5)(2x - 1)$

6. Factor $18x^2 - 6x - 24$.

6. Note: There is a greatest common monomial factor of 6. First factor out the GCF of 6.

   $$18x^2 - 6x - 24 = 6(3x^2 - x - 4)$$

   The trinomial $3x^2 - x - 4$ now has a prime

number as the coefficient of the $x^2$.

1. $c = -4$, middle term $= -1x$
   $(3x\quad)(x\quad)$

2. $+4, -1$
   $-4, +1$
   $+2, -2$
   Note: $-2, +2$ is the same as $+2, -2$. $\therefore$ Testing both pairs of numbers is not necessary.

3. $(3x + \overset{4x}{4})(x \underset{-3x}{-1})$

   $4x + (-3x) = 1x \neq -1x$

   $(3x \overset{-x}{-1})(x \underset{+12x}{+4})$

   $-x + (12x) = 11x \neq -1x$

   $(3x \overset{-4x}{-4})(x \underset{3x}{+1})$

   $-4x + 3x = -1x = -1x \checkmark$

Factors of $3x^2 - x - 4 = (3x - 4)(x + 1)$

Factors of
$$18x^2 - 6x - 24 = 6(3x^2 - x - 4)$$
$$= 6(3x - 4)(x + 1)$$

---

**When $ax^2 + bx + c$ is unfactorable**

---

Trinomials of the form $ax^2 + bx + c$ are not always factorable. **If none of the procedures work leave the trinomial is $ax^2 + bx + c$ form.**

7. Can $x^2 + 2x - 7$ be factored?

7. $c = -7$, middle term $= +2x$

1. $(x\quad)(x\quad)$  The trinomial has $a = 1$.

2. $+7, -1$  Identify the factor pairs of $c$.
   $-7, +1$

3. Insert and test.

   $(x + \overset{+7x}{7})(x \underset{-1x}{-1})$

   $7x + (-1x) = 6x \neq 2x$

   $(x \overset{-7x}{-7})(x \underset{+1x}{+1})$

   $-7x + 1x = -6x \neq 2x$

None of the possibilities work.

$$x^2 + 2x - 7 \text{ is not factorable.}$$

## Combined Factoring or Factor Completely

Some expressions can be factored into more than two factors. To explore this possibility:

> ● Factor out any Greatest Common Factor (GCF).
> ● Factor the resulting polynomial, if possible.
> ● Factor any resulting polynomial factors, if possible.
> ● Express the original polynomial as a product of its factors.

**Examples:**

8. Factor $3x^4 - 48$.

**Solutions:**

8. $3x^4 - 48$    There is a GCF of 3
   $= 3(x^4 - 16)$

   $x^4 - 16$ is the difference of two squares
   $x^2 - y^2 = (x + y)(x - y)$
   $\therefore x^4 - 16 = (x^2 + 4)(x^2 - 4)$

   $3(x^4 - 16)$
   $= 3(x^2 + 4)(x^2 - 4)$

   $x^2 - 4$  is the difference of two squares
   $\therefore x^2 - 4 = (x + 2)(x - 2)$

   $3(x^2 + 4)(x^2 - 4)$
   $= 3(x^2 + 4)(x + 2)(x - 2)$

## Practice Problems 14.4

1. Factor $x^2 - 7x + 10$.

   A) $(x - 10)(x - 1)$  B) $(x + 10)(x + 1)$
   C) $(x - 5)(x - 2)$  D) $(x + 2)(x + 5)$
   E) $(x - 2)(x + 5)$

2. Factor completely $3x^2 - 27$.

   A) $3(x^2 - 9)$  B) $3(x^2 + 3)(x^2 - 3)$
   C) $3(x - 3)(x - 3)$  D) $3(x + 3)(x - 3)$
   E) $3x(x - 9)$

3. Factor $3x^2 - 4x + 4$.

   A) $(3x + 2)(x - 2)$  B) $(3x - 2)(x + 2)$
   C) $(3x - 2)(3x + 2)$  D) $(3x + 2)(x + 2)$
   E) $(3x - 2)(x - 2)$

4. Factor $15y^2 - 5y$.

   A) $(5y - 1)(3y + 5)$  B) $(5y)(3y)$
   C) $(5y)(3y - 1)$  D) $(5y - y)(3y + 5)$
   E) $(3y - 1)(5y + y)$

5. Factor completely $2x^3 - 16$.

   A) $2(x^3 - 8)$  B) $2(x + 4)(x^2 - 4)$
   C) $2(x^2 + 4)(x - 4)$  D) $2(x^2 + 2x + 4)$
   E) $2(x - 2)(x^2 + 2x + 4)$

6. Factor completely $4x^2 + 12x - 16$.

   A) $(4x - 4)(x + 4)$  B) $4(x + 4)(x - 1)$
   C) $4(x - 4)(x + 1)$  D) $(4x - 16)(x + 1)$
   E) $(x + 4)(x - 4)$

7. Factor completely $8x^8 - 8$.

  A) $8(x^8 - 1)$  B) $8(x^4 - 8)(x^4 + 1)$
  C) $8(x^6 - 8)(x^2 + 1)$
  D) $8(x^4 + 1)(x^2 + 1)(x + 1)(x - 1)$
  E) $8(x^4 - 1)(x^2 - 1)(x + 1)(x - 1)$

8. Factor $4x^2 - 8x + 3$.

  A) $(2x - 3)(2x - 1)$  B) $(4x - 3)(x - 1)$
  C) $(2x - 1)(x - 3)$  D) $(4x - 1)(x - 3)$
  E) $4x(x - 8 + 3x^2)$

9. Factor $7x^2 - 13x - 2$.

  A) $(7x - 2)(x + 1)$  B) $(7x + 1)(x - 2)$
  C) $(7x^2)(13x - 2)$
  D) $(7x - 6)(x - 7)$  E) $(7x + 1)(x + 2)$

10. Factor $x^2 - 5x + 6$.

  A) $(x + 6)(x - 1)$  B) $(x - 6)(x + 1)$
  C) $(x + 3)(x - 2)$  D) $(x - 3)(x - 2)$
  E) $(x - 3)(x + 2)$

11. Factor $2x^2 + x - 3$.

  A) $(2x - 3)(x + 1)$  B) $(2x + 1)(x - 3)$
  C) $(2x + 3)(x - 1)$  D) $(2x - 1)(x + 3)$
  E) $2(x - 3)(x + 1)$

12. Factor completely $4x^3 - 49x$.

  A) $4(x^3 - 12x)$  B) $4x(x^2 - 12)$
  C) $(2x - 7)(2x^2 + 7)$  D) $x(4x^2 - 49)$
  E) $x(2x + 7)(2x - 7)$

Solutions on page 259

# 14.5 Applications of Factoring I (Multiplying and Dividing Polynomial Fractions)

- Express the problem as a series of fraction multiplications.

$$\frac{a}{b} \cdot \frac{c}{d} \cdot \frac{e}{f} \cdots$$

- If there is division, invert the fraction after the division sign and multiply it by the first fraction.

$$\left. \begin{array}{c} \dfrac{a}{b} \div \dfrac{x}{y} \\[2em] \dfrac{\dfrac{a}{b}}{\dfrac{x}{y}} \end{array} \right\} \Rightarrow \frac{a}{b} \cdot \frac{y}{x}$$

- Factor all the polynomials which can be factored.
- Reduce like factors and opposite factors to $\frac{1}{1}$ and $\frac{-1}{1}$, respectively.

☞ Like Factors: Every term in each factor has a correspondence with the same sign.
Opposite Factors: Every term in each factor has a correspondence with the opposite sign.

$$\frac{x+1}{x+1} \Rightarrow \frac{x \text{ is "+", 1 is "+"}}{x \text{ is "+", 1 is "+"}} = \frac{1}{1} = 1 \qquad \frac{x-1}{1-x} \Rightarrow \frac{x \text{ is "+", 1 is "−"}}{x \text{ is "−", 1 is "+"}} = \frac{-1}{1} = -1$$

$$\frac{-1-x}{1-x} \Rightarrow \frac{1 \text{ is "−", } x \text{ is "−"}}{1 \text{ is "+", } x \text{ is "+"}} = \frac{-1}{1} = -1 \qquad \frac{x-1}{x-1} \Rightarrow \frac{x \text{ is "+", 1 is "−"}}{x \text{ is "+", 1 is "−"}} = \frac{1}{1} = 1$$

$$\frac{x+1}{x-1} \Rightarrow \left. \frac{x \text{ is "+", 1 is "+"}}{x \text{ is "+", 1 is "−"}} \right\} \text{ cannot be reduced}$$

$$\frac{-x+1}{1-x} \Rightarrow \frac{x \text{ is "−", 1 is "+"}}{x \text{ is "−", 1 is "+"}} = \frac{1}{1} = 1$$

⚡ $$\frac{x^2+2}{2} \neq x^2$$

$$\neq x^2 + 1$$

⚡ $$\frac{3x^2 + 2x}{3x^2} \neq 2x$$

$$\neq 1 + 2x$$

● Multiply and simplify the remaining factors.

$$\frac{a}{b} \cdot \frac{c}{d} \cdot \frac{e}{f} = \frac{ace}{bdf}$$

**Examples:**

1. Multiply $\dfrac{x^2+x}{x^2-6x-7} \cdot \dfrac{3x-21}{x^2-x}$.

2. Multiply $\dfrac{x^2-16}{3-3x} \cdot \dfrac{3x^2-3}{x^2-3x-4}$.

3. Divide $\dfrac{a^3+6a^2}{a^2+12a+36} \div \dfrac{6a-36}{a^2-36}$.

**Solutions:**

1. $\dfrac{\overset{(1)}{\cancel{x}}\overset{(1)}{(\cancel{x+1})}}{(\cancel{x+1})(x\cancel{+7})} \cdot \dfrac{3\overset{(1)}{(\cancel{x+7})}}{\cancel{x}(x-1)} = \dfrac{3}{x-1}$

2. $\dfrac{(x+4)\overset{(1)}{(\cancel{x+4})}}{\cancel{3}(1\cancel{-x})} \cdot \dfrac{\overset{(1)}{\cancel{3}}(\overset{(1)}{\cancel{x-1}})\overset{(-1)}{(\cancel{x-1})}}{(\cancel{x+1})(\cancel{x+4})} = -(x+4)$

Note: $\dfrac{x-1}{1-x} \Rightarrow \dfrac{x \text{ is "+", 1 is "−"}}{x \text{ is "−", 1 is "+"}} = \dfrac{-1}{1}$.

3. Division by $\dfrac{6a-36}{a^2-36}$ should be changed to multiplication by $\dfrac{a^2-36}{6a-36}$

$$\dfrac{a^3+6a^2}{a^2+12a+36} \cdot \dfrac{a^2-36}{6a-36}$$

$$= \dfrac{a^2(\overset{(1)}{\cancel{a+6}})}{(\cancel{a+6})(\cancel{a+6})} \cdot \dfrac{\overset{(1)}{(\cancel{a+6})}\overset{(1)}{(\cancel{a-6})}}{6(\cancel{a-6})}$$

$$= \dfrac{a^2}{6}$$

4. Simplify $\dfrac{x^2 - 6x - 16}{3x^2 + 30x + 48} \div \dfrac{2x^2 - 128}{x^2 + 16x + 64}$.

4. $\dfrac{x^2 - 6x - 16}{3x^2 + 30x + 48} \div \dfrac{2x^2 - 128}{x^2 + 16x + 64}$

Division should first be changed to multiplication

$$= \dfrac{x^2 - 6x - 16}{3x^2 + 30x + 48} \cdot \dfrac{x^2 + 16x + 64}{2x^2 - 128}$$

$$= \dfrac{\overset{(1)}{(x \not{+} 2)} \overset{(1)}{(x \not{+} 8)}}{3(x \not{+} 2)(x \not{+} 8)} \cdot \dfrac{\overset{(1)}{(x \not{+} 8)} \overset{(1)}{(x \not{+} 8)}}{2(x \not{+} 8)(x \not{+} 8)} = \dfrac{1}{6}$$

## Practice Problems 14.5

1. Simplify $\dfrac{x+3}{2x+4} \cdot \dfrac{x^2 - x - 6}{2x + 6}$.

   A) $\dfrac{4}{x-3}$    B) $\dfrac{x-3}{4}$    C) $\dfrac{1}{x-3}$

   D) $x - 3$    E) $\dfrac{1}{4}$

3. Simplify $\dfrac{x^2 + 3x + 2}{x^2 - x - 2} \cdot \dfrac{x^2 + x - 6}{x^2 - x - 12}$.

   A) $\dfrac{x+2}{x-4}$    B) $\dfrac{x-4}{x+2}$    C) $-\dfrac{1}{2}$

   D) $\dfrac{1}{2}$    E) $\dfrac{(x+1)(x+2)}{(x+3)(x-4)}$

2. Simplify $\dfrac{a^2 + 8a + 15}{a^2 + 5a} \div \dfrac{a^2 - 9}{a^2 + 2a - 3}$.

   A) $\dfrac{a-1}{a}$    B) $\dfrac{a(a-3)}{(a-1)(a+3)}$    C) $\dfrac{a}{a-1}$

   D) $\dfrac{a-1}{a+1}$    E) $\dfrac{(a+3)(a-1)}{a(a-3)}$

4. Simplify $\dfrac{a+2}{2a} \cdot \dfrac{2a-4}{2+a} \cdot \dfrac{a^2 - 4}{12 - 6a} \div \dfrac{2+a}{12a^2}$.

   A) $-2(a-2)$    B) $2a(a-2)$

   C) $-2a(a-2)$    D) $(a+2)(a-2)$

   E) $\dfrac{-2}{2+a}$

Solutions on page 261

## 14.6 Applications of Factoring II (Solving Equations)

To solve an equation by factoring:

- Rearrange the equation into $f(x) = 0$ form, $ax^2 + bx + c = 0$ if it's a quadratic.
- Factor the equation completely (if possible).
- Set each factor separately equal to zero and solve.

This procedure is predicated on the property: $lmn = 0$ if and only if $l = 0$ or $m = 0$ or $n = 0$.

**Examples:**

1. Solve $(x - 2)(x + 3) = 0$.

2. Solve $(x - 2)(x + 3) = 6$.

3. Find the solution set of the equation
   $x^3 - 2x + x = 0$.

**Solutions:**

1.  $\quad (x - 2)(x + 3) = 0$

$$x - 2 = 0 \mid x + 3 = 0$$
$$x = 2 \quad \mid \quad x = -3$$
$$x = 2 \text{ or } x = -3$$

2. Since the equation $\neq 0$, it must be rearranged.

$$(x - 2)(x + 3) = 6$$
$$x^2 + x - 6 = 6$$
$$x^2 + x - 12 = 0$$
$$(x - 3)(x + 4) = 0$$
$$x - 3 = 0 \mid x + 4 = 0$$
$$x = 3 \mid x = -4$$
$$x = 3 \text{ or } x = -4$$

3. $\qquad\qquad x^3 - 2x^2 + x = 0$

$$x(x^2 - 2x + 1) = 0 \text{ GCF: } x$$
$$x(x - 1)(x - 1) = 0$$
$$x = 0 \mid x - 1 = 0 \mid x - 1 = 0$$
$$x = 1 \mid \quad x = 1$$

$$x = 0 \text{ or } x = 1 \text{ (double root)}$$

$$\{0, 1\}$$

## Practice Problems 14.6

1. Solve for $x$: $x^2 - 4 = 0$.

   A) $x = 4$ or $x = 1$  B) $x = 2$ only
   C) $x = 2$ or $x = -2$  D) $x = -2$ only
   E) $x = -4$ or $x = 1$

2. Solve for $x$: $x^2 + 9 = 10x$.

   A) $x = 1$ or $x = 9$  B) $x = -1$ or $x = -9$
   C) $x = 1$ or $x = -9$  D) $x = -1$ or $x = 9$
   E) $x = 1$ only

3. Find the solution set of the equation

   $2n^3 - 2n = 0$

   A) $\{0\}$  B) $\{1\}$  C) $\{-1\}$
   D) $\{1, -1\}$  E) $\{0, 1, -1\}$

4. Find the solution set for the equation

   $8x^2 + 10x - 3 = 0$

   A) $\{0, 3\}$  B) $\left\{\dfrac{1}{4}, \dfrac{-3}{2}\right\}$  C) $\left\{\dfrac{-3}{8}, \dfrac{-1}{4}\right\}$

   D) $\left\{\dfrac{-1}{4}, \dfrac{3}{2}\right\}$  E) $\left\{\dfrac{3}{8}, \dfrac{1}{4}\right\}$

Solutions on page 261

# 15. Solving Equations

This section reviews a multiplicity of solution techniques as applied to equations or systems of equations. Included are solution techniques for:

**Equalities**

| Type of Equation or System | | Solution Technique | | |
| --- | --- | --- | --- | --- |
| | | Algebra | Graphic | Determinants |
| 15.1 | Linear | ✓ | | |
| 15.2 | Proportions | ✓ | | |
| 15.3 | Literal | ✓ | | |
| 15.4 | Linear-Linear | ✓ | ✓ | ✓ |
| 15.5 | Quadratic | ✓ | ✓ | |
| 15.6 | Linear-Quadratic | ✓ | ✓ | |
| 15.7 | Quadratic-Quadratic | ✓ | ✓ | |
| 15.8 | Radical | ✓ | | |
| 15.9 | Absolute Value | ✓ | | |
| 15.10 | 3 Simultaneous Linear | ✓ | | ✓ |

**Inequalities**

| | | | Algebra | Graphic |
| --- | --- | --- | --- | --- |
| 15.11 | Linear | 1 Variable | ✓ | Graphic Solution Representation |
| | | 2 Variables | | ✓ |
| 15.12 | Linear-Linear | | | ✓ |
| 15.13 | Quadratic | 1 Variable | | Graphic Solution Representation |
| | | 2 Variables | | ✓ |
| 15.14 | Quadratic-Linear Quadratic-Quadratic | | | ✓ |

192

# 15.1 Solving Linear Equations

The technique for solving a linear equation with one variable is outlined below.

| Step | |
|---|---|
| 1 | Remove fractions or decimals by multiplication. For fractions the multiplier is the lowest common denominator (LCD). For decimals it is the smallest power of 10 which clears all the decimal places. $$2\left(\frac{x}{2} - 3 = x\right) \Rightarrow x - 6 = 2x \quad 100(.3x + 1 = .01) = 30x + 100 = 1$$ |
| 2 | Remove all the parentheses by using the distributive law of multiplication.  Don't forget to distribute minus signs. $$7 - (x + 3) = 7 - x - 3$$ |
| 3 | Collect all terms containing the unknown (for which you are solving) on the same side of the equal sign. All terms *not* containing the unknown should be brought to the *other* side of the equation.  Usually the terms containing the unknown are brought to the left side of the equation.  Remember, whenever a term crosses from one side of the equal sign to the other it changes sign. $$2x - 3 = x - 2 \quad \Rightarrow \quad 2x - x = -2 + 3$$ |
| 4 | Determine the coefficient of the unknown either by combining similar terms or factoring when terms cannot be combined. Similarly combine all common terms on the other side of the equation. $$3x + 2x = 12 + 7 \Rightarrow 5x = 19$$ $$ax + bx = 12 - 1 \Rightarrow x(a + b) = 11$$ |
| 5 | Divide both sides of the equation by the coefficient of the unknown. $$\frac{5x}{5} = \frac{19}{5} \quad \bigg| \quad \frac{x(a + b)}{a + b} = \frac{11}{a + b}$$ $$x = \frac{19}{5} \quad \bigg| \quad x = \frac{11}{a + b}$$ |

 An equation is balanced to the left and right of the equal sign. You can apply any legitimate mathematical operation to either side of the equation provided:

1. it is applied to the side as a whole.
2. it is applied to the other side as a whole, to maintain the balance.

**Examples:**

1. Solve $3x - 15 = x + 1$ for $x$.

2. Solve $\dfrac{2x}{5} + 9 = \dfrac{x}{3} + 10$.

3. Solve $2(x + 1) - 10 = 7 - (x + 3)$.

4. Solve $.4x + .16 = 1.36$.

**Solutions:**

1. Steps 1 and 2 do not apply.

$$3x - 15 = x + 1$$

Step 3: Collect the variable terms on one side of the equation:

$$3x - x = 1 + 15$$

Step 4: Combine common terms:

$$2x = 16$$

Step 5: Divide by the coefficient of the variable:

$$x = 8$$

2. Step 1: Clear all fractions:

$$15 \left( \frac{2x}{5} + 9 = \frac{x}{3} + 10 \right) \Rightarrow 6x + 135 = 5x + 150$$

Step 2 does not apply.

Steps 3, 4, 5:

$$6x + 135 = 5x + 150$$
$$6x - 5x = 150 - 135$$
$$x = 15$$

3. Step 1 does not apply.

$$2(x + 1) - 10 = 7 - (x + 3)$$

Step 2: Remove parentheses:

$$2x + 2 - 10 = 7 - x - 3$$

Steps 3, 4, 5:

$$2x - 8 = 4 - x$$
$$2x + x = 4 + 8$$
$$\frac{3x}{3} = \frac{12}{3}$$
$$x = 4$$

4. $100(.4x + .16 = 1.36)$
$$40x + 16 = 136$$
$$40x = 120$$
$$x = 3$$

# Practice Problems 15.1

1. Solve for $x$: $3x + 5 = 5x - 3$.

   A) 4    B) 1   C) $\frac{1}{8}$   D) 0   E) $-1$

2. Solve for $d$: $\frac{d}{2} + 6 = 4$.

   A) 4   B) $-4$   C) 2   D) $-2$   E) $-1$

3. Solve for $y$: $x - .2 = 1.8$.

   A) .2   B) 1.6   C) 2   D) 16   E) 20

4. Solve for $b$: $4(b - 1) - 3 = 17$.

   A) $-4$   B) 2.5   C) 4   D) 6   E) $\frac{21}{4}$

5. Solve for $n$: $\frac{2n}{3} - \frac{5n}{12} = \frac{5}{4}$.

   A) $-\frac{5}{3}$   B) $\frac{5}{3}$   C) $\frac{40}{9}$   D) $\frac{1}{2}$   E) 5

6. Solve for $a$: $5 + 3(a + 2) = 14$.

   A) 1   B) $\frac{3}{2}$   C) $-\frac{1}{4}$   D) $-\frac{1}{2}$   E) $-1$

Solutions on page 261

## 15.2 Solving Proportions

**Proportions** are ratios that are equal. There are two common formats:

$$\underset{\text{extremes}}{\overset{\overset{\text{means}}{\frown}}{a : b = c : d}} \quad \text{or} \quad \frac{a}{b} = \frac{c}{d}$$

The symbol ":" is read "is to" and the symbol "=" is read "as". Therefore $2 : 3 = x : 4$ is read "Two *is to* three *as* $x$ *is to* four".
The positions $a$ and $d$ are called the extremes.
The positions $b$ and $c$ are called the means.

---

The procedure for solving an equation in the format of a proportion is quite simple.

*Step*

1   Convert the proportion to $\frac{a}{b} = \frac{c}{d}$ format.

$$2 : 3 = x : 7 \Rightarrow \frac{2}{3} = \frac{x}{7}$$

2   Cross multiply and solve as a linear equation.

$$\frac{2}{3} = \frac{x}{7} \Rightarrow 3x = 14$$
$$x = \frac{14}{3}$$

---

 A formal statement of the procedure used in step 2 is:
"The product of the means equals the product of the extremes."

**Examples:**

1. Solve for $n$: $\dfrac{n}{6} = \dfrac{11}{2}$.

2. Solve for $x$: $\dfrac{x+8}{5} = \dfrac{x+2}{3}$.

**Solutions:**

1. Cross multiply:

$$\frac{n}{6} = \frac{11}{2}$$
$$2n = 66$$
$$n = 33$$

2. Crossmultiply:

$$\frac{x+8}{5} = \frac{x+2}{3}$$
$$3(x + 8) = 5(x + 2)$$
$$3x + 24 = 5x + 10$$
$$2x = 14$$
$$x = 7$$

## Practice Problems 15.2

1. Solve for $a$: $\dfrac{3a+1}{4} = \dfrac{5}{2}$.

A) 4   B) 3   C) 2   D) $\dfrac{19}{6}$   E) 1

2. Solve for $m$: $\dfrac{m}{4} = \dfrac{5}{2}$.

A) .1   B) 1   C) 10   D) 100   E) 0

3. Solve for $x$:

$$\frac{x+1}{8} = \frac{2x+1}{4}$$

A) 3   B) $-3$   C) $\dfrac{1}{3}$   D) $-\dfrac{1}{3}$   E) 2

4. Solve for $l$:

$$\frac{3l-1}{5l+2} = \frac{2}{7}$$

A) 1   B) $-\dfrac{3}{2}$   C) $\dfrac{3}{11}$   D) $\dfrac{-3}{11}$   E) $\dfrac{7}{11}$

Solutions on page 262

## 15.3 Literal Equations

Literal equations are equations in which many or all of the constants are expressed in terms of letters rather than numbers. It is paramount to keep track of which letter you are solving for.

The procedure for solving linear literal equations is identical to that for linear equations or proportions as applicable.

**Examples:**

1. Solve for $x$: $ax + b = c$.

2. Solve for $x$: $\dfrac{1}{x} + a = 2b$.

3. Solve for $n$: $\dfrac{cn + 1}{x} = \dfrac{2y}{3}$.

**Solutions:**

1. Collect all the terms containing the unknown on one side of the equation. Divide by the coefficient of the unknown:

$$ax + b = c$$

$$\frac{ax}{a} = \frac{c - b}{a}$$

$$x = \frac{c - b}{a}$$

2. Clear the fraction:

$$x\left(\frac{1}{x} + a = 2b\right) \Rightarrow 1 + ax = 2bx$$

Bring all $x$ terms to one side of the equation:

$$1 + ax = 2bx$$

$$ax - 2bx = -1$$

Factor out a single $x$:

$$x(a - 2b) = -1$$

Divide by the coefficient of the $x$:

$$\frac{x(a - 2b)}{(a - 2b)} = \frac{-1}{a - 2b}$$

$$x = \frac{-1}{a - 2b} \quad \text{or} \quad \frac{1}{2b - a}$$

☞ If $x = \dfrac{-1}{a - 2b}$, multiplying the top and bottom of the fraction by $-1$ yields

$$x = \frac{-1}{(a - 2b)}\frac{(-1)}{(-1)} = \frac{1}{2b - a}$$

3. $\dfrac{cn + 1}{x} = \dfrac{2y}{3}$

Cross multiply:

$$= 3(cn + 1) = 2xy$$

Bring all $n$ terms to one side of the equation:

$$3cn + 3 = 2xy$$

$$3cn \quad = 2xy - 3$$

Divide by the coefficient of the unknown:

$$\frac{3cn}{3c} = \frac{2xy - 3}{3c}$$

$$n = \frac{2xy - 3}{3c}$$

# Practice Problems 15.3

1. Solve for $a$: $\dfrac{cx + 1}{a} = \dfrac{b}{3}$

A) $\dfrac{ab - 1}{3}$  B) $\dfrac{3cx + 1}{b}$  C) $\dfrac{3b}{cx + 1}$

D) $\dfrac{3(cx + 1)}{b}$  E) $3c$

2. Solve for $x$: $3(x + 2) = a(b + x)$

A) $\dfrac{ab - 6}{3 - a}$  B) $\dfrac{ab + 2}{2}$  C) $\dfrac{ab - 2}{2}$

D) $\dfrac{ab + 6}{a - 3}$  E) $\dfrac{3 - a}{ab - 6}$

3. Solve for $t$: $a = \dfrac{t + b}{t - c}$

A) $\dfrac{b}{c}$  B) $\dfrac{ac}{b}$  C) $\dfrac{b + ac}{a - 1}$

D) $\dfrac{c + b}{a - 1}$  E) $\dfrac{ac + b}{a + 1}$

4. Solve for $n$: $\dfrac{1}{a} = \dfrac{1}{n} + \dfrac{1}{b}$

A) $a + b$  B) $\dfrac{b - a}{ab}$  C) $\dfrac{ab}{b - a}$

D) $1$  E) $ab$

Solutions on page 262

## 15.4 Linear-Linear Systems

The solution set to a linear-linear system is the set of values of the variables which satisfy both equations simultaneously. These are also called simultaneous equations. Graphically the solution set is where the two graphs intersect. There are three types of linear systems:

- **Consistent:** There is one solution

- **Inconsistent:** There are no solutions. Parallel lines are an example.

- **Dependent:** Infinite solutions. The two lines are coincident. They overlap.

There are two algebraic solution techniques:

- **Elimination**
- **Substitution**

Both are discussed and demonstrated here. This is followed by the graphical approach and lastly a solution by determinants is presented.

● **Elimination:** Multiply one or both of the equations by suitable constants in order to make the coefficients of one of the variables the same—but opposite in sign. Remember that multiplying *all* terms in an equation by the same constant does not upset the balance of the equation. One of the unknowns can then be eliminated by adding the two equations. When working with simultaneous equations, always be sure to have the terms containing the variables on one side of the equation and the remaining terms on the other side.

Once one of the variables is eliminated solve the linear equation. Using the value for the found variable, substitute this value back into one of the original equations to find the value of the other variable.

**Example 1:** Solve: $3x + y = 4$
$\qquad\qquad\qquad\quad x - 2y = 6$

Plan: Eliminate $y$ $\qquad\qquad 3x + y = 4$
$\qquad\qquad\qquad\qquad\qquad\quad x - 2y = 6$

Multiply the first $\qquad 2(3x + y = 4) \Rightarrow 6x + 2y = 8$
equation by 2:

Add the two equations: $\quad 6x + 2y = 8$
$\qquad\qquad\qquad\qquad\qquad \dfrac{\quad x - 2y = 6}{}$
$\qquad\qquad\qquad\qquad\qquad 7x \quad\;\; = 14$
$\qquad\qquad\qquad\qquad\qquad\; x \quad\;\; = 2$

Substitute $x = 2$ $\qquad\qquad 3x + y = 4$
into one of the $\qquad\qquad\; 3(2) + y = 4$
original equations: $\qquad\qquad\;\; 6 + y = 4$
$\qquad\qquad\qquad\qquad\qquad\qquad\;\; y = -2$

Solution: $x = 2$, $y = -2$ or written as an ordered pair $(2, -2)$

**Example 2:** Solve: $2x - 3y = 11$
$\qquad\qquad\qquad\qquad\quad 3x - 4y = -9$

Plan: Eliminate $x$ $\qquad\qquad\qquad 2x - 3y = 11$
$\qquad\qquad\qquad\qquad\qquad\qquad \dfrac{3x + 4y = -9}{}$

Multiply the first $\qquad\qquad -3(2x - 3y = 11)$
equation by $-3$ $\qquad\qquad\quad \dfrac{2(3x + 4y = -9)}{}$
and the second by 2:

Add the equations: $\qquad\qquad -6x + 9y = -33$
$\qquad\qquad\qquad\qquad\qquad\qquad\; \dfrac{6x + 8y = -18}{}$
$\qquad\qquad\qquad\qquad\qquad\qquad\; 17y = -51$
$\qquad\qquad\qquad\qquad\qquad\qquad\quad\; y = -3$

Substitute $y = -3$ $\qquad\qquad\; 2x - 3y = 11$
into one of the $\qquad\qquad\qquad 2x - 3(-3) = 11$
original equations: $\qquad\qquad\;\; 2x + 9 = 11$
$\qquad\qquad\qquad\qquad\qquad\qquad\quad 2x = 2$
$\qquad\qquad\qquad\qquad\qquad\qquad\quad\;\; x = 1$

Solution:

$$x = 1, y = -3$$

or written as an ordered pair:

$$(1, -3)$$

> ● **Substitution:** Solve the easiest of the two equations for the value of either variable. Substitute this value into the second equation as appropriate. Solve the linear equation. Using the value for the found variable substitute this value back into one of the original equations to find the value of the other variable.

**Example 3:** Solve: $3x + y = 4$
$$x - 2y = 6$$

Plan: Solve the
second equation for $x$

$$x - 2y = 6$$
$$x = 6 + 2y$$

Substitute $x = 6 + 2y$ into the first equation at each appearance of $x$:

$$3x + y = 4$$
$$3(6 + 2y) + y = 4$$
$$18 + 6y + y = 4$$
$$7y = -14$$
$$y = -2$$

Substitute $y = -2$ into one of the original equations:

$$x - 2y = 6$$
$$x - 2(-2) = 6$$
$$x + 4 = 6$$
$$x = 2$$

Solution:

$$x = 2, y = -2$$

or written as an ordered pair:

$$(2, -2)$$

**Example 4:** Solve: $2x - 3y = 11$
$$3x + 4y = -9$$

Plan: Solve the
first equation for $x$

$$2x - 3y = 11$$
$$2x = 11 + 3y$$
$$x = \frac{11 + 3y}{2}$$

Substitute $x = \dfrac{11 + 3y}{2}$ into the second equation at each appearance of $x$:

$$3x + 4y = -9$$
$$3\left(\frac{11 + 3y}{2}\right) + 4y = -9$$

Multiply the entire equation by 2 to clear the fractions

$$2\left(\frac{33 + 9y}{2} + 4y = -9\right)$$
$$33 + 9y + 8y = -18$$
$$17y = -51$$
$$y = -3$$

Substitute $y = -3$ into one of the original equations:

$$2x - 3y = 11$$
$$2x - 3(-3) = 11$$
$$2x + 9 = 11$$
$$2x = 2$$
$$x = 1$$

Solution:

$$x = 1, y = -3$$

or written as an ordered pair:

$$(1, -3)$$

● **Graphic:** Graph each of the lines on the same set of axes. Any graphical technique can be used, e.g., $y = mx + b$, a chart, or intercepts. The intersection of the two graphs represents the solution.

**Example 5:**

Solve: $3x + y = 4$
$x - 2y = 6$

---

Plan: Graph using $y = mx + b$

$3x + y = 4$

Slope-intercept $\quad y = -3x + 4$

$m = \text{slope} = \dfrac{-3}{1}\; \begin{matrix}\leftarrow\text{change in } y, \text{3 down}\\\leftarrow\text{change in } x, \text{1 to right}\end{matrix}$

$b = y\text{-intercept} = 4$

---

$x - 2y = 6$

$-2y = -x + 6$

$y = \dfrac{x}{2} - 3$

$m = \text{slope} = \dfrac{1}{2}\; \begin{matrix}\leftarrow\text{change in } y, \text{1 up}\\\leftarrow\text{change in } x, \text{2 to right}\end{matrix}$

$b = y\text{-intercept} = -3$

**Example 6:**

$2x - 3y = 11$
$3x + 4y = -9$

Plan: Graph by developing a table of values

$2x - 3y = 11$

| $x$ | $y$ |
|---|---|
| 1 | $-3$ |
| 4 | $-1$ |
| $-2$ | $-5$ |

$3x + 4y = -9$

| $x$ | $y$ |
|---|---|
| $-3$ | 0 |
| 1 | $-3$ |
| $-7$ | 3 |

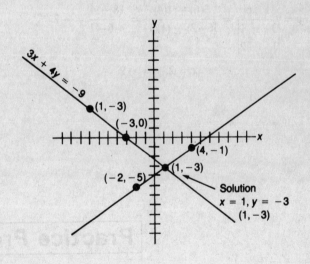

● **Determinants:** Align the equations in the format

$$ax + by = c$$
$$dx + ey = f$$

By Cramer's Rule:

$$x = \frac{\begin{vmatrix} c & b \\ f & e \end{vmatrix}}{\begin{vmatrix} a & b \\ d & e \end{vmatrix}} \qquad y = \frac{\begin{vmatrix} a & c \\ d & f \end{vmatrix}}{\begin{vmatrix} a & b \\ d & e \end{vmatrix}}$$

As a review: $\begin{vmatrix} a & c \\ b & d \end{vmatrix} = ad - bc$

**Example 7:**

Solve: $3x + y = 4$
$\qquad x - 2y = 6$

$$x = \frac{\begin{vmatrix} 4 & 1 \\ 6 & -2 \end{vmatrix}}{\begin{vmatrix} 3 & 1 \\ 1 & -2 \end{vmatrix}} = \frac{4(-2) - (6)(1)}{3(-2) - (1)(1)} = \frac{-8 - 6}{-6 - 1} = 2$$

$$y = \frac{\begin{vmatrix} 3 & 4 \\ 1 & 6 \end{vmatrix}}{\begin{vmatrix} 3 & 1 \\ 1 & -2 \end{vmatrix}} = \frac{3(6) - 1(4)}{3(-2) - (1)(1)} = \frac{18 - 4}{-6 - 1} = -2$$

$$x = 2, y = -2$$

**Example 8:**

Solve: $2x - 3y = 11$
$\qquad 3x + 4y = -9$

$$x = \frac{\begin{vmatrix} 11 & -3 \\ -9 & 4 \end{vmatrix}}{\begin{vmatrix} 2 & -3 \\ 3 & 4 \end{vmatrix}} = \frac{11(4) - (-3)(-9)}{2(4) - (-3)(3)}$$

$$= \frac{44 - 27}{8 + 9} = \frac{17}{17} = 1$$

$$y = \frac{\begin{vmatrix} 2 & 11 \\ 3 & -9 \end{vmatrix}}{\begin{vmatrix} 2 & -3 \\ 3 & 4 \end{vmatrix}} = \frac{2(-9) - 3(11)}{2(4) - (-3)(3)}$$

$$= \frac{-18 - 33}{8 + 9} = \frac{-51}{17} = -3$$

$$x = 1, y = -3$$

# Practice Problems 15.4

1. Solve by substitution:

$$y - x = -1$$
$$2y + x = 4$$

A) (2, 1)  B) (1, 2)  C) (−2, −1)
D) (−1, −2), E(−1, −1)

2. Solve by elimination:

$$3x + 4y = 7$$
$$x - 4y = 1$$

A) $(4, \frac{1}{2})$  B) $(\frac{1}{2}, 4)$  C) $(2, \frac{1}{4})$
D) $(\frac{1}{4}, -2)$  E) $(\frac{1}{4}, 2)$

3. Solve by determinants:

$$3x - 2y = -1$$
$$2x + 3y = 8$$

A) (1, 2)  B) (−1, 2)  C) (1, −2)
D) (−1, −2)  E) (1, $\frac{1}{2}$)

6. Solve by determinants:

$$4x - y = 10$$
$$x + y = 5$$

A) (3, 2)  B) (4, 4)  C) (4, 1)
D) (6, −1)  E) (−3, 2)

4. Solve by substitution:

$$2x - y = 6$$
$$x + y = 9$$

A) (−5, −4)  B) (−5, 4)  C) (5, −4)
D) (5, 4)  E) ($\frac{1}{5}$, $\frac{1}{4}$)

7. Solve graphically:

$$x + y = 3$$
$$2x - y = 6$$

A) (0, 3)  B) (0, 0)  C) (3, 3)
D) (3, 0)  E) (4, −7)

5. Solve by elimination:

$$y - 4 = x$$
$$5x - 20 = y$$

A) (4, 8)  B) (1, 5)  C) (−4, −8)
D) (6, 10)  C) (5, 5)

8. Solve by substitution:

$$x + y = 12$$
$$y = 3x$$

A) (1, 3)  B) (9, 3)  D) (3, 9)
D) ($\frac{1}{3}$, 1)  E) (−1, 1)

Solutions on page 263

## 15.5 Quadratic Equations

There are three algebraic techniques for solving a quadratic equation.

**Factoring**—the easiest, if it can be applied.
**Completing the square**—useful to obtain $(x - h)^2$ and $(y - k)^2$ terms for graphing.
**Quadratic formula**—always works.

The solutions (roots, zeros) of a quadratic equation are the values of the variable which make the equation equal to zero. The roots are graphically represented as the $x$-intercepts if $y = f(x)$. There are always two roots, not necessarily real.

**Factoring**

Find the roots of $f(x) = x^2 - 5x - 6$ **by factoring**.

1. To find the roots set $f(x) = 0$.
2. Factor.
3. Set each factor equal to 0.
4. Solve for $x$.

$$x^2 - 5x - 6 = 0$$
$$(x - 6)(x + 1) = 0$$
$$x - 6 = 0 \mid x + 1 = 0$$
$$x = 6 \mid x = -1$$

The roots are 6, −1.

## Completing the Square

Find the roots of $f(x) = x^2 - 5x - 6$ **by completing the square**.

$$x^2 - 5x - 6 = 0$$

$$x^2 - 5x = 6$$

$$x^2 - 5x + \left(\frac{-5}{2}\right)^2 = 6 + \left(\frac{-5}{2}\right)^2$$

$$\left(x - \frac{5}{2}\right)^2 = \frac{49}{4}$$

$$\sqrt{\left(x - \frac{5}{2}\right)^2} = \pm\sqrt{\frac{49}{4}}$$

$$x - \frac{5}{2} = \pm\frac{7}{2}$$

1. To find the roots set $f(x) = 0$.
2. Algebraically move the numerical term to the other side of the equation.
3. Divide the coefficient of the $x$ term by 2 and add this number squared to both sides of the equation.
4. a) Express the left side of the equation as $(x + a)^2$, where $a$ is the value from step 3 found by dividing the coefficient of the $x$ by 2.
   b) Express the right side as a single numerical value.
5. Take the square root of each side of the equation.
6. Separate the equation into two parts and solve for $x$.

$$x - \frac{5}{2} = \frac{7}{2} \qquad \Big| \qquad x - \frac{5}{2} = \frac{-7}{2}$$

$$x = 6 \qquad \Big| \qquad x = -1$$

The roots are $-1$ and 6.

## Quadratic Formula

Find the roots of $f(x) = x^2 - 5x - 6$ **using the quadratic formula**.

### Quadratic Formula

$$f(x) = ax^2 + bx + c \qquad\qquad f(y) = ay^2 + by + c$$

$$x = \frac{-b \pm \sqrt{b^2 - 4ac}}{2a} \qquad\qquad y = \frac{-b \pm \sqrt{b^2 - 4ac}}{2a}$$

$$x^2 - 5x - 6 = 0$$

$$a = 1 \quad b = -5 \quad c = -6$$

$$x = \frac{-(-5) \pm \sqrt{(-5)^2 - 4(1)(-6)}}{2(1)}$$

1. To find the roots set $f(x) = 0$.
2. Determine the values of $a$, $b$ and $c$.
3. Apply the formula.
4. Solve for the values of the variable.

$$= \frac{5 \pm \sqrt{25 + 24}}{2} = \frac{5 \pm \sqrt{49}}{2}$$

$$= \frac{5 \pm 7}{2}$$

$$x = \frac{5 + 7}{2} = 6 \qquad x = \frac{5 - 7}{2} = -1$$

The roots are 6 and $-1$.

# Graphing

The roots are where the equation crosses the x axis if $y = f(x)$.
Find the roots of $f(x) = x^2 - 5x - 6$ using graphing.

$$f(x) = x^2 - 5x - 6$$

$$x = \frac{-b}{2a} = \frac{5}{2} = \frac{5}{2}$$

1. Arrange the equation into the form $f(x) = ax^2 + bx + c$.
2. Determine the axis of symmetry $x = \frac{-b}{2a}$.
3. Establish a table of values with $x = -\frac{b}{2a}$ the center of 7 values. That is, 3 values above and 4 values below. The values above and below should only be integers.
4. The roots are the x-intercepts.

| $x$ | $x^2 - 5x - 6$ | $y$ |
|---|---|---|
| 0 | $0^2 - 5(0) - 6$ | $-6$ |
| 1 | $1^2 - 5(1) - 6$ | $-10$ |
| 2 | $2^2 - 5(2) - 6$ | $-12$ |
| $\frac{5}{2}$ | $(\frac{5}{2})^2 - 5(\frac{5}{2}) - 6$ | $-\frac{49}{4}$ |
| 3 | $3^2 - 5(3) - 6$ | $-12$ |
| 4 | $4^2 - 5(4) - 6$ | $-10$ |
| 5 | $5^2 - 5(5) - 6$ | $-6$ |

# Practice Problems 15.5

1. Find the roots of $f(x) = 2x^2 + 3x - 2$ (use factoring).

A) $\frac{1}{2}$ and $-2$  B) 2 and $-2$
C) $\frac{1}{2}$ and $-\frac{1}{2}$  D) $\frac{1}{2}$ and 2
E) $-\frac{1}{2}$ and 2

2. Find the roots of $x^2 - 4x - 7$ (use completing the square).

A) $2 \pm \sqrt{7}$  B) $2 \pm \sqrt{15}$  C) $2 \pm \sqrt{11}$
D) $-2$ and 2  E) $\pm\sqrt{11}$

3. Find the roots of $x^2 + 4x + 9$ (use the quadratic formula).

A) $2 \pm \sqrt{5}$    B) $-2 \pm i\sqrt{5}$    C) $2 \pm \sqrt{2}$

D) $2 \pm 5i\sqrt{2}$    E) $5 \pm 3\sqrt{2}$

4. Find the roots of $x^2 - 4x + 4$ (use graphing).

A) Double root at 2    B) 2, $-2$    C) 2, 3

D) 4, 1    E) 0, 4

Solutions on page 264

## 15.6 Linear-Quadratic Systems

These systems are solved algebraically by substitution or by graphing. There are **usually two solutions, which can be real or imaginary**.

---

- **Substitution:** Solve the linear equation for either variable, whichever is easiest. Substitute the value into the quadratic for each appearance of the variable. Solve the resulting quadratic equation by factoring, quadratic formula, or completing the square. For each value obtained find the value of the other variable by substituting into the linear equation.

---

 If there are two solutions, they can be:
- 2 real and different, no imaginary
- 2 imaginary and no real
- 2 real and equal, no imaginary
  (only one solution is visible on a graph)

Typical graph

| 2 solutions | 2 solutions | 1 solution (2 the same) | 0 real solutions (2 imaginary) |

**Example 1:**

Solve: $x^2 - 4x - y - 5 = 0$

$$y = 3x - 11$$

The linear equation is already solved for $y$. Substitute into the quadratic:

$$x^2 - 4x - (3x - 11) - 5 = 0$$
$$x^2 - 4x - 3x + 11 - 5 = 0$$
$$x^2 - 7x + 6 = 0$$
$$(x - 6)(x - 1) = 0$$

$$x - 6 = 0 \quad | \quad x - 1 = 0$$
$$x = 6 \quad | \quad x = 1$$

For each $x$ value determinine the corresponding $y$:

| $x = 6$ | $x = 1$ |
|---|---|
| $y = 3x - 11$ | $y = 3(1) - 11$ |
| $y = 3(6) - 11$ | $= 3 - 11$ |
| $y = 7$ | $= -8$ |
| $\therefore x = 6, y = 7$ | $\therefore x = 1, y = -8$ |
| $(6, 7)$ | $(1, -8)$ |

**Example 2:**

$$x^2 + y^2 - 10y - 24 = 0$$
$$y = x - 2$$

The linear equation is already solved for $y$. Substitute into the quadratic:

$$x^2 + (x - 2)^2 - 10(x - 2) - 24 = 0$$
$$x^2 + x^2 - 4x + 4 - 10x + 20 - 24 = 0$$
$$\frac{2x^2}{2} \frac{-14x}{2} = \frac{0}{2}$$
$$x^2 - 7x = 0$$
$$x(x - 7) = 0$$

$$x = 0 \quad | \quad x - 7 = 0$$
$$| \quad x = 7$$

For each $x$-value determine the corresponding $y$ value:

| $y = x - 2$ | $y = x - 2$ |
|---|---|
| $y = 0 - 2$ | $y = 7 - 2$ |
| $y = -2$ | $y = 5$ |
| $x = 0, \quad y = -2$ | $x = 7, y = 5$ |
| $(0, -2)$ | $(7, 5)$ |

# Graphing

> Graph the equations on the same set of axis. The real solutions are at the intersections of the graphs.

Solve: $(x - 2)^2 + (y + 3)^2 = 9$
$$y = 2x - 3$$

$$(x - 2)^2 + (y + 3)^2 = 9$$

This is a circle with center $(2, -3)$ and radius $= 3$.

$$y = 2x - 3$$

This is a line with $m = $ slope $= \dfrac{2 \leftarrow \text{up}}{1 \leftarrow \text{right}}$ and $y$-intercept of $-3$.

2 real solutions indicated by the $X$'s

# Practice Problems 15.6

1. Solve: $y = 5 + 2x$
   $x^2 + y^2 = 25$

   A) (1, 7) and (0, 5)
   B) (−4, −3) and (1, 7)
   C) (−4, −3) and (0, 5)
   D) (3, 4) and (−3, −4)
   E) (2, 9) and (2, $\sqrt{21}$)

2. Solve: $x^2 + y^2 - 4x - 2y + 1 = 0$
   $y + x = 1$

   A) ($\frac{1}{2}$, $\frac{1}{2}$) and (0, 1)
   B) (1, 0) and (3, −2)
   C) ($\frac{1}{4}$, $\frac{3}{4}$) and (−2, 3)
   D) (0, 1) and ($\frac{1}{2}$, $\frac{1}{2}$)
   E) (0, 1) and (2, −1)

3. Solve: $y = 2x^2 - 5x + 5$
   $y - x = 5$

   A) (7, 2) and (4, −1)
   B) (0, 5) and (3, 8)
   C) (0, 5) and (−1, 4)
   D) (−3, −8) and (1, −6)
   E) ($\frac{5}{2}$, $\frac{1}{2}$) and (5, 0)

4. Solve: $y = -x^2 + 4x - 3$
   $y + 1 = x$

   A) (1, 0) and (2, 1)
   B) (3, 2) and (−1, 0)
   C) (0, −1) and (2, 3)
   D) (1, 2) and (0, 1)
   E) (5, 4) and (−6, −7)

Solutions on page 265

## 15.7 Quadratic-Quadratic Systems

In general, either substitution or elimination can be used to solve these systems. Sometimes depending on the particular equations involved only one of these methods may be practical. These systems can have 4, 3, 2, 1 or 0 real solutions.

4 real solutions

3 real solutions

2 real solutions

1 real solution

0 real solutions

- **Elimination:** Multiply one or both of the equations by suitable constants in order to make the coefficients of one of the variables the same—but opposite in sign. This unknown can be eliminated by adding the two equations. Remember, when using elimination be sure the variable terms are all on one side of the equation and the numerical or constant terms are on the other.

  Once one of the variables is eliminated solve the remaining equation. Using the values for the solved variable substitute these values back into one of the original equations to find the corresponding values of the eliminated variable.

● **Substitution:** Solve the easiest of the two equations for the value of either variable or either variable squared. Substitute this value into the other equation and solve. Using the values for the solved variable, substitute these values back into one of the original equations to find the corresponding values of the other variable.

**Example 1:**

Solve by elimination:

$$2x^2 = 6 + 3y^2$$
$$x^2 = 17 - 2y^2$$

Rearrange the equations.

$$2x^2 - 3y^2 = 6$$
$$x^2 + 2y^2 = 17$$

Multiply the second equation by $-2$ to eliminate the $x^2$.

$$-2(x^2 + 2y^2 = 17) \Rightarrow -2x^2 - 4y^2 = -34$$

$$\begin{array}{r} -2x^2 - 4y^2 = -34 \\ 2x^2 - 3y^2 = 6 \\ \hline -7y^2 = -28 \\ y^2 = 4 \\ y = \pm 2 \end{array}$$

For each value of $y$ find the corresponding $x$ values.

| $y = 2$ | $y = -2$ |
|---|---|
| $x^2 = 17 - 2y^2$ | $x^2 = 17 - 2y^2$ |
| $x^2 = 17 - 2(2)^2$ | $x^2 = 17 - 2(-2)^2$ |
| $x^2 = 17 - 8 = 9$ | $x^2 = 17 - 8 = 9$ |
| $x = \pm 3$ | $x = \pm 3$ |
| $x = 3, y = 2$ | $x = 3, y = -2$ |
| $x = -3, y = 2$ | $x = -3, y = -2$ |

**4 solutions**

**Example 2:**

Solve by substitution:

$$2x^2 = 6 + 3y^2$$
$$x^2 = 17 - 2y^2$$

The second equation is solved for $x^2$. Substitute into first equation:

$$2x^2 = 6 + 3y^2$$
$$2(17 - 2y^2) = 6 + 3y^2$$
$$34 - 4y^2 = 6 + 3y^2$$
$$7y^2 = 28$$
$$y^2 = 4$$
$$y = \pm 2$$

For each value of $y$ find the corresponding $x$ values:

| $y = 2$ | $y = -2$ |
|---|---|
| $x^2 = 17 - 2y^2$ | $x^2 = 17 - 2y^2$ |
| $x^2 = 17 - 2(2)^2$ | $x^2 = 17 - 2(-2)^2$ |
| $x^2 = 17 - 8 = 9$ | $x^2 = 17 - 8 = 9$ |
| $x = \pm 3$ | $x = \pm 3$ |
| $x = 3, y = 2$ | $x = 3, y = -2$ |
| $x = -3, y = 2$ | $x = -3, y = -2$ |

• **Graphing:** Plot or sketch each of the curves. Their points of intersection are the real solutions.

Solve graphically: $2x^2 = 6 + 3y^2$
$x^2 = 17 - 2y^2$

$$2x^2 = 6 + 3y^2 \Rightarrow 2x^2 - 3y^2 = 6 \Rightarrow \frac{x^2}{3} - \frac{y^2}{2} = 1 \quad \text{Hyperbola}$$

$$x^2 = 17 - 2y^2 \Rightarrow x^2 + 2y^2 = 17 \Rightarrow \frac{x^2}{17} + \frac{y^2}{\frac{17}{2}} = 1 \quad \text{Ellipse}$$

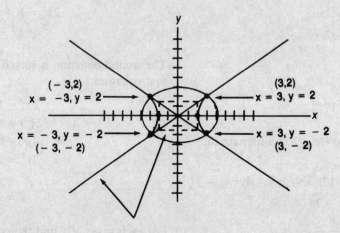

4 Solutions

# Practice Problems 15.7

1. How many real solutions are there for the quadratic system: $4x^2 + y^2 = 20$
$x^2 + 4y^2 = 20$

A) 0   B) 1   C) 2   D) 3   E) 4

2. How many real solution are there for the quadratic system: $y^2 + 2x = -1$
$x^2 - y^2 = 16$

A) 0   B) 1   C) 2   D) 3   E) 4

3. How many real solutions are there for the quadratic system: $x^2 + y = 5$
$x^2 + y^2 = 1$

A) 0   B) 1   C) 2   D) 3   E) 4

4. How many real solutions are there for the quadratic system: $xy = 12$
$x^2 + y^2 = 16$

A) 0   B) 1   C) 2   D) 3   E) 4

Solutions on page 266

## 15.8 Radical Equations

In solving equations containing radicals it is important to get the radical term alone on one side of the equation. Then square both sides of the equation to eliminate the radical sign. Solve the resulting equation.

☞ $\sqrt{x}\cdot\sqrt{x} = (\sqrt{x})^2 = x \quad x \geq 0$

$\sqrt{4} = 2$ only, $\sqrt{4} \neq -2$

Remember that all solutions to radical equations **must be checked**, as squaring both sides sometimes results in extraneous solutions. It is possible for the solution set of a radical equation to be a null or empty set.

Remember also in squaring each side of an equation, do not make the mistake of simply squaring each term; the entire side of the equation must be multiplied by itself.

⚡ $(x+2)^2 \neq x^2 + 2^2, \quad (x+2)^2 = (x+2)(x+2) = x^2 + 4x + 4$

If more than one radical term appears in the equation eliminate one radical term as described above. Then get the second radical term alone and repeat the process. These types of equations involve double squaring operations.

**Examples:**

1. Solve for $x$: $\sqrt{x-3} = 4$

2. Solve for $x$: $\sqrt{x^2-7} - 1 = x$

**Solutions:**

1. Get the $\sqrt{\phantom{x}}$ term alone, then square both sides.

$$\sqrt{x-3} = 4$$
$$(\sqrt{x-3})^2 = (4)^2$$
$$x - 3 = 16$$
$$x = 19$$

*Check*

$$\sqrt{x-3} \stackrel{?}{=} 4$$
$$\sqrt{19-3} \stackrel{?}{=} 4$$
$$\sqrt{16} \stackrel{?}{=} 4$$
$$4 \stackrel{?}{=} 4 \checkmark$$
1 solution
$$x = 4$$

2. Get the $\sqrt{\phantom{x}}$ term alone, then square both sides.

$$\sqrt{x^2-7} - 1 = x$$
$$\sqrt{x^2-7} = x + 1$$
$$(\sqrt{x^2-7})^2 = (x+1)^2 = (x+1)(x+1)$$
$$x^2 - 7 = x^2 + 2x + 1$$
$$2x + 8 = 0$$
$$x = -4$$

*Check*

$$\sqrt{x^2-7} - 1 = x$$
$$\sqrt{(-4)^2-7} - 1 \stackrel{?}{=} -4$$
$$\sqrt{16-7} - 1 \stackrel{?}{=} -4$$
$$\sqrt{9} - 1 \stackrel{?}{=} -4$$
$$3 - 1 \neq -4$$

No solution
{ }

3. Solve for $x$: $\sqrt{x+2} + \sqrt{x-3} = 5$

3. First get one of $\sqrt{\phantom{x}}$ by itself and square the equation.

$$\sqrt{x+2} + \sqrt{x-3} = 5$$
$$\sqrt{x+2} = 5 - \sqrt{x-3}$$
$$(\sqrt{x+2})^2 = (5 - \sqrt{x-3})^2$$
$$x+2 = 25 - 10\sqrt{x-3} + (x-3)$$

Rearrange and get the $\sqrt{\phantom{x}}$ by itself and square the equation.

$$10\sqrt{x-3} = 20$$
$$\sqrt{x-3} = 2$$
$$(\sqrt{x-3})^2 = (2)^2$$
$$x-3 = 4$$
$$x = 7$$

*Check*

$$\sqrt{x+2} + \sqrt{x-3} = 5$$
$$\sqrt{7+2} + \sqrt{7-3} \stackrel{?}{=} 5$$
$$\sqrt{9} + \sqrt{4} \stackrel{?}{=} 5$$
$$3 + 2 \stackrel{?}{=} 5$$
$$5 \stackrel{?}{=} 5 \checkmark$$

1 solution
$$x = 7$$

# Practice Problems 15.8

1. Solve for $x$: $\sqrt{x^2+3} = x+1$

   A) $\pm 1$   B) $+1$ only   C) $-1$ only
   D) No solution   E) 2

2. Solve for $x$: $\sqrt{x+18} + 2 = x$

   A) 7 and $-2$   B) $-2$ only   C) 7 only
   D) No solution   E) $-7$ and $-2$

3. If $\sqrt{x+1} + 2 = 0$, what is the solution set for all real values of $x$?

   A) $\{-1\}$   B) $\{2\}$   C) $\{-1, 2\}$
   D) $\{3\}$   E) $\{\ \}$

4. The equation $\sqrt{2x} - \sqrt{x+1} = 0$ has

   A) 1 as its only root
   B) 0 as its only root
   C) 0 and 1 as its only roots
   D) 0, 1 and 2 as its roots
   E) no solution

Solutions on page 267

## 15.9 Absolute Value Equations

Absolute value equations are of the format $|x| = a$. To solve these equations convert the equation into:

$$|x| = a$$
$$\swarrow \quad \searrow$$
$$x = -a \qquad x = a$$

Solve each equation separately. All possible solutions **must be checked** into the original equation. It is possible for the solution set to an absolute value equation to be the empty or null set.

**Examples:**

1. Solve for $x$: $|x - 3| = 9$

2. Solve for $x$: $|2x - 3| = x + 1$

**Solutions:**

1.
$$|x - 3| = 9$$
$$\swarrow \quad \searrow$$

$$x - 3 = -9 \quad | \quad x - 3 = 9$$
$$x = -6 \quad | \quad x = 12$$

*Check*

$$|x - 3| = 9 \quad | \quad |x - 3| = 9$$
$$|-6 - 3| \overset{?}{=} 9 \quad | \quad |12 - 3| \overset{?}{=} 9$$
$$|-9| \overset{?}{=} 9 \quad | \quad |9| = 9$$
$$9 = 9 \checkmark \quad | \quad 9 = 9 \checkmark$$

2 solutions:
$$x = -6, x = 12$$

2.
$$|2x - 3| = x + 1$$
$$\swarrow \quad \searrow$$

$$2x - 3 = -(x + 1) \quad | \quad 2x - 3 = x + 1$$
$$2x - 3 = -x - 1 \quad | \quad x = 4$$
$$3x = +2 \quad |$$
$$x = \tfrac{2}{3} \quad | \quad x = 4$$

*Check*

$$|2x - 3| = x + 1 \quad | \quad |2x - 3| = x + 1$$
$$|2(\tfrac{2}{3}) - 3| \overset{?}{=} \tfrac{2}{3} + 1 \quad | \quad |2(4) - 3| \overset{?}{=} 4 + 1$$
$$|-\tfrac{5}{3}| \overset{?}{=} \tfrac{5}{3} \quad | \quad |5| \overset{?}{=} 5$$
$$\tfrac{5}{3} = \tfrac{5}{3} \quad | \quad \sqrt{5} = 5 \checkmark$$

2 solutions:
$$x = \tfrac{5}{3}, x = 4$$

3. Solve for $x$: $|x + 4| = -2$

3.
$$|x + 4| = -2$$

$$x + 4 = -(-2) \quad | \quad x + 4 = -2$$
$$x = -2 \quad | \quad x = -6$$

*Check*

$$|x + 4| = -2 \quad | \quad |x + 4| = -2$$
$$|-2 + 4| \stackrel{?}{=} -2 \quad | \quad |-6 + 4| \stackrel{?}{=} -2$$
$$|2| \stackrel{?}{=} -2 \quad | \quad |-2| \stackrel{?}{=} -2$$
$$2 \neq -2 \quad | \quad 2 \neq -2$$

No solutions

# Practice Problems 15.9

Solutions on page 268

1. The solution set of $|x - 1| = 3$ is

A) $\{-2, 4\}$  B) $\{2, -4\}$  C) $\{-2\}$
D) $\{4\}$  E) $\{ \}$

3. The solution set of $|x^2 - 5x| = 6$ is

A) $\{ \}$  B) $\{2, 3\}$  C) $\{-1, 6\}$
D) $\{3, 6\}$  E) $\{-1, 2, 3, 6\}$

2. What is the solution set of the equation

$|x + 7| = 2x - 1$

A) $\{-2\}$  B) $\{8\}$  C) $\{-2, 8\}$
D) $\{2, -8\}$  E) $\{ \}$

4. What is the solution set of the equation

$|x - 7| = 2x - 10$

A) $\{\frac{17}{3}\}$  B) $\{3\}$  C) $\{-3\}$
D) $\{\frac{17}{3}, 3\}$  E) $\{ \}$

## 15.10 Three Simultaneous Equations

● **Elimination:** This type of system is composed of three variables. Select two different pairs of equations and, using the elimination method, eliminate the same variable from each pair of equations. The result is two linear equations with two variables. Using any of the linear-linear system methods, solve for both variables. Using the two found variables, substitute into any of the three original equations to find the value of the third variable.

**Example:**

1. Solve the system of equations:

$$3x + y - 4z = -2$$
$$x + y + 2z = -1$$
$$x + 2y + 6z = -2$$

**Solution:**

1. Selecting equations 1 with 2 and 1 with 3.

$$3x+y-4z = -2 \qquad\qquad 3x+y-4z = -2$$
$$x+y+2z = -1 \qquad\qquad x+2y+6z = -2$$

Multiply first equation by $-1$ and add it to the second equation

Multiply first equation by $-2$ and add it to the second equation

$$-3x-y+4z = 2 \qquad\qquad -6x-2y+8z = 4$$
$$\underline{x+y+2z = -1} \qquad\qquad \underline{x+2y+6z = -2}$$

$$\boxed{-2x \quad +6z = 1} \qquad\qquad \boxed{-5x \quad +14z = 2}$$

$$-2x + 6z = 1$$
$$-5x + 14z = 2$$

Multiply first equation by 5 and second equation by $-2$ to eliminate $x$

$$5(-2x + 6z = 1)$$
$$-2(-5x + 14z = 2)$$
$$\Downarrow$$
$$-10x + 30z = 5$$
$$\underline{10x - 28z = -4}$$
$$2z = 1$$

$$\boxed{z = \tfrac{1}{2}}$$

Substitute $z = \tfrac{1}{2}$ into one of the two 2-variable equations.

$$-2x + 6z = 1$$
$$-2x + 6(\tfrac{1}{2}) = 1$$
$$-2x + 3 = 1$$
$$-2x = -2$$
$$\boxed{x = 1}$$

Substitute $x = 1$, $z = \tfrac{1}{2}$ into one of the three 3-variable equations.

$$x + y + 2z = -1$$
$$1 + y + 2(\tfrac{1}{2}) = -1$$
$$y + 2 = -1$$

$$\boxed{y = -3}$$

$$x = 1, \; y = -3, \; z = \tfrac{1}{2}$$

- **Determinants:** Arrange the equation in the format:

$$ax + by + cz = \mathbf{d}$$
$$ex + fy + gz = \mathbf{h}$$
$$jx + ky + lz = \mathbf{m}$$

If any term is missing insert the variable with a zero coefficient.

$$x = \frac{\begin{vmatrix} d & b & c \\ h & f & g \\ m & k & l \end{vmatrix}}{\begin{vmatrix} a & b & c \\ e & f & g \\ j & k & l \end{vmatrix}} \qquad y = \frac{\begin{vmatrix} a & d & c \\ e & h & g \\ j & m & l \end{vmatrix}}{\begin{vmatrix} a & b & c \\ e & f & g \\ j & k & l \end{vmatrix}} \qquad z = \frac{\begin{vmatrix} a & b & d \\ e & f & h \\ j & k & m \end{vmatrix}}{\begin{vmatrix} a & b & c \\ e & f & g \\ j & k & l \end{vmatrix}}$$

Review:

$$j \cdot f \cdot c + k \cdot g \cdot a + l \cdot e \cdot b$$

$$\begin{vmatrix} a & b & c \\ e & f & g \\ j & k & l \end{vmatrix} \begin{matrix} a & b \\ e & f \\ j & k \end{matrix} \quad f = [a \cdot f \cdot l + b \cdot g \cdot y + c \cdot e \cdot k] - [j \cdot f \cdot c + k \cdot g \cdot a + l \cdot e \cdot b]$$

$$a \cdot f \cdot l + b \cdot g \cdot j + c \cdot e \cdot k$$

**Example:**

2. Using determinants solve the system of equations:

$$3x + y - 4z = -2$$
$$x + y + 2z = -1$$
$$x + 2y + 6z = -2$$

**Solutions:**

$$x = \frac{\begin{vmatrix} -2 & 1 & -4 \\ -1 & 1 & 2 \\ -2 & 2 & 6 \end{vmatrix}}{\begin{vmatrix} 3 & 1 & -4 \\ 1 & 1 & 2 \\ 1 & 2 & 6 \end{vmatrix}} = \frac{\begin{vmatrix} -2 & 1 & -4 \\ -1 & 1 & 2 \\ -2 & 2 & 6 \end{vmatrix} \begin{matrix} -2 & 1 \\ -1 & 1 \\ -2 & 2 \end{matrix}}{\begin{vmatrix} 3 & 1 & -4 \\ 1 & 1 & 2 \\ 1 & 2 & 6 \end{vmatrix} \begin{matrix} 3 & 1 \\ 1 & 1 \\ 1 & 2 \end{matrix}}$$

$$= \frac{(-12 - 4 + 8) - (8 - 8 - 6)}{(18 + 2 - 8) - (-4 + 12 + 6)}$$

$$= \frac{-2}{-2} = 1$$

$$y = \frac{\begin{vmatrix} 3 & -2 & -4 \\ 1 & -1 & 2 \\ 1 & -2 & 6 \end{vmatrix}}{\begin{vmatrix} 3 & 1 & -4 \\ 1 & 1 & 2 \\ 1 & 2 & 6 \end{vmatrix}} = \frac{\begin{vmatrix} 3 & -2 & -4 \\ 1 & -1 & 2 \\ 1 & -2 & 6 \end{vmatrix} \begin{matrix} 3 & -2 \\ 1 & -1 \\ 1 & -2 \end{matrix}}{\begin{vmatrix} 3 & 1 & -4 \\ 1 & 1 & 2 \\ 1 & 2 & 6 \end{vmatrix} \begin{matrix} 3 & 1 \\ 1 & 1 \\ 1 & 2 \end{matrix}}$$

$$= \frac{(-18 - 4 + 8) - (4 - 12 - 12)}{-2}$$

$$= \frac{6}{-2} = 3$$

$$z = \begin{vmatrix} 3 & 1 & -2 \\ 1 & 1 & -1 \\ 1 & 2 & -2 \end{vmatrix} = \begin{vmatrix} 3 & 1 & -2 \\ 1 & 1 & -1 \\ 1 & 2 & -2 \end{vmatrix} \begin{matrix} 3 & 1 \\ 1 & 1 \\ 1 & 2 \end{matrix}$$

$$\frac{\begin{vmatrix} 3 & 1 & -4 \\ 1 & 1 & 2 \\ 1 & 2 & 6 \end{vmatrix}}{\begin{vmatrix} 3 & 1 & -4 \\ 1 & 1 & 2 \\ 1 & 2 & 6 \end{vmatrix} \begin{matrix} 3 & 1 \\ 1 & 1 \\ 1 & 2 \end{matrix}}$$

$$= \frac{(-6-1-4)-(-2-6-2)}{-2}$$

$$= \frac{-1}{-2} = \frac{1}{2}$$

 Note all three denominators are the same. Only calculate the value once.

# Practice Problems 15.10

1. Solve by elimination:

$$x + y + 2z = 1$$
$$2x + 3y + 4z = 4$$
$$3x - y - z = 2$$

A) $x=1, y=2, z=-1$   B) $x=2, y=2, z=\frac{3}{2}$
C) $x=4, y=6, z=4$   D) $x=1, y=1, z=\frac{1}{4}$
E) $x=-1, y=2, z=1$

3. Solve by determinants:

$$2x + 2y - z = 10$$
$$x + 3y + 3z = 4$$
$$5x + z = 3$$

A) $x=1, y=1, z=1$   B) $x=1, y=3, z=-2$
C) $x=1, y=2, z=3$   D) $x=-1, y=2, z=-3$
E) $x=1, y=2, z=-3$

2. Solve by elimination:

$$2x - 3y + 2z = 26$$
$$x - 2y - 3z = -5$$
$$x + y - z = -7$$

A) $x=4, y=3, z=-14$   B) $x=0, y=5, z=-2$
C) $x=-1, y=-2, z=-1$   D) $x=1, y=1, z=1$
E) $x=2, y=-4, z=5$

4. Solve by determinants:

$$y + 2z = -6$$
$$2x + z = 11$$
$$x + 2y = 16$$

A) $x=0, y=12, z=3$   B) $x=8, y=4, z=-5$
C) $x=5, y=0, z=1$   D) $x=4, y=-5, z=8$
E) $x=4, y=6, z=0$

Solutions on page 270

## 5.11 Linear Inequalities

There are two types of **linear inequalities** which we will review in this book.

**1 variable**—these lend themselves to algebraic solution and a demonstration of the solution on a number line.

**2 variables**—these lend themselves to graphic representation and represent a 'space' on the coordinate plane.

---

**1 variable:** The rules to solve a single variable inequality are the same as those for an equality with one exception:

**Any multiplication or division by a negative number reverses the direction of the inequality.**

The solution can be represented on a number line.

---

**Examples:**

1. Find the solution set for $2x - 1 \geqslant 5$ and represent the solution on a number line.

**Solutions:**

1.
$$2x - 1 \geqslant 5$$
$$2x \geqslant 6$$
$$x \geqslant 3$$

Solid because $x \geqslant 3$, not $x > 3$

 ● **represents a particular number is included,** e.g., $x \geqslant 3$, $x \leqslant 3$, $x = 3$.
○ **represents a particular value is not included,** e.g., $x > 3$, $x < 3$.

2. If $8 - 2x > 4$, represent the solution set on a number line.

2.
$$8 - 2x > 4$$
$$-2x > -4$$
$$x < 2$$

Division by $-2$ reverses the direction of the inequality.

Hollow because $x < 2$, not $x \leqslant 2$

**2 variables:** The equation $ax + by = c$ is the boundary between the space $ax + by > c$ and $ax + by < c$. The boundary itself is included if $ax + by \geqslant c$ or $ax + by \leqslant c$ and is shown by a solid line (to show inclusion). The boundary is excluded if $ax + by > c$ or $ax + by < c$ and is shown as a dotted line (to show exclusion).

$ax + by \geqslant c$, $ax + by > c$, $ax + by \leqslant c$ and $ax + by < c$ represent a space on the coordinate plane.

**Examples:**

3. Represent $y + 2x \geqslant 4$.

**Solutions:**

3. First graph the boundary $y + 2x = 4$. Since the original problem shows $\geqslant$ the boundary is solid (inclusion).

$$y + 2x = 4$$

| x | y |
|---|---|
| 0 | 4 |
| 1 | 2 |
| 2 | 0 |

Test a sample point on one side of the boundary line. For example, test the origin $(0, 0)$:

$$y + 2x \geqslant 4$$

Test point $(0, 0)$:

$$0 + 2(0) \overset{?}{\geqslant} 4$$

$$0 \overset{?}{\geqslant} 4 \quad \text{False}$$

$\therefore$ The test point is *not* in the space.

Shade the side of the line *not* including the test point.

4. Represent $x < 4$.

4. First graph the boundary $x = 4$. Since the original equation involves only $<$, the boundary is dotted (excluded).

Test a sample point on one side of the boundary line. For example, test point $(-1, 3)$:

$$x < 4$$

Test point $(-1, 3)$:

$$-1 < 4 \quad \text{True}$$

∴ The test point *is* in the sample space.
Shade the side of the line including the test point.

# Practice Problems 15.11

1. Graphically represent the solution set to

   $3x + 1 > 7$

   A)  ⟶  0   2  (closed dot at 2, arrow right)   B)  ⟶  0   2  (open dot at 2, arrow right)

   C)  ⟵  0   2  (open dot at 2, arrow left)   D)  ⟵  0   2  (closed dot at 2, arrow left)

   E)  0  ——  (closed dot)

2. Graphically represent the solution set to

   $1 - 3x \leqslant -5$

   A)  ⟶ (closed dot, arrow right)   B)  ⟶ (open dot, arrow right)

   C)  ⟵ (open dot, arrow left)   D)  ⟵ (closed dot, arrow left)

   E)  —— (closed dot)

3. Represent $x - y < 0$

4. Represent $2x + y \leqslant -6$

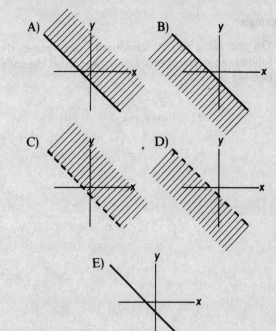

Solutions on page 271

## 15.12 Linear-Linear Inequalities

> The solution set is the coordinate space in the region overlapped by the two linear graphs. Graph each inequality on the same set of axes and notate the overlap.

**Example:**

1. On the same set of coordinate axes, graph the following system of inequalities and label the solution set $A$:

$$2y \geqslant x - 6$$
$$x + y < 2$$

**Solution:**

1. Graph the boundary $2y = x - 6$ as a solid line:

| x | y |
|---|---|
| 0 | -3 |
| 2 | -2 |
| 4 | -1 |

Test point $(0, 0)$:

$$2y \geqslant x - 6$$
$$2(0) \geqslant 0 - 6$$
$$0 \geqslant -6 \quad \text{True}$$

Shade the side of the line including the test point.

---

Graph the boundary $x + y = 2$ as a dashed line:

| x | y |
|---|---|
| 0 | 2 |
| 1 | 1 |
| 2 | 0 |

Test point $(0, 0)$:

$$x + y < 2$$
$$0 + 0 < 2$$
$$0 < 2 \quad \text{True}$$

Shade the side of the boundary including the test point.

---

The solution set $A$ is the overlapping region.

# Practice Problems 15.12

1. On the same set of axes, graph the following system of inequalities and label the solution set $S$:

$$y > x + 1$$
$$x \leqslant -1$$

2. On the same set of axes, graph the following system of inequalities and label the solution set $N$:

$$y < x - 2$$
$$y \geqslant -2x + 4$$

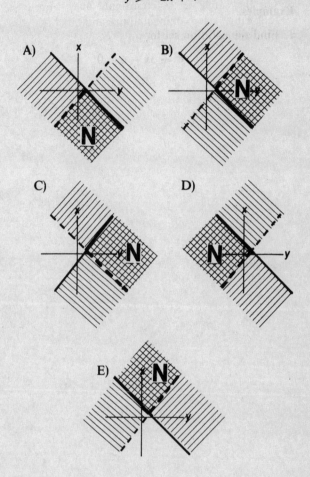

Solutions on page 272

## 15.13 Quadratic Inequalities

There are two types of quadratic inequalities that we will review in this book.

1. **variable**—These lend themselves to an algebraic solution and/or analysis on a number line. Using a number analysis, find the values which make the quadratic function equal to 0. Analyze the intervals between these critical values.

2. **variables**—These lend themselves to graphic representation and represent a 'space' on the coordinate plane.

**1 variable:** Arrange the quadratic function into $ax^2 + bx + c \geqslant 0, > 0, < 0, \leqslant 0$ form. Set the quadratic function equal to 0. The values which satisfy this equality are critical values in a number line analysis. Analyze the interval between these critical values to determine if the inequality is valid. The critical values are included if $\leqslant$ or $\geqslant$; they are excluded if $>$ or $<$. Critical values are represented by ● for inclusion, ○ for exclusion.

**Examples:**

1. Find the solution set for

$$x^2 + 3x - 4 \geqslant 0$$

**Solutions:**

1. To find the critical values set $f(x) = 0$.

$$x^2 + 3x - 4 = 0$$
$$(x + 4)(x - 1) = 0$$
$$x + 4 = 0 \quad | \quad x - 1 = 0$$
$$x = -4 \qquad \quad x = 1$$

The critical values are $-4$ and $1$. They are included because of the $\geqslant$ in the original problem.

$$x^2 + 3x - 4 \geqslant 0$$

Test the interval $x < -4$, e.g., $x = -10$:

$(-10)^2 + 3(-10) - 4$
$66 \geqslant 0$ True $\checkmark$

Darken $x < -4$ on the number line.

Test the interval $-4 < x < 1$, e.g., $x = 0$:
$0^2 + 3(0) - 4 \geqslant 0$
$\qquad -4 \geqslant 0$ False

Do *not* darken the number line in this interval.

Test the interval $x > 1$, e.g., $x = 10$:
$10^2 + 3(10) - 4 \geqslant 0$
$\qquad 126 \geqslant 0$ True $\checkmark$

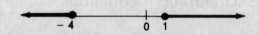

Darken $x > 1$ on the number line.

2. Find the solution set for

$$x^2 - 4 < 0$$

2. To find the critical values set $f(x) = 0$

$$x^2 - 4 = 0$$
$$(x + 2)(x - 2) = 0$$
$$x = -2 \,|\, x = 2$$

The critical values are $x=2$, $x=-2$.
These values are excluded because of the $<$ in the original problem.
Test the interval $x < -2$, e.g., $x = -10$:

$$x^2 - 4 < 0 \quad 10^2 - 4 < 0 \quad \text{False}$$

Do *not* darken the number line in this interval.

Test the interval $-2 < x < 2$, e.g., $x = 0$:

$$x^2 - 4 < 0 \quad 0^2 - 4 < 0 \quad \text{True}$$

Darken the number line in this interval.
Test the interval $x > 2$, e.g., $x = 10$:

$$x^2 - 4 < 0 \quad 10^2 - 4 < 0 \quad \text{False}$$

Do *not* darken the number line in this interval.

- **2 variables:** The quadratic equation with an $=$ instead of the $\leqslant$, $\geqslant$, $>$ or $<$ represents the boundary between the $>$ and $<$ regions. The boundary itself is included if $\geqslant$ or $\leqslant$ is represented (the boundary is solid). The boundary is excluded if $<$ or $>$ is represented (the boundary is dashed). The quadratic inequality is basically the "space" on the coordinate axes inside or outside the shape with the boundary either included or excluded as appropriate.

**Examples:**

3. Represent $x^2 + y^2 < 4$.

**Solutions:**

3. First graph $x^2 + y^2 = 4$. This is a circle with center $(0, 0)$ and radius $= 2$. The boundary is dashed as the original equation contains a $<$.

Test point $(0, 0)$:

$$x^2 + y^2 < 4$$
$$0^2 + 0^2 < 4$$
$$0 < 4 \quad \text{True}$$

The test point is in the required space. Shade in the inside of the circle.

4. Represent $\dfrac{x^2}{9} - \dfrac{y^2}{4} \leqslant 1$.

4. First sketch $\dfrac{x^2}{9} - \dfrac{y^2}{4} = 1$. This is a hyperbola. The boundary is solid as the original equation has a $\leqslant$.

Test point (0, 0):

$$\dfrac{x^2}{9} - \dfrac{y^2}{4} \leqslant 1$$

$$\dfrac{0^2}{9} - \dfrac{0^2}{4} \leqslant 1$$

$$0 \leqslant 1 \quad \text{True}$$

∴ The test point is in the required space. Shade the space including the test point.

# Practice Problems 15.13

1. Graphically represent the solution set to
$$x^2 - 5x + 6 < 0$$

3. Represent $\dfrac{x^2}{9} + \dfrac{y^2}{16} \geqslant 1$.

2. Graphically represent the solution set to
$$(x + 3)(x - 1) \leqslant 0$$

4. Represent $x^2 + y^2 > 16$.

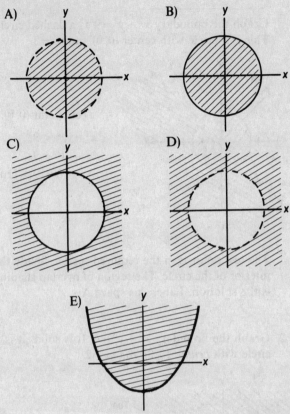

A)   B)

C)   D)

E)

Solutions on page 273

## 15.14 Quadratic-Linear and Quadratic-Quadratic Inequality Systems

> The solution set is the coordinate space in the region overlapped by the two graphs. Graph each inequality on the same set of axes and notate the overlapped region.

**Examples:**

1. On the same set of coordinate axes graph the following system of inequalities and label the solution set $N$.

$$x - y > 0$$
$$x^2 + y^2 < 9$$

**Solutions:**

1. Graph the boundary $x - y = 0$ as a dashed line

| $x$ | $y$ |
|-----|-----|
| 0 | 0 |
| 1 | 1 |
| 2 | 2 |

Test point (6, 0):

$$x - y > 0$$
$$6 - 0 > 0$$
$$6 > 0 \quad \text{True}$$

Shade the side of the boundary including the test point.

---

Graph the boundary $x^2 + y^2 = 9$ as a dashed circle. This is a circle with center $(0, 0)$, radius $= 3$.

Test point $(0, 0)$:

$$x^2 + y^2 < 9$$
$$0^2 + 0^2 < 9$$
$$0 < 9 \quad \text{True}$$

The test point *is* in the required region. Shade the interior of the circle. The region of overlap shading is the solution. Label this space $N$.

2. On the same set of coordinate axes graph the following system of inequalities.

$$x^2 + y^2 \geqslant 4$$
$$x^2 + y^2 \leqslant 9$$

2. Graph the boundary $x^2 + y^2 = 4$. It is solid. It is a circle with *center*: $(0, 0)$, *radius* $= 2$.

Test point $(0, 0)$:

$$x^2 + y^2 \geqslant 4$$
$$0^2 + 0^2 \geqslant 4$$
$$0 \geqslant 4 \quad \text{False}$$

The test point is not in the required space. Shade outside the circle.

---

Graph the boundary $x^2 + y^2 = 9$. It is solid. It is a circle with *center*: $(0, 0)$, *radius* $= 3$.

Test point $(0, 0)$:

$$x^2 + y^2 \leqslant 9$$
$$0^2 + 0^2 \leqslant 9$$
$$0 \leqslant 9 \quad \text{True}$$

The test point *is* in the required space. Shade inside the circle.

The overlapping region is the solution set. It is the space between the circles including the boundaries.

# Practice Problems 15.14

1. On the same set of axes, graph the solution set to the system of inequalities given below.

$$x \leqslant 1$$
$$y < x^2 + 3x$$

2. Graphically display the solution set to the following system of inequalities:

$$\frac{x^2}{9} + \frac{y^2}{4} < 1$$

$$\frac{x^2}{9} - \frac{y^2}{4} \geqslant 1$$

A)

B)

A)

B)

C)

D)

C)

D)

E)

E)

No solution set

Solutions on page 274

# 16. Polynomials

## 16.1 Fundamental Facts

**Monomial**—An algebraic expression with one term. Monomials are either a constant, a variable or the product of a constant and one or more variables. Examples of monomials are:

$$5 \quad -5 \quad x \quad 5x \quad 5x^2 \quad 5x^2y \quad xyz \quad xy^2z^3$$

**Polynomial**—The sum or difference of monomials. Examples include:

$$5 + x \quad x^2 + 3x + 2 \quad 3x^5 + 2x^4 - 3x^3 + xy - 6 \quad x^{75} - 1$$

—Two-term polynomials are called binomials, e.g., $x^2 - 1$.
—Three-term polynomials are called trinomials, e.g., $x^2 - 3x + 1$.

**Degree of a monomial**—The sum of the exponential values of the variable factors. For example,
$3x^3y$ is a 4th degree monomial. The variable $x$ has an exponential value of 3. The variable $y$ has an exponential value of 1. Total: $3 + 1 = 4$

$6a^2b^2c^3$ is a 7th degree monomial. The variable $a$ has an exponential value of 2. The variable $b$ has an exponential value of 2. The variable $c$ has an exponential value of 3. Total: $2 + 2 + 3 = 7$

**Degree of a polynomial**—Equal in value to the degree of its highest monomial term

**Simplifying a polynomial**—Combining like terms. For example,

$$3xy - 2xy = xy \quad 7x^4 + 4x^4 = 11x^4$$

**Standard form of a polynomial**—Arranging the terms in descending order of degree. Highest degree terms first. For example,

$$P(x) = x^5 - 3x^4 + 3x^3 + x^2 + x - 1$$

## Addition of polynomials

Addition can be done either horizontally or vertically.

---

**Addition:**

- Arrange the polynomials in standard form.
- Associate like terms.
- Add like terms by adding their coefficients.

---

**Example:**

1. Add $6x^3 + 4x^2 + 2x - 7$ and $8x^2 - 3x + 4$.

**Solution:**

1. The polynomials are already in standard form.

*Horizontal Addition*

$$(6x^3 + 4x^2 + 2x - 7) + (8x^2 - 3x + 4)$$

$$= 6x^3 + \underbrace{4x^2 + 8x^2} + \underbrace{2x - 3x} \underbrace{-7 + 4}$$

$$= 6x^3 + \quad +12x^2 \qquad -x \qquad -3$$

$$= 6x^3 + 12x^2 - x - 3$$

*Vertical Addition*

$$6x^3 + 4x^2 + 2x - 7$$
$$\underline{\phantom{6x^3 + }8x^2 - 3x + 4}$$
$$6x^3 + 12x^3 - x - 3$$

# Subtraction of polynomials

Subtraction is the addition of the opposite (used when subtracting vertically) or the distribution of the minus sign (used when subtracting horizontally).

---

**Subtraction**

- Arrange the polynomials in standard form.
- For vertical calculations change all the signs of the polynomial being subtracted and then add.
- For horizontal calculations distribute the minus sign.
- Associate like terms.
- Add like terms by adding their coefficients.

---

**Example:**

2. From $6x^3 + 4x^2 - 2x - 7$ subtract $8x^2 - 3x - 4$.

**Solution:**

2. Polynomials are already in standard form.

*Horizontal Subtraction*

$$(6x^3 + 4x^2 - 2x - 7) - (8x^2 - 3x - 4)$$

$$6x^3 + 4x^2 - 2x - 7 - 8x^2 + 3x + 4$$

$$6x^3 + \underbrace{4x^2 - 8x^2} \underbrace{-2x + 3x} \underbrace{-7 + 4}$$
$$6x^3 \quad -4x^2 \qquad +x \qquad -3$$

$$6x^3 - 4x^2 + x - 3$$

*Vertical Subtraction*

$$6x^3 + 4x^2 - 2x - 7$$
$$-\qquad 8x^2 - 3x - 4$$

$$\Downarrow$$

$$6x^3 + 4x^2 - 2x - 7$$
$$+\quad \underline{-8x^2 + 3x + 4}$$
$$6x^3 - 4x^2 + \phantom{x} x - 3$$

# Multiplying polynomials

Multiplication is best done horizontally.

---

**Multiplication:**

### *A polynomial by a monomial*

- **Distribute** the monomial to each term of the polynomial.

$$a(b + c + d) = ab + ac + ad$$

- **Associate** and add like terms.

### *A binomial by a binomial*

- **Distribute** each term of the first binomial to each term of the second binommial.

$$(a + b)(c + d) = ac + ad + bc + bd$$

This procedure is sometimes called **FOIL**. Referring to the terms of the binomials, multiply the First terms, Outside terms, Inside and Last terms.

$$(a + b)(c + d) = \frac{ac}{\text{First}} + \frac{ad}{\text{Outside}} + \frac{bc}{\text{Inside}} + \frac{bd}{\text{Last}}$$

- **Associate** and **add** like terms.

### *A polynomial by a polynomial*

- **Distribute** each term of the first polynomial to each term of the second polynomial.

$$(a + b + c)(x + y + z) = a(x + y + z) + b(x + y + z) + c(x + y + z) \text{ or}$$
$$= ax + ay + az + bx + by + bz + cx + cy + cz$$

- **Associate** and **add** like terms.

---

☞ Recall $x^a \cdot x^b = x^{a+b}$ and $x \cdot y = xy$.

**Examples:**

3. $6x(3x^2 - 2x - 4xy + 5) =$

**Solutions:**

3. $6x(3x^2 - 2x - 4xy + 5)$

$$= 18x^3 - 12x^2 - 24x^2y + 30x$$

4. Simplify $3x^2 + 2(x+1)(x-3) - (x+1)(x-1)$.

4.
$$3x^2 + 2\underbrace{(x+1)(x-3)} \quad \underbrace{-(x+1)(x-1)}$$
$$= 3x^2 + 2(x^2 - 3x + x - 3) - (x^2 - x + x - 1)$$
$$= 3x^2 + 2\underbrace{(x^2 - 2x - 3)} - \underbrace{(x^2 - 1)}$$
$$= 3x^2 + 2x^2 - 4x - 6 - x^2 + 1$$
$$= \underbrace{3x^2 + 2x^2 - x^2}_{4x^2} \underbrace{-4x}_{-4x} \underbrace{-6+1}_{-5}$$
$$= 4x^2 - 4x - 5$$

5. Multiply $(x^2 - 3x + 1)(x^2 - 2x - 3)$.

5.
$$(x^2 - 3x + 1)(x^2 - 2x - 3)$$
$$= x^2(x^2 - 2x - 3) - 3x(x^2 - 2x - 3) + 1(x^2 - 2x - 3) \text{ or}$$
$$= x^4 - 2x^3 - 3x^2 - 3x^3 + 6x^2 + 9x + x^2 - 2x - 3$$
$$= x^4 \underbrace{-2x^3 - 3x^3}_{-5x^3} \underbrace{-3x^2 + 6x^2 + x^2}_{+4x^2} \underbrace{+9x - 2x}_{+7x} -3$$
$$= x^4 - 5x^3 + 4x^2 + 7x - 3$$

# Division of polynomials

Division is best expressed in fraction form.

---

**Division:**

*A polynomial by a polynomial*

- Divide each term of the polynomial by the monomial

$$\frac{ab + ac}{a} = \frac{ab}{a} + \frac{ac}{a} = b + c$$

- This is done similiar to long division for numbers

| *Steps* | **Example** |
|---|---|
| | divide $x^2 - 5x - 7$ by $x - 1$ |
| | $x - 1 \overline{\smash{)}x^2 - 5x - 7}$ |
| 1. **Divide** the left term of the divisor into the left term of the dividend to obtain the first term of the quotient. Place the result in the quotient over the corresponding term of the dividend. | 1. divide $x$ into $x^2 \Rightarrow x$ $$\begin{array}{r} x \phantom{{}-5x-7} \\ x-1\overline{\smash{)}x^2 - 5x - 7} \end{array}$$ |

---

2. **Multiply** the divisor by the first term of the quotient and

   **Subtract** this result from the dividend

   ☞ subtraction is done by changing the signs of the polynomial being subtracted and then add

2. $x(x-1)=x^2-x$

$$
\begin{array}{r}
x \quad\quad\quad \\
x-1{\overline{\smash{\big)}\,-x^2-5x-7}} \\
x^2-x \quad\quad\quad
\end{array}
$$

☞
$$
\begin{array}{r}
x \quad\quad\quad\quad \\
x-1{\overline{\smash{\big)}\,x^2-5x-7}} \\
+ \quad\quad\quad\quad \\
\underline{-x^2+x \quad\quad\quad} \\
-4x-7
\end{array}
$$

3. **Divide** the left term of the divisor into the left term of the **new** dividend to obtain the next term of the quotient.

3. divide $x$ into $-4x \Rightarrow -4$

$$
\begin{array}{r}
x-4 \quad\quad\quad \\
x-1{\overline{\smash{\big)}\,x^2-5x-7}} \\
\underline{-x^2+\ \ x \quad\quad\quad} \\
-4x-7
\end{array}
$$

4. **Multiply** the divisor by the new term of the quotient and

   **Subtract** this result from the new dividend (Add the opposite)

4. $-4(x-1)=-4x+4$

$$
\begin{array}{r}
x-4 \quad\quad\quad \\
x-1{\overline{\smash{\big)}\,x^2-5x-7}} \\
\underline{-x^2+\ \ x \quad\quad\quad}
\end{array}
$$

$$
\begin{array}{r}
-4x-7 \\
-4x+4
\end{array}
$$

becomes

$$
\begin{array}{r}
x-4 \quad\quad\quad \\
x-1{\overline{\smash{\big)}\,x^2-5x-7}} \\
\underline{-x^2+\ \ x \quad\quad\quad} \\
-4x-7 \\
+ \quad\quad\quad\quad
\end{array}
$$

5. **Continue** this process until the subtraction has a remainder of 0 or is of a lower power than the divisor.

   If the remainder $=0$, the quotient is complete
   If the remainder $\neq 0$, express as a fraction
   $\dfrac{\text{remainder}}{\text{divisor}}$

$$
\begin{array}{r}
\underline{4x-4} \\
-11
\end{array}
$$

5. The remainder $-11$ is a lower power than the divisor $x-1$

   Solution: $x-4+\dfrac{-11}{x-1}$

$$
\underset{\text{Divisor}}{\Big\lfloor}\overline{\underset{\text{Dividend}}{\text{Quotient}+\dfrac{\text{Remainder}}{\text{Divisor}}}}
\quad \text{or} \quad
\dfrac{\text{Dividend}}{\text{Divisor}}=\text{Quotient}+\dfrac{\text{Remainder}}{\text{Divisor}}
$$

Replace any missing degree terms in the dividend before dividing. For example, if the dividend is $x^2-1$ use $x^2+0x-1$; if it is $x^3-3x+2$ use $x^3+0x^2-3x+2$.

**Examples:**

6. Divide $\dfrac{3x^3 + 6x^2 + 3x}{3x}$.

7. Divide $2x^3 + x - 1$ by $x + 3$.

**Solutions:**

6. $\dfrac{3x^3 + 6x^2 + 3x}{3x}$

$$= \frac{3x^3}{3x} + \frac{6x^2}{3x} + \frac{3x}{3x}$$

$$= x^2 + 2x + 1$$

 **Common error:**

$$\frac{3x^3 + 6x^2 + 3x}{3x} = x^2 + 2x$$

Note: $\dfrac{3x}{3x} = 1$ not $0$

7.

$$x + 3 \overline{\smash{)}2x^3 + 0x^2 + x - 1}$$

Insert the missing term:

$$\begin{array}{r} 2x^2 \\ x + 3 \overline{\smash{)}2x^3 + 0x^2 + x - 1} \end{array} \qquad \begin{array}{r} 2x^2 \\ x \overline{\smash{)}2x^3} \end{array}$$

Multiply $2x^2(x + 3)$ and subtract from the dividend:

$$\begin{array}{r} 2x^2 \\ x + 3 \overline{\smash{)}-2x^3 + 0x^2 + x - 1} \\ \underline{2x^3 + 6x^2} \end{array} \quad \Rightarrow$$

$$\begin{array}{r} 2x^2 \\ x + 3 \overline{\smash{)}2x^3 + 0x^2 + x - 1} \\ + \underline{-2x^3 - 6x^2} \\ -6x^2 + x - 1 \end{array}$$

$$\begin{array}{r} -6x \\ x \overline{\smash{)}-6x^2} \end{array}$$

$$\begin{array}{r} 2x^2 - 6x \\ x + 3 \overline{\smash{)}\;2x^3 + 0x^2 + x - 1} \\ + \underline{-2x^3 - 6x^2} \\ -6x^2 + x - 1 \end{array}$$

Multiply $-6x(x + 3)$ and subtract from the new dividend:

$$\begin{array}{r} 2x^2 - 6x \\ x + 3 \overline{\smash{)}\;2x^3 + 0x^2 + x - 1} \\ + \underline{-2x^3 - 6x^2} \\ -6x^2 + x - 1 \\ - \underline{-6x^2 - 18x} \end{array}$$

$$x + 3 \overline{\smash{\big)}\,\begin{aligned} &\phantom{x+3\,)\,}2x^2 - 6x \\ &2x^3 + 0x^2 + x - 1 \end{aligned}}$$

$$+\ \underline{-2x^3 - 6x^2}$$
$$-6x^2 + x - 1$$
$$+\ \ \underline{\mathbf{6x^2 + 18x}}$$
$$\mathbf{19x - 1}$$

$$x \overline{\smash{\big)}\,\begin{aligned} &\phantom{x\,)\,}19 \\ &19x \end{aligned}}$$

$$x + 3 \overline{\smash{\big)}\,\begin{aligned} &\phantom{x+3\,)\,}2x^2 - 6x + \mathbf{19} \\ &2x^3 + 0x^2 + x - 1 \end{aligned}}$$
$$+\ \underline{-2x^3 - 6x^2}$$
$$-6x^2 + x - 1$$
$$+\ \ \underline{6x^2 + 18}$$
$$19x - 1$$

Multiply $19(x + 3)$ and subtract from the new dividend:

$$x + 3 \overline{\smash{\big)}\,\begin{aligned} &\phantom{x+3\,)\,}2x^2 - 6x + 19 \\ &2x^3 + 0x^2 + x - 1 \end{aligned}}$$
$$+\ \underline{-2x^3 - 6x^2}$$
$$-6x^2 + x - 1$$
$$+\ \ \underline{6x^2 + 18x}$$
$$-19x - 1$$
$$\mathbf{19x + 57}$$

$$x + 3 \overline{\smash{\big)}\,\begin{aligned} &\phantom{x+3\,)\,}2x^2 - 6x + 19 \\ &2x^3 + 0x^2 + x - 1 \end{aligned}}$$
$$+\ \underline{-2x^3 - 6x^2}$$
$$-6x^2 + x - 1$$
$$+\underline{6x^2 + 18x}$$
$$-19x - 1$$
$$+\ \ \underline{\mathbf{19x - 57}}$$
$$\mathbf{-58}$$

$$\text{Quotient} = 2x^2 - 6x + 19 + \frac{-58}{x + 3}$$

# Synthetic division

If a polynomial is divided by a binomial of the form $x - a$, synthetic division simplifies the division process.

---

**Synthetic Division:**

### *Division of a polynomial by $x - a$*

### **Example**

Divide $2x^3 + x + 1$ by $x + 3$

*Steps*

● Insert any missing terms.

$$x + 3 \,\overline{\smash{\big)}\, 2x^3 + 0x^2 + x - 1}$$

● Represent the coefficients of the dividend. Place the value of $a$ in the bracket on the left side.

● Bring down the first number.

● Multiply the brought down number by the bracket number and place under the second coefficient number.

$-3(2) = -6$

● Add the new number to the coefficient number and bring down the new sum.

● Multiply the new brought down number by the bracket number and place under the 3rd coefficient number.

$$
\begin{array}{r|rrrr}
-3 & 2 & 0 & 1 & -1 \\
   &   & -6 & 18 &  \\
\hline
   & 2 & -6 & 19 &
\end{array}
$$

● Add the new number to the coefficient number and bring down the new sum.

$-3(-6) = 18$

Continue until all places are filled.

$$
\begin{array}{r|rrrr}
-3 & 2 & 0 & 1 & -1 \\
   &   & -6 & 18 & -57 \\
\hline
   & 2 & -6 & 19 & -58
\end{array}
$$

$-3(19) = -57$

Box the last number. This represents the remainder. The numbers to the left of the remainder represent the quotient (with each term one power less than the original dividend.

$$
\begin{array}{r|rrrr}
-3 & 2 & 0 & 1 & -1 \\
   &   & -6 & 18 & -57 \\
\hline
   & 2 & -6 & 19 & \boxed{-58}
\end{array}
$$

$$\qquad\quad\ \underset{x^2}{\uparrow}\ \ \underset{x}{\uparrow}\ \ \underset{\#}{\uparrow}$$

Quotient $= 2x^2 - 6x + 19 + \dfrac{-58}{x + 3}$

**Examples:**

8. Divide $x^2 - 5x - 7$ by $x - 1$.

**Solutions:**

8. $a$  $x^2$    $x$    #
   ↓   ↓     ↓     ↓

   $\underline{1|}$  1  $-5$  $-7$

   　　　　　  1  $-4$

   　　　1  $-4$  $\boxed{-11}$
   　　　↑　　↑
   　　　$x$　　#

$$\text{Quotient} = x - 4 + \frac{-11}{x-1}$$

9. Divide $x^2 - 5x + 6$ by $x - 5$.

9. $a$  $x^2$    $x$    #
   ↓   ↓     ↓     ↓

   $\underline{5|}$  1  $-5$  $+6$

   　　　　　  5   0

   　　　1   0  $\boxed{6}$
   　　　↑　　↑
   　　　$x$　　#

$$\text{Quotient} = x + \frac{6}{x-5}$$

# Practice Problems 16.1

1. Multiply $(2y - 5)(y - 4)$.

   A) $3y + 20$   B) $-7y - 9$
   C) $2y^2 + 20$   D) $2y^2 - 9y + 20$
   E) $2y^2 - 13y + 20$

2. From $6x^2 - 8x + 6$ subtract $x^2 - 3x - 6$.

   A) $3x^2 - 5x$   B) $5x^2 - 11x + 12$
   C) $5x^2 - 5x + 12$   D) $-5x^2 + 5x - 12$
   E) $5x - 5$

3. Divide $9x^6 + 3x^2$ by $3x^2$.

   A) $3x^3$   B) $3x^4 + 1$   C) $3x^3 + 1$
   D) $3x^4$   E) $6x^4 + 3x^2$

4. $(x - 2)(x^2 + 4x - 4) =$

   A) $2x^3 + 5x + 8$   B) $x^2 + 5x - 6$
   C) $x^3 + 8$   D) $x^3 - 8$
   E) $x^3 + 2x^2 - 12x + 8$

5. $(1 + a^3b^3)(1 - a^3b^3) =$

   A) $1 - a^6b^6$   B) 2   C) $1 - a^9b^9$
   D) $1 - 2a^3b^3$   E) $1 - 2a^6b^6 - a^9b^9$

6. $x^3 + 3x^2 - 6x - 3(2x + 3) + 2x(-3x + 1) =$

   A) $-8x^3 + 5x^2 - 6x - 9$
   B) $x^3 - 3x^2 - 10x - 9$
   C) $x^3 - 3x^2 + 2x - 9$
   D) $x^3 - 2x^2 - 2x + 1$
   E) $x^8 + 4$

7. Divide $x^3 + 2x^2 - 4x - 8$ by $x + 2$.

   A) $x^2 + 2$   B) $x^2 - 3x + 2$   C) $x^2 - 4$
   D) $x - 2$   E) $x^2 + 2x - 2$

8. Divide $x^2 + 9x - 10$ by $x + 10$.

   A) $x - 1$   B) $x + 1$   C) $x - 9$
   D) $x + 9$   E) $x - 10$

Solutions on page 275

# 16.2 Asymptotes

**Asymptotes**—These are lines which a graph approaches (but does not get to) at some finite value of $x$ or when $x = \pm \infty$.

There are three types of asymptotes: **Vertical**
**Horizontal**
**Slant**

| Given: $f(x) = \dfrac{g(x)}{h(x)}$ | | | |
|---|---|---|---|
| **Type of Asymptote** | **Criterion** | **Equation of Asymptote** | **Graphic Illustration** |
| **Vertical** | $h(a) = 0$ | $x = a$ | |
| **Horizontal** | $\lim\limits_{x \to \infty} f(x) = a$  $a \neq \infty$ | $y = a$ | |
| **Slant** | $\dfrac{g(x)}{h(x)} = Q(x) + R$  $g(x)$ is of a higher degree than $h(x)$ | $y = Q(x)$ | |

**Examples:**

1. Find any vertical, horizontal or slant asymptotes for $f(x) = \dfrac{4x}{x-3}$.

**Solutions:**

1. $f(x) = \dfrac{g(x)}{h(x)} = \dfrac{4x}{x-3}$

- $h(x) = 0$, $x - 3 = 0$, $x = 3$ There is a vertical asymptote at $x = 3$.

- $\lim\limits_{x \to \infty} \dfrac{4x}{x-3} = \lim\limits_{x \to \infty} \dfrac{\dfrac{4x}{x}}{\dfrac{x}{x} - \dfrac{3}{x}}$

$= \lim\limits_{x \to \infty} \dfrac{4}{1 - \dfrac{3}{x}}$

$= 4$

There is a horizontal asymptote at $y = 4$.

- There are no slant asymptotes; the degree of $g(x)$ is not greater than the degree of $h(x)$.

2. Find any horizontal, vertical or slant asymptotes of

$$f(x) = \frac{x^2+1}{x-1}$$

2. $f(x) = \dfrac{g(x)}{h(x)} = \dfrac{x^2+1}{x-1}$

$h(x)=0$, $x-1=0$, $x=1$ There is a vertical asymptote at $x=1$.

$$\lim_{x\to\infty} \frac{x^2+1}{x-1} = \lim_{x\to\infty} \frac{\dfrac{x^2}{x^2}+\dfrac{1}{x^2}}{\dfrac{x}{x^2}-\dfrac{1}{x^2}} = \lim_{x\to\infty} \frac{1+\dfrac{1}{x^2}}{\dfrac{1}{x}-\dfrac{1}{x^2}}$$

$$= \frac{1+0}{0-0} = \frac{1}{0} = \infty$$

No horizontal asymptote

Since $\lim_{x\to\infty} f(x) = \infty$, there is no horizontal asymptote.

Divide $\dfrac{x^2+1}{x-1}$ using synthetic division.

$$
\begin{array}{c c c c}
 & x^2 & x & \# \\
 & \downarrow & \downarrow & \downarrow \\
\underline{1\rfloor} & 1 & 0 & 1 \\
 & & 1 & 1 \\
\hline
 & 1 & 1 & \boxed{2} \\
 & \uparrow & \uparrow & \\
 & x & \# &
\end{array}
$$

$$\frac{x^2+1}{x-1} = x+1+\frac{2}{x-1}$$

There is a slant asymptote of $y=x+1$.

# Practice Problem 16.2

1. Determine any vertical asymptote(s) for

$$f(x) = \frac{x^3+3x^2+4x-3}{x}.$$

A) $x=1$  B) $y=0$  C) $x=0$
D) $y=1$  E) No vertical asymptote

2. Determine any horizontal asymptote(s) for

$$f(x) = \frac{x^2}{x^2+4}$$

A) $x=1$  B) $x=-2$  C) $x=2$
D) $y=1$  E) $y=-2$

3. Determine the vertical asymptote(s) for

$$f(x) = \frac{x^2}{x^2-4}$$

A) $y=1$  B) $x=2$ only  C) $x=-2$ only
D) $x=\pm2$  E) No vertical asymptotes

4. Determine the slant asymptote(s) for

$$x^2-yx+y+2=0$$

A) $y=2$  B) $x=2$  C) $y=x+1$
D) $x=y+1$  E) No slant asymptote

Solutions on page 276

# 16.3 Properties of Polynomials

| $P(x) = a_nx^n + a_{n-1}x^{n-1} \cdots + a_1x + a_0$ | |
|---|---|
| **Roots or Zeros of a Polynomial** | The values of $x$ for which $P(x)=0$. |
| **The Number of Roots** | A polynomial of degree $n$ has $n$ roots, some of which may be real or complex, and some of which may have a multiplicity greater than 1. |
| **Fundamental Theorem of Algebra** | Every polynomial of degree $\geq 1$ has at least one real or complex root. |
| **Remainder Theorem** | If $\dfrac{P(x)}{x-a} = Q(x)+R$    The remainder $R = P(a)$. |
| **Factor Theorem** | If $\dfrac{P(x)}{x-a} = Q(x)+0$ or $P(a)=0$, then $(x-a)$ is a factor of $P(x)$ and $a$ is a root. |
| **Imaginary Roots** | If $P(x)$ has real coefficients and if $a + bi$ is a root, then $a-bi$ is a root. |
| **Irrational Roots** | If $P(x)$ has rational coefficients and if $a + \sqrt{b}$ is root, then $a - \sqrt{b}$ is a root. |
| **Rational Root Theorem** | If $a_n$ to $a_0$ are integral coefficients, and if $\dfrac{p}{q}$ reduced to lowest terms is a rational number and a root of $P(x)$, then $p$ is a factor of $a_0$ and $q$ is a factor of $a_n$. |
| **Descartes' Rule of Sign** | If $P(x)$ has real coefficients and $a_0 \neq 0$, the number of positive real roots is equal to the number of sign variations of $P(x)$ or less than that number by a multiple of two. The number of negative real roots is equal to the number of sign variations of $P(-x)$ or less than that number by a multiple of two. |
| **Upper and Lower Bounds** | If $P(x)$ has real coefficients and $a_n$ is positive, there is a lower limit to the roots, that is, no root is smaller than $L$ if for $L \leq 0$ the third row in the synthetic division of $\dfrac{P(x)}{x-L}$ has values that alternate in sign. (Zero is considered $+$ or $-$ as required.) There is an upper limit to the roots, that is, no root is larger than $U$ if for $U \geq 0$ the third row of the synthetic division of $\dfrac{P(x)}{x-U}$ has only non-negative numbers. |

| The Sums and Products of the Roots | |
|---|---|
| | $\dfrac{a_{n-1}}{a_n}$ = Sum of the roots |
| | $-\dfrac{a_{n-2}}{a_n}$ = Sum of the roots taken two at a time |
| | $\dfrac{a_{n-3}}{a_n}$ = Sum of the roots taken three at a time |
| | $-\dfrac{a_{n-4}}{a_n}$ = Sum of the roots taken four at a time |
| | $\vdots \qquad \vdots$ |
| | $\dfrac{a_0}{n}$ (if $n$ is even) $\left.\right\}$ Sum of the roots taken $n$ at a time or |
| | $-\dfrac{a_0}{n}$ (if $n$ is odd) $\left.\right\}$ Product of the roots |

---

## Procedure for **locating the roots (zeros) of a polynomial** with real/rational coefficients

| Step | Description | Rule Utilized |
|---|---|---|
| 1 | Display the possible rational roots in the form $\dfrac{p}{q}$. | Rational Root Theorem |
| 2 | Eliminate impossible roots. For example, there may be no positive roots. | Descartes' Rule of Signs |
| 3 | Test possible values, looking for $P(a)=0$. Then $a$ is a root and $(x-a)$ is a factor. | Factor Theorem |
| 4 | Upon finding a root, depress the equation by removing the factor and lowering the degree of the equation. | Synthetic Division |
| 5 | Continue to search for roots using the depressed equation. Remember: Look for upper and lower bounds. : A root may have multiplicity $>1$. : Try the easier numbers first. | |
| 6 | If possible, continue locating roots until a quadratic equation is achieved. Solve the quadratic equation to obtain the last two roots/factors. | Quadratic Formula |

**Examples:**

1. Find the roots (zeros) of

$$P(x) = x^5 + 4x^4 - 4x^3 - 34x^2 - 45x - 18$$

**Solutions:**

1. List the possible roots: $\dfrac{\text{factors of } 18}{\text{factors of } 1}$

$$\frac{p}{q} = \frac{\pm 18,\ \pm 9,\ \pm 6,\ \pm 3,\ \pm 2,\ \pm 1}{\pm 1}$$

Descarte's rule of signs: 1 sign change ∴ 1 positive root

$$P(-x) = -x^5 + x^4 + 4x^3 - 34x^2 + 45x - 18$$

4 sign changes

|  | + | − | imag. |
|---|---|---|---|
| ∴ | 1 | 4 | 0 |
|  | 1 | 2 | 2 |
|  | 1 | 0 | 4 |

Look for the positive root:
$$P(x) = x^5 + 4x^4 - 4x^3 - 34x^2 - 45x - 18$$
$$P(1) = 1 + 4 - 4 - 34 - 45 - 18 \neq 0$$
$$P(2) = 32 + 64 - 32 - 136 - 90 - 18 \neq 0$$
$$P(3) = 243 + 324 - 108 - 306$$
$$\qquad\quad - 135 - 18 = 0$$

$$x^5 + 4x^4 - 4x^3 - 34x^2 - 45x - 18$$

```
3| 1   4   -4   -34   -45   -18
       3   21    51    51    18
   1   7   17    17     6    |0|
```
$$(x - 3)(x^4 + 7x^3 + 17x^2 + 17x + 6)$$

There are no more positive roots. Try for any negative roots using the depressed equation:

$$D_1(x) = x^4 + 7x^3 + 17x^2 + 17x + 6$$
$$D_1(-1) = 1 - 7 + 17 - 17 + 6 = 0 \ \checkmark$$

```
-1| 1   +7   +17   +17   +6
        -1    -6   -11   -6
    1    6    11     6   |0|
```
$$(x - 3)(x + 1)(x^3 + 6x^2 + 11x + 6)$$

Using the depressed equation $x^3 + 6x^2 + 11x + 6$:

$$D_2(x) = x^3 + 6x^2 + 11x + 6$$
$$D_2(-1) = -1 + 6 - 11 + 6 = 0 \ \checkmark$$

(−1 was used again looking for a possible multiple root.)

```
-1| 1   +6   +11   +6
        -1    -5   -6
    1    5     6   |0|
```
$$(x - 3)(x + 1)(x + 1)(x^2 + 5x + 6)$$
$$\underbrace{\qquad\qquad}_{(x + 3)(x + 2)}$$

∴ $P(x) = (x - 3)(x + 1)(x + 1)(x + 3)(x + 2)$
Roots are 3, −1, −2, −3.
−1 is a double root.

# Practice Problems 16.3

1. What is the multiplicity of the root $+1$ for the equation $x^3 - 3x^2 + 3x - 1 = 0$?

   A) 4  B) 3  C) 2  D) 1  E) 0

2. The equation $2x^3 + 3x^2 - 11x - 6 = 0$ has how many fractional roots?

   A) 4  B) 3  C) 2  D) 1  E) 0

3. What is the remainder when $x^{98} + 3$ is divided by $x - 1$?

   A) $(98)^{98}$  B) 98  C) 3  D) 2  E) 4

4. What is the sum of the roots of the equation $4x^3 - 2x^2 - x + 7 = 0$?

   A) 2  B) $\frac{1}{2}$  C) $\frac{7}{4}$  D) $\frac{4}{7}$  E) $-\frac{1}{2}$

5. What is the highest multiplicity of a root of the equation $x^3 - 3x^2 + 4 = 0$?

   A) 4  B) 3  C) 2  D) 1  E) 0

6. How many roots of $x^4 - 16 = 0$ are not real?

   A) 4  B) 3  C) 2  D) 1  E) 0

7. Which of the following is not a possible rational root of $f(x) = 4x^3 + 6x^2 + 8x + 9$?

   A) $-3$  B) $-\frac{3}{2}$  C) $-\frac{3}{4}$  D) 2  E) 9

8. If a polynomial with real/rational coefficients has one root $3 + 2i$ and another $1 - 3i$ what is the lowest degree equation possible?

   A) 5th  B) 4th  C) 3rd  D) 2nd
   E) 1st

Solutions on page 277

## 16.4 Sketching a Polynomial

$$f(x) = \frac{g(x)}{h(x)}$$

| Step | Details | Comment |
|---|---|---|
| 1 | Locate $y$-intercept | Let $x = 0$, solve for $y$ |
| 2 | Locate $x$-intercepts (roots) | Let $y = 0$, solve for $x$ |
| 3 | Note any single, double, triple roots, etc. | <br>Factors of the form $(x - a)$ cross the $x$-axis at $a$ as shown by either the solid or dotted line. |

Factors of the form $(x-a)^2$, double roots at $x=a$, do not cross the axis at $a$ they are tangent to it as shown by the solid or dotted curve.

Factors of the form $(x-a)^3$, triple roots cross the $x$-axis as shown by either the solid or dotted line.

Factors of the form $(x-a)^n$, where $n$ is even, vary as they **touch** the $x$-axis by their degree of flatness.

Factors of the form $(x-a)^n$, where $n$ is odd, vary as they **cross** the $x$-axis by their degree of flatness

---

**4** Determine when the graph is above and below the $x$-axis.

Test the region between the roots and between the lowest root and $-\infty$, and the largest root and $+\infty$. A number line can prove useful. In each interval test one value of $x$ to determine if $f(x)$ is $+$ or $-$.

In this case there are four intervals to test. If there are vertical asymptotes use these as

|  | | additional points on the number line to achieve additional intervals. |
| --- | --- | --- |
| | | |
| 5 | Evaluate asymptotes:<br>Vertical<br><br>Horizontal<br><br><br>Slant | $h(a)=0$   $x=a$ is a vertical asymptote<br><br>$\lim\limits_{x\to\infty} f(x)=a$   $y=a$ is a horizontal asymptote<br><br>$f(x)=\dfrac{g(x)}{h(x)}=Q(x)+R$<br><br>$y=Q(x)$ is a slant asymptote |
| 6* | Determine the relative min/max points.<br>Test the critical values. | Use the first derivative $=0$ to determine possible min/max points.<br>Determine if the points are/are not min/max points. |
| 7* | Determine the points of inflection.<br>Determine concavity. | Use the second derivative $=0$ to determine possible points of inflection.<br>Determine if the curve is concave or convex, etc. |

*These are further topics discussed in a calculus context which aid in sketching the graph of $f(x)$.

**Examples:**

1. Sketch the graph of

$$f(x)=x^3-7x^2+11x-5$$

**Solutions:**

1. $y=x^3-7x^2+11x-5$

Let $x=0$

$y=0^3-7(0)^2+11(0)-5=-5$

The $y$-intercept is $(0,\,-5)$

---

Let $y=0$

$0=x^3-7x^2+11x-5$

$\dfrac{p}{q}=\dfrac{\pm 5,\ \pm 1}{\pm 1}$

$P(x)=x^3-7x^2+11x-5$
$P(1)=1^3-7(1)^2+11(1)-5=0\ \checkmark$

$$\underline{1|}\quad 1 \quad -7 \quad 11 \quad -5$$
$$\phantom{\underline{1|}\quad 1\quad}\ 1 \quad -6 \quad\ \ 5$$
$$\overline{\phantom{\underline{1|}}\quad 1 \quad -6 \quad\ \ 5 \quad\ \underline{|0}}$$

$P(x)=(x-1)(x^2-6x+5)$
$\phantom{P(x)}=(x-1)(x-5)(x-1)$
$\phantom{P(x)}=(x-1)^2(x-5)$

There is a root at $x=5$ and a double root at $x=1$.

---

There are no asymptotes since there is no division.

Test the intervals between the roots.

Test the intervals:

$-\infty$ to 1    $f(x)="-"$
1 to 5    $f(x)="-"$
5 to $\infty$    $f(x)="+"$

a "$-$" indicates the graph is below the $x$-axis in this interval.

a "$+$" indicates the graph is above the $x$-axis in this interval.

The shaded area on the number line above shows the regions the graph passes through.

Recall that a single root at $x=5$ passes through the axis as a straight line.

Recall: a double root at $x=1$ indicates the curve is tangent to the axis.

2. **Sketch the graph of**

$$f(x)=x^3+x^2-5x+3$$

(Use the first and second derivatives to aid in sketching.)

2. $y=x^3+x^2-5x+3$

Let $x=0$
$y=0^3+0^2-5(0)+3=3$
The $y$-intercept is (0, 3)

Let $y=0$
$0=x^3+x^2-5x+3$
$\dfrac{p}{q}=\dfrac{\pm 3, \pm 1}{\pm 1}$
$P(x)=x^3+x^2-5x+3$

$$P(1) = 1^3 + 1^2 - 5(1) + 3 = 0 \checkmark$$

$$
\begin{array}{r|rrrr}
1\!| & 1 & 1 & -5 & 3 \\
  &   & 1 & 2 & -3 \\
\hline
  & 1 & 2 & -3 & \underline{|0} \\
\end{array}
$$

$$= (x-1)(x^2 + 2x - 3)$$
$$= (x-1)(x+3)(x-1)$$
$$= (x-1)^2(x+3)$$

There is a root at $x = -3$ and a double root at $x = 1$.

There are no asymptotes since there is no division.

Test the intervals between the roots.

Evaluate the intervals:

$-\infty$ to $-3$    $f(x) = "-"$
$-3$ to $1$      $f(x) = "+"$
  $1$ to $\infty$      $f(x) = "+"$

"$-$" means the graph is below the $x$-axis.
"$+$" means the graph is above the $x$-axis.
Also recall that the single root at $x = -3$ crosses the $x$ axis as a line at $x = -3$.
The double root at $x = 1$, is tangent to the axis.

*Using calculus:

$$P'(x) = 3x^2 + 2x - 5 = 0$$
$$(3x + 5)(x - 1) = 0$$
$$x = -\tfrac{5}{3} \quad x = 1$$
$$y = \tfrac{256}{27} \quad y = 0$$
$$\left(-\tfrac{5}{3}, \tfrac{256}{27}\right) \quad (1, 0)$$

$$P''(x) = 6x + 2 = 0$$
$$x = -\tfrac{1}{3}$$
$$\left(-\tfrac{1}{3}, \tfrac{128}{27}\right)$$

is a point of inflection

$P''(1) = 6(1) + 2 = 8 \quad \therefore (1, 0)$ is a minimum point
$P''(-\tfrac{5}{3}) = 6(-\tfrac{5}{3}) + 2 = -8 \quad \therefore \left(-\tfrac{5}{3}, \tfrac{256}{27}\right)$ is a maximum point.

# Practice Problems 16.4

1. Sketch a graph of $f(x) = \dfrac{x^2}{x^2 - 16}$.

2. Sketch a graph of $f(x) = x^3 - 4x^2 - 3x + 18$.

Solutions on page 278

# Cumulative Review Test

1. If $5^{3x-1} = 25$, find $x$.

   A) 1   B) $\frac{2}{3}$   C) $\frac{3}{2}$   D) $-1$   E) $\frac{1}{3}$

2. The graph of the equation $x^2 + 3x + 4y^2 = 36$ is a(n)

   A) line   B) parabola   C) ellipse
   D) circle   E) hyperbola

3. Factor completely $6x^2 - 7x - 3$.

   A) $(6x-1)(x+3)$   B) $(3x+1)(2x-3)$
   C) $(3x-3)(2x+1)$   D) $3(2x-1)(2x+1)$
   E) $3(x-1)(2x+1)$

4. If the measures of the three angles of a triangle are represented by $x$, $2x-20$ and $3x-10$, then the triangle is

   A) right   B) obtuse   C) acute
   D) equilateral   E) isosceles

5. Solve the following system of equations
   for $x$:   $4x - 3y = 8$
   $\qquad\qquad 2x + y = -1$

   A) $-2$   B) $-1$   C) 0   D) $\frac{1}{2}$   E) $\frac{3}{4}$

6. If $x^4 - 7x^3 + 9x^2 - 4 = 0$, find the sum of the roots.

   A) 9   B) $-9$   C) 7   D) $-7$   E) 4

7. Factor $y^{2n} - 16$.

   A) $(y^n+4)(y^n-4)$   B) $(y^n+1)(y^n-16)$
   C) $(y+4)(y^{2n-1}-4)$   D) $2(y^n-8)$
   E) $2(y+4)(y-4)$

8. The graph of the equation $y = x^2 + 3x + k$ passes through the point $(2, 0)$. Find $k$.

   A) $-\frac{3}{2}$   B) 2   C) $-2$   D) $-10$
   E) 10

9. Find the equation of the line which passes through the points $(1, 6)$ and $(-1, -2)$.

   A) $y-6 = 4(x-1)$   B) $y+6 = -4(x+1)$
   C) $y-2 = -4(x-1)$   D) $y-1 = 4(x-6)$
   E) $y-6 = -\frac{1}{4}(x-1)$

10. Express $6.5 \times 10^4$ as an integer.

    A) .00065   B) .0065   C) 6.54
    D) 65000   E) 65400

11. Express $\sqrt{-81} + \sqrt{-25} + 3\sqrt{-49}$ in terms of $i$.

    A) 35   B) $35i$   C) $-35$
    D) $-35i$   E) $21i$

12. Solve for $x$: $2 + 3\sqrt{x} = 8$.

    A) 0   B) 2   C) 4   D) 6   E) 8

13. Factor $12x^2 - 5x - 3$.

    A) $(3x+1)(4x-3)$   B) $3(x-1)(4x+1)$
    C) $4(3x-1)(x-1)$   D) $(12x+1)(x-3)$
    E) $(6x-1)(2x+3)$

14. If $f(x) = \dfrac{x-3}{x+2}$, find $f^{-1}(x)$.

    A) $\dfrac{x+2}{x-3}$   B) $\dfrac{x+3}{x-2}$   C) $\dfrac{2x+3}{x+1}$

    D) $\dfrac{2x+3}{1-x}$   E) $\dfrac{3x+2}{x+1}$

15. What is the locus of points 5 units from a given line and 3 units from a point on the line?

    A) 0 points   B) 1 point   C) 2 points
    D) 3 points   E) 4 points

16. Solve for $r$: $S = \dfrac{rl-a}{r-1}$.

    A) $\dfrac{S-a}{l}$   B) $\dfrac{S-a}{S-l}$   C) $\dfrac{l-a}{S}$

    D) $\dfrac{S-1}{l+1}$   E) $\dfrac{S-l}{S-a}$

250

17. Find the roots of $f(x) = 3x^2 + 2x - 6$.

    A) $-1 \pm \sqrt{19}$   B) $-2 \pm \sqrt{38}$   C) $\dfrac{-2 \pm \sqrt{19}}{3}$

    D) $\dfrac{-1 \pm \sqrt{19}}{3}$   E) $\dfrac{-2 \pm \sqrt{19}}{2}$

18. If $1 + i$ and $1 - i$ are the roots of the quadratic equation $x^2 + bx + c$, what is the value of $c$?

    A) 0   B) 1   C) 2   D) 3   E) 4

19. The solution set for the equation $2x^3 - 3x^2 - 23x + 12 = 0$ is

    A) $\{-4, \frac{1}{2}, -3\}$   B) $\{-4, -2, -1\}$
    C) $\{4, 2, -1\}$   D) $\{3, -1, \frac{1}{2}\}$
    E) $\{4, \frac{1}{2}, -3\}$

20. The solution set for the equation $3 + |x - 1| = 2x$ is

    A) $\{2\}$   B) $\{2, \frac{4}{3}\}$   C) $\{\frac{4}{3}\}$
    D) $\{\ \}$   E) $\{2, -\frac{4}{3}\}$

Solutions on page 279

# Diagnostic Chart for Cumulative Review Test

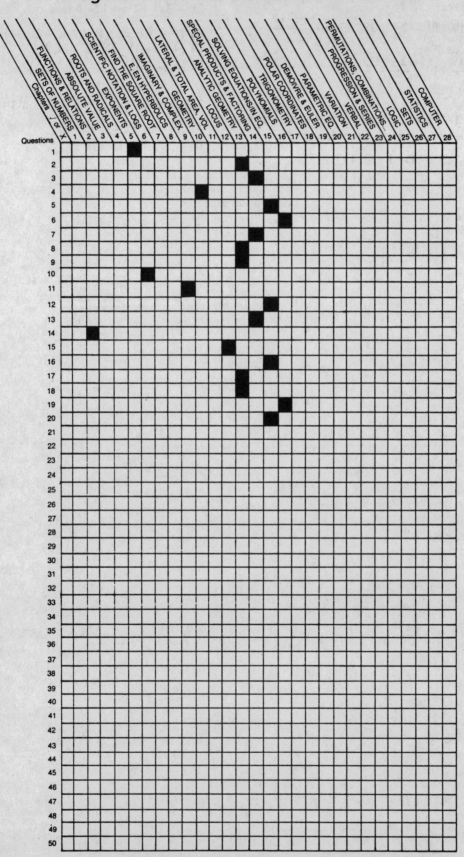

# Solutions to Problems in Unit Four

## Practice Problems 13.1

**1. (A)** $N(-3, -10), L(4, -4)$

$$m_{NL} = \frac{\Delta y}{\Delta x} = \frac{-4-(-10)}{4-(-3)}$$

$$= \frac{-4+10}{4+3} = \frac{6}{7}$$

**2. (B)** $N(-7, 0), L(-4, -2)$

$$d_{NL} = \sqrt{(x_2-x_1)^2 + (y_2-y_1)^2}$$

$$= \sqrt{[-4-(-7)]^2 + [-2-0]^2}$$

$$= \sqrt{(-4+7)^2 + (-2)^2}$$

$$= \sqrt{3^2+(-2)^2} = \sqrt{9+4} = \sqrt{13}$$

**3. (E)**

$$m_{RS} = \frac{\Delta y}{\Delta x} = \frac{y_2-y_1}{x_2-x_1} = \frac{-2-2}{-5-5} = \frac{-4}{-10}$$

$$= \frac{2}{5}$$

**4. (C)**

$$M_{JK} = \left(\frac{x_1+x_2}{2}, \frac{y_1+y_2}{2}\right) = \left(\frac{-6+(-8)}{2}, \frac{-2+4}{2}\right)$$

$$= \left(\frac{-14}{2}, \frac{2}{2}\right) = (-7, 1)$$

**5. (B)**

$x$ midpoint $= 3 = \dfrac{8+x}{2}$

$6 = 8+x, \quad x = -2$

$y$ midpoint $= 2 = \dfrac{3+y}{2}$

$4 = 3+y, \quad y = 1$

$\quad B(-2, 1)$

**6. (A)** The three points $N$, $J$, and $L$ are colinear if $m_{NJ} = m_{NL} = m_{JL}$.

$N(-2, -11)\ J(0, -7)\ \ L(4, 1)$

$$m_{NJ} = \frac{y_2-y_1}{x_2-x_1} = \frac{-7-(-11)}{0-(-2)} = \frac{-7+11}{2}$$

$$= \frac{4}{2} = 2$$

$$m_{JL} = \frac{y_2-y_1}{x_2-x_1} = \frac{1-(-7)}{4-(0)} = \frac{1+7}{4}$$

$$= \frac{8}{4} = 2$$

$$m_{NL} = \frac{y_2-y_1}{x_2-x_1} = \frac{1-(-11)}{4-(-2)} = \frac{1+11}{4+2}$$

$$= \frac{12}{6} = 2$$

The points are colinear.

**7. (D)**

$r$ is the distance from the center to point $S$.

$$d = \sqrt{(x_1-x_2)^2 + (y_2-y_1)^2}$$

$$= \sqrt{(2-6)^2 + (4-1)^2} = \sqrt{(-4)^2 + (3)^2}$$

$$= \sqrt{16+9} = \sqrt{25} = 5$$

**8. (D)** Perpendicular lines have slopes that are negative reciprocals.

$m_1 = -\dfrac{5}{2}$, then $m_2 = +\dfrac{2}{5}$

 Negative reciprocals are two numbers with
a) opposite signs
b) one upside down of the other
c) multiply to $-1$

# Practice Problems 13.2

1. **(D)** First find the slope of the given line:

$$m = \tfrac{1}{2}$$

A line perpendicular to the given line has a slope which is the negative reciprocal.

$$\therefore m = -\frac{2}{1}$$

Using $m = -2$ and the point $(-1, 5)$

$$y - 5 = -2(x - (-1))$$
$$y - 5 = -2(x + 1)$$

2. **(C)** $N(5, 0)$   $L(0, 5)$

These are the $x$- and $y$-intercepts.

$$\frac{x}{5} + \frac{y}{5} = 1 \quad \text{or} \quad x + y = 5$$

3. **(C)** Rearrange the equation into $y = mx + b$ form:

$$7 + 3y = -4x$$
$$3y = -4x - 7$$
$$y = -\tfrac{4}{3}x - \tfrac{7}{3}$$
$$m = -\tfrac{4}{3}$$

4. **(A)** $N(0, 3)$   $L(1, 4)$

$$m = \frac{\Delta y}{\Delta x} = \frac{4-3}{1-0} = \frac{1}{1} = 1$$

$$y - 3 = 1(x - 0)$$

To find the $x$-intercept, let $y = 0$:

$$0 - 3 = 1(x)$$
$$-3 = x$$
$$(-3, 0)$$

5. **(B)** First find the slope of the given line:

$$2x - 3y = 6$$

$$\frac{-3y}{-3} = \frac{-2x}{-3} + \frac{6}{-3}$$

$$y = \frac{2}{3}x - 2$$

$$m = \frac{2}{3}$$

A line parallel to the given line has the same slope.

$$\therefore m = \frac{2}{3} \text{ and using the point } (1, 3)$$

$$(y - 3) = \frac{2}{3}(x - 1)$$

☞ If a point $(x_1, y_1)$ is on a line, then the coordinates $x_1$ and $y_1$, when substituted into the equation of the line, will satisfy it.

6. **(C)** Test the five answer choices:

$$2y = 4 - x$$

A) $(2, 1)$   $2(1) = 4 - (-2)$
$$2 = 2 \checkmark$$

B) $(-4, 4)$   $2(4) = 4 - (-4)$
$$8 = 8 \checkmark$$

C) $(-2, -3)$   $2(-3) = 4 - (-2)$
$$-6 \neq 6 \text{ X}$$

D) $(0, 2)$   $2(2) = 4 - 0$
$$4 = 4 \checkmark$$

E) $(4, 0)$   $2(0) = 4 - 4$
$$0 = 0 \checkmark$$

7. **(E)** To be colinear $m_{NJ} = m_{JL} = m_{NL}$:

$$m_{NJ} = \frac{6-0}{3-1} = \frac{6}{2} = \frac{3}{1}$$

$$m_{NL} = \frac{k-0}{-4-1} = \frac{k}{-5}$$

$$\frac{3}{1} = \frac{k}{-5} \qquad k = -15$$

8. **(D)** $y = 3x + b$

If a point $(-3, 1)$ is on a line, it can be substituted into the equation of the line to make a true sentence:

$$1 = 3(-3) + b$$
$$1 = -9 + b$$
$$b = 10$$

# Practice Problems 13.4

1. **(D)** $r_1 = 1$   $r_2 = 2$
   Factors were $(x-1)(x-2)=0$
   $$x^2 - 3x + 2 = 0$$

2. **(D)** Real and equal roots of Discriminant $= 0$.
   $$x^2 + 4x + c = 0$$
   $$\therefore a = 1 \quad b = 4 \quad c = 4$$
   $$b^2 - 4ac = 0$$
   $$4^2 - 4(1)(c) = 0$$
   $$16 = 4c$$
   $$c = 4$$

3. **(A)** $y = x^2 - 4x$
   Complete the square:
   $$y + (-2)^2 = x^2 - 4x + (-2)^2$$
   $$y + 4 = (x-2)^2$$
   $$(x-2)^2 = y + 4$$

   $$\begin{cases} V:(2, -4) \\ p = 1 \\ F: (2, -3) \end{cases}$$

4. **(B)** $x^2 = 12(y-4)$

   $$\begin{cases} V: (0, 4) \\ 4p = 12 \quad p = 3 \end{cases}$$

5. **(C)**
   $$x^2 - 4x - 5 = 0$$
   $$(x-5)(x+1) = 0$$
   $$x - 5 = 0 \;\big|\; x + 1 = 0$$
   $$x = 5 \;\big|\; x = -1$$
   $$\{-1, 5\}$$

6. **(A)** $f(\text{root}) = 0$   $r_1 = 2$
   $$y = x^2 + kx - 2$$
   $$y = (2)^2 + k(2) - 2 = 0$$
   $$4 + 2k - 2 = 0$$
   $$2k = -2$$
   $$k = -1$$

7. **(B)** A parabola tangent to the $x$-axis has equal roots. The Discrimant $= 0$.
   $$kx^2 - 4x + k = 0$$
   $$a = k \quad b = -4 \quad c = k$$
   $$b^2 - 4ac = 0$$
   $$(-4)^2 - 4(k)(k) = 0$$
   $$16 - 4k^2 = 0$$
   $$4k^2 = 16$$
   $$k^2 = 4$$
   $$k = \pm 2$$

8. **(B)** $2x^2 - 5x - 4 = 0$
   $$a = 2 \quad b = -5 \quad c = -4$$
   $$x = \frac{-b \pm \sqrt{b^2 - 4ac}}{2a}$$
   $$= \frac{-(-5) \pm \sqrt{(-5)^2 - 4(2)(-4)}}{2(2)}$$
   $$= \frac{5 \pm \sqrt{25 + 32}}{4}$$
   $$= \frac{5 \pm \sqrt{57}}{4}$$

# Practice Problems 13.5

1. **(E)** $(x-h)^2 + (y-k)^2 = r^2$   *center*: $(-1, 1)$
   $$(x+1)^2 + (y-1)^2 = 3^2 \quad r = 3$$
   $$(x+1)^2 + (y-1)^2 = 9$$

2. **(B)**
   $$x^2 + y^2 = r^2 \quad \text{*center*: } (0, 0)$$
   $$x^2 + y^2 = 8 \quad r = \sqrt{8} = 2\sqrt{2}$$

**3. (A)** Complete the square.

$$x^2 - 6x + (-3)^2 + y^2 + 4y + (+2)^2$$
$$= -10 + (-3)^2 + (2)^2$$
$$(x-3)^2 + (y+2)^2 = 3$$
*center:* $(3, -2)$   $r = \sqrt{3}$

**4. (E)** *center:* $(-1, -1)$   *P:* $(4, -5)$

$$(x-h)^2 + (y-k)^2 = r^2$$
$$(x+1)^2 + (y+1)^2 = r^2$$

Substitute the point on the circle:

$$(4+1)^2 + (-5+1)^2 = r^2$$
$$5^2 + (-4)^2 = r^2$$
$$25 + 16 = r^2$$
$$r^2 = 41$$
$$(x+1)^2 + (y+1)^2 = 41$$

# Practice Problems 13.6

**1. (B)**

form: $\dfrac{y^2}{a^2} + \dfrac{x^2}{b^2} = 1$

$c$ = distance focus to center = 2
$a$ = distance vertex to center = 6

$$b^2 = a^2 - c^2 = 36 - 4 = 32$$

$$\frac{y^2}{36} + \frac{x^2}{32} = 1$$

**2. (E)** To set the equation equal to 1, divide each term by 3

$$\frac{\dfrac{(x-1)^2}{4}}{3} + \frac{\dfrac{(y-2)^2}{3}}{3} = \frac{3}{3}$$

$$\frac{(x-1)^2}{12} + \frac{(y-2)^2}{9} = 1$$

$a$ = distance from center to major vertex
$\quad = \sqrt{12}$ or $2\sqrt{3}$

**3. (C)**

$$\frac{4x^2}{36} + \frac{9y^2}{36} = \frac{36}{36} \quad \text{Divide each term by 36.}$$

$$\frac{x^2}{9} + \frac{y^2}{4} = 1 \quad \begin{matrix} a = 3 \\ b = 2 \end{matrix}$$

Length of latus rectum

$$= \left| \frac{2b^2}{a} \right| = \frac{2(2)^2}{3} = \frac{8}{3}$$

**4. (A)** Rewrite the given equation in the form:

$$\frac{x^2}{a^2} + \frac{y^2}{b^2} = 1$$

☞  $\dfrac{a}{b} = \dfrac{1}{\dfrac{b}{a}}$,   e.g.,  $\dfrac{25}{9} = \dfrac{1}{\dfrac{9}{25}}$

$$\frac{25x^2}{9} + 4y^2 = 1$$

$$\frac{x^2}{\frac{9}{25}} + \frac{y^2}{\frac{1}{4}} = 1$$

$$a = \sqrt{\frac{9}{25}} = \frac{3}{5}, \quad b = \sqrt{\frac{1}{4}} = \frac{1}{2}$$

$$b^2 + c^2 = a^2$$

$$\tfrac{1}{4} + c^2 = \tfrac{9}{25}$$

$$c^2 = \frac{9}{25} - \frac{1}{4} = \frac{26 - 25}{100} = \frac{11}{100}$$

$$c = \frac{\sqrt{11}}{10}$$

## Practice Problems 13.7

1. **(B)** $a=5$   $b=3$

    *latus rectum* length $=\dfrac{2b^2}{a}=\dfrac{2(3)^2}{5}=\dfrac{18}{5}$

2. **(B)** center: $(1, -11)$

    $a=\sqrt{2},\quad b=\sqrt{2}$

    $b^2+a^2=c^2$

    $2+2=c^2$

    $c=2$

The foci are each a distance 2 from the center.

$F_1: (3, -11)\quad F_2: (-1, -11)$

3. **(A)** Complete the square

    $16(x^2+6x+(3)^2)-9(y^2-4y+(-2)^2)$
    $$=36+16(3)^2-9(-2)^2$$

    $16(x+3)^2-9(y-2)^2=144$

    Divide each term by 144 to make the equation equal to 1

    $$\frac{16(x+3)^2}{144}-\frac{9(y-2)^2}{144}=\frac{144}{144}$$

    $$\frac{(x+3)^2}{9}-\frac{(y-2)^2}{16}=1$$

    center: $(-3, 2)$   $a=3,\quad b=4$

    Asymptotes:   $y=\pm\frac{4}{3}(x+3)+2$

    $\therefore y=\frac{4}{3}(x+3)+2\Rightarrow 3y-4x-18=0$

    $\phantom{\therefore} y=-\frac{4}{3}(x+3)+2\Rightarrow 3y+4x+6=0$

4. **(C)** $xy=-6$

    Make a table of values.

$xy=-6$

| $x$ | $y$ |
|---|---|
| 1 | $-6$ |
| 6 | $-1$ |
| $-1$ | 6 |
| $-6$ | 1 |
| 3 | $-2$ |
| 2 | $-3$ |
| $-2$ | 3 |
| $-3$ | 2 |

II and IV only

## Practice Problems 13.8

1. **(E)** The $x^2$ and $y^2$ terms have the same sign, different coefficients.

    Ellipse

2. **(A)** The $x^2$ and $y^2$ terms have the same sign and the same coefficients.

    Circle

3. **(D)** The $x^2$ and $y^2$ terms have different signs.

    Hyperbola

4. **(B)** One variable is squared, the other is linear.

    Parabola

# Practice Problems 14.1

1. **(A)** Using the model

$$(x-y)(x-y)=x^2-2xy+y^2$$

where

$$x=3f \text{ and } y=6g^2$$

$$(3f-6g^2)(3f+6g^2)=(3f)^2-2(3f)(6g^2)$$
$$+(6g^2)^2$$
$$=9f^2-36fg^2+36g^4$$

2. **(A)** Using the model

$$(x-y)(x+y)=x^2-y^2$$

where $x=5m^3$ and $y=3$

$$(5m^3-3)(5m^3+3)=(5m^3)^2-(3)^2$$
$$=25m^6-9$$

3. **(D)** Using the model

$$(x-y)(x^2+xy+y^2)=x^3-y^3$$

where $x=5$ and $y=3x$

$$(5-3x)(25+15x+9x^2)=(5)^3-(3x)^3$$
$$=125-27x^3$$

4. **(B)** Using the model

$$(x+y)(x-y)=x^2-y^2$$

where $x=7\sqrt{a}$ and $y=3\sqrt{b}$

$$(7\sqrt{a}+3\sqrt{b})(7\sqrt{a}-3\sqrt{b})=(7\sqrt{a})^2-(3\sqrt{b})^2$$
$$=49a-9b$$

# Practice Problems 14.2

1. **(B)** 9 is the largest common numerical factor.
   $x^2$ is the highest common power of $x$.
   There is no common power of $y$.

   1. The GCF $=9x^2$

   2. $\dfrac{9x^2-9x^2y}{9x^2}=\dfrac{9x^2}{9x^2}-\dfrac{9x^2y}{9x^2}=1-y$

   3. $9x^2-9x^2y=9x^2(1-y)$

2. **(C)** 3 is the largest common numerical factor.
   $m^2$ is the highest common power of $m$.
   $n^2$ is the highest common power of $n$.

   1. The GCF $=3m^2n^2$

   2. $\dfrac{3m^2n^2+3m^3n^3}{3m^2n^2}=\dfrac{3m^2n^2}{3m^2n^2}+\dfrac{3m^3n^3}{3m^2n^2}$
      $$=1+mn$$

   3. $3m^2n^2+3m^3n^3=3m^2n^2(1+mn)$

3. **(A)** There is no common numerical factor.
   $a$ is the highest common power of $a$.
   There is no common power of $b$.

   1. The GCF $=a$

   2. $\dfrac{4a^2+20a+5a^3b}{a}=\dfrac{4a^2}{a}+\dfrac{20a}{a}+\dfrac{5a^3b}{a}$
      $$=4a+20+5a^2b$$

   3. $4a^2+20a+5a^3b=a(4a+20+5a^2b)$

4. **(D)** 3 is the greatest common numerical factor.
   $n$ is the highest common power of $n$.
   $l$ is the highest common power of $l$.

   1. The GCF $=3nl$

   2. $\dfrac{3n^2l^2+6nl+9n^3l^3}{3nl}=\dfrac{3n^2l^2}{3nl}+\dfrac{6nl}{3nl}+\dfrac{9n^3l^3}{3nl}$
      $$=nl+2+3n^2l^2$$

   3. $3n^2l^2+6nl+9n^3l^3=3nl(nl+2+3n^2l^2)$

# Practice Problems 14.3

1. **(E)** Using the format $x^2-y^2$, where $x=3a$ and $y=4$,

$$x^2-y^2=(x+y)(x-y)$$
$$9a^2-16=(3a+4)(3a-4)$$

2. **(B)** Using the format $x^2-y^2$, where $x=4$ and $y=3a$,

$$x^2-y^2=(x+y)(x-y)$$
$$16-9a^2=(4+3a)(4-3a)$$

3. **(D)** Using the format $x^2+2xy+y^2$, where $x=a$ and $y=5$,

$$x^2+2xy+y^2=(x+y)(x+y)$$
$$a^2+10a+25=(a+5)(a+5)$$

4. **(D)** Using the format $x^3-y^3$, where $x=3a$ and $y=2$,

$$x^3-y^3=(x-y)(x^2+xy+y^2)$$
$$27a^3-8=(3a-2)(9a^2+6a+4)$$

5. **(B)** Using the format $x^3+y^3$, where $x=3a$ and $y=2$,

$$x^3+y^3=(x+y)(x^2-xy+y^2)$$
$$27a^3+8=(3a+2)(9a^2-6a+4)$$

6. **(C)** Using the format $x^2-2xy+y^2$, where $x=3a$ and $y=2$,

$$x^2-2xy+y^2=(x-y)(x-y)$$
$$9a^2-12a+4=(3a-2)(3a-2)$$

## Practice Problems 14.4

1. **(C)** $(x\quad)(x\quad)\qquad a=1$

$c=10$, middle term $=-7x$

$+10, +1$
$-10, -1$
$+5, +2$
$-5, -2$

---

$(x+10)(x+1)$ $\quad 10x+1x=11x\neq-7x$ $\quad$ [ $10x$ over $1x$ ]

---

$(x-10)(x-1)$ $\quad -10x-1x=-11x\neq-7x$ $\quad$ [ $-10x$ over $-x$ ]

---

$(x+5)(x+2)$ $\quad 5x+2x=7x\neq-7x$ $\quad$ [ $5x$ over $2x$ ]

---

$(x-5)(x-2)$ $\quad -5x-2x=-7x=-7x\ \checkmark$ $\quad$ [ $-5x$ over $-2x$ ]

Factors are: $(x-5)(x-2)$

2. **(D)** $3x^2-27=3(x^2-9)$ $\quad$ GCF: 3
$$=3(x+3)(x-3)$$
$x^2-9$ is the difference of two squares.

3. **(A)** $(3x\quad)(x\quad)$

$a=3$ a prime number
$c=4$, middle term $=-4x$

$+4, +1$
$-4, -1$
$+2, -2$

---

$(3x+4)(x+1)$ $\quad 4x+3x=7x\neq-4x$ $\quad$ [ $4x$ over $3x$ ]

---

$(3x+1)(x+4)$ $\quad x+12x=13x\neq-4x$ $\quad$ [ $x$ over $12x$ ]

---

$(3x-4)(x-1)$ $\quad -4x+(-3x)=-7x\neq-4x$ $\quad$ [ $-4x$ over $-3x$ ]

---

$(3x-1)(x-4)$ $\quad -x+(-12x)=-13x\neq-4x$ $\quad$ [ $-x$ over $-12x$ ]

---

$(3x+2)(x-2)$ $\quad 2x+(-6x)=-4x=-4x\ \checkmark$ $\quad$ [ $+2x$ over $-6x$ ]

Factors are: $(3x+2)(x-2)$

4. **(C)** $15y^2-5y=5y(3y-1)$ $\quad$ GCF: $5y$

5. **(E)** $2x^3-16=2(x^3-8)$ $\quad$ GCF: 2

$x^3-8$ is the difference of two cubes $x^3-y^3$, where $x=x$, $y=2$.

$$2x^3-16=2(x-2)(x^2+2x+4)$$

6. **(B)**

$4x^2+12x-16$ $\quad$ GCF: 4
$4(x^2+3x-4)$ $\quad$ $a$ is 1
$4(x\quad)(x\quad)$
$c=-4$, middle term $=+3x$

$+4, -1$
$-1, +4$
$-2, +2$

---

$(x+4)(x-1)$ $\quad 4x+(-x)=3x=3x\ \checkmark$ $\quad$ [ $4x$ over $-x$ ]

Factors are: $4(x+4)(x-1)$

7. **(D)**

$8x^8 - 8$    GCF: 8

$= 8(x^8 - 1)$

$x^8 - 1$ is the difference of two squares.

$= 8(x^4 + 1)(x^4 - 1)$

$x^4 - 1$ is the difference of two squares.

$= 8(x^4 + 1)(x^2 + 1)(x^2 - 1)$

$x^2 - 1$ is the difference of two squares.

$= 8(x^4 + 1)(x^2 + 1)(x + 1)(x - 1)$

8. **(A)** $a = 4$ a non-prime number

$\left.\begin{array}{l} 4, 1 \\ 2, 2 \end{array}\right\}$ factors of $a$

$(4x \quad )(x \quad )$

$c = 3$, middle term $= -8x$

$\left.\begin{array}{l} 3, \ 1 \\ -3, \ -1 \end{array}\right\}$ factors of $c$

$(4\underset{4x}{\underline{x + 3}})(\overset{3x}{x + 1})$

$3x + 4x = 7x \neq -8x$

$(4\underset{12x}{\underline{x + 1}})(\overset{x}{x + 3})$    $x + 12x = 13x \neq -8x$

$(4\underset{-4x}{\underline{x - 3}})(\overset{-3x}{x - 1})$    $-3x + (-4x) = -7x \neq -8x$

$(4\underset{-12x}{\underline{x - 1}})(\overset{-x}{x - 3})$    $-x + (-12x) = -13x \neq -8x$

$(2\underset{2x}{\underline{x + 3}})(\overset{6x}{2x + 1})$    $6x + 2x = 8x \neq -8x$

$(2\underset{-2x}{\underline{x - 3}})(\overset{-6x}{2x - 1})$    $-6x + (-2x) = -8x = -8x \ \checkmark$

Factors are: $(2x - 3)(2x - 1)$

9. **(B)** $(7x \quad )(x \quad )$

$a = 7$ a prime number

$c = -2$, middle term $= -13x$

$-2, \ +1$

$+2, \ -1$

$(7\underset{7x}{\underline{x - 2}})(\overset{-2x}{x + 1})$    $-2x + 7x = 5x \neq -13x$

$(7\underset{-14x}{\underline{x + 1}})(\overset{x}{x - 2})$    $x + (-14x) = -13x = -13x \ \checkmark$

Factors are: $(7x + 1)(x - 2)$

10. **(D)** $(x \quad )(x \quad )$    $a = 1$

$c = 6$, middle term $= -5x$

$\begin{array}{rr} 6 & 1 \\ -6 & -1 \\ 3 & 2 \\ -3 & -2 \end{array}$

$(\underset{x}{\underline{x + 6}})(\overset{6x}{x + 1})$    $6x + 1x = 7x \neq -5x$

$(\underset{-x}{\underline{x - 6}})(\overset{-6x}{x - 1})$    $-6x - x = -7x \neq -5x$

$(\underset{2x}{\underline{x + 3}})(\overset{3x}{x + 2})$    $3x + 2x = 5x \neq -5x$

$(\underset{-2x}{\underline{x - 3}})(\overset{-3x}{x - 2})$    $-3x - 2x = -5x \neq -5x \ \checkmark$

Factors are: $(x - 3)(x - 2)$

11. **(C)** $(2x \quad )(x \quad )$

$a = 2$ a prime number

$c = -3$, middle term $= +1x$

$-3, \ 1$

$3, \ -1$

$(2\underset{2x}{\underline{x - 3}})(\overset{-3x}{x + 1})$    $-3x + 2x = -x \neq x$

$(2\underset{-6x}{\underline{x + 1}})(\overset{x}{x - 3})$    $x + (-6x) = -5x \neq x$

$(2\underset{-2x}{\underline{x + 3}})(\overset{3x}{x - 1})$    $3x + (-2x) = x = x \ \checkmark$

Factors are: $(2x + 3)(x - 1)$

12. **(E)** $4x^3 - 49x$    GCF: $x$

$= x(4x^2 - 49)$

$4x^2 - 49$ is the difference of two squares.

$= x(2x + 7)(2x - 7)$

Factors are: $x(2x + 7)(2x - 7)$

## Practice Problems 14.5

**1. (B)**

$$\frac{x+3}{2x+4} \cdot \frac{x^2-x-6}{2x+6}$$

$$= \frac{\overset{(1)}{\cancel{x+3}}}{2\cancel{(x+2)}} \cdot \frac{(x-3)\overset{(1)}{\cancel{(x+2)}}}{2\cancel{(x+3)}} = \frac{x-3}{4}$$

**2. (E)**

$$\frac{a^2+8a+15}{a^2+5a} \div \frac{a^2-9}{a^2+2a-3}$$

$$= \frac{a^2+8a+15}{a^2+5a} \cdot \frac{a^2+2a-3}{a^2-9}$$

$$= \frac{(a+3)\overset{(1)}{\cancel{(a+5)}}}{a\cancel{(a+5)}} \cdot \frac{(a+3)(a-1)}{\cancel{(a+3)}(a-3)}$$

$$= \frac{(a+3)(a-1)}{a(a-3)}$$

**3. (A)**

$$\frac{x^2+3x+2}{x^2-x-2} \cdot \frac{x^2+x-6}{x^2-x-12}$$

$$= \frac{(x+2)\overset{(1)}{\cancel{(x+1)}}}{\cancel{(x+1)}\cancel{(x-2)}} \cdot \frac{\overset{(1)}{\cancel{(x-2)}}\overset{(1)}{\cancel{(x+3)}}}{\cancel{(x+3)}(x-4)}$$

$$= \frac{x+2}{x-4}$$

**4. (C)**

$$\frac{a+2}{2a} \cdot \frac{2a-4}{2+a} \cdot \frac{a^2-4}{12-6a} \div \frac{2+a}{12a^2}$$

$$= \frac{a+2}{2a} \cdot \frac{2a-4}{2+a} \cdot \frac{a^2-4}{12-6a} \cdot \frac{12a^2}{2+a}$$

$$= \frac{\overset{(1)}{\cancel{a+2}}}{\cancel{2a}} \cdot \frac{2(a-2)}{\cancel{a+2}} \cdot \frac{(\overset{(1)}{\cancel{a+2}})(\overset{(-1)}{\cancel{a-2}})}{\cancel{6}(2-a)} \cdot \frac{\overset{(2)}{\cancel{12}}\overset{(a)}{a^2}}{2+a}$$

$$= -2a(a-2)$$

## Practice Problems 14.6

**1. (C)**

$$x^2-4=0$$
$$(x+2)(x-2)=0$$
$$x+2=0 \mid x-2=0$$
$$x=-2 \mid x=2$$
$$x=-2 \text{ or } 2$$

**2. (A)**

$$x^2+9=10x$$
$$x^2-10x+9=0$$
$$(x-1)(x-9)=0$$
$$x-1=0 \mid x-9=0$$
$$x=1 \mid x=9$$
$$x=1 \text{ or } x=9$$

**3. (E)** $2n^3-2n=0$

$$2n(n^2-1)=0 \qquad \text{GCF: } 2n$$
$$2n(n+1)(n-1)=0$$
$$2n=0 \mid n+1=0 \mid n-1=0$$
$$n=0 \mid n=-1 \mid n=1$$
$$\{0, 1, -1\}$$

**4. (B)**

$$8x^2+10x-3=0$$
$$(4x-1)(2x+3)=0$$
$$4x-1=0 \mid 2x+3=0$$
$$x=\tfrac{1}{4} \mid x=-\tfrac{3}{2}$$
$$\{\tfrac{1}{4}, -\tfrac{3}{2}\}$$

## Practice Problems 15.1

**1. (A)** $3x+5=5x-3$

$$5+3=5x-3x$$
$$2x=8$$
$$x=4$$

**2. (B)** $2\left(\dfrac{d}{2}+6=4\right)$

$$d+12=8$$
$$d=-4$$

3. **(C)** $10(x-.2=1.8)$
$10x-2=18$
$10x=20$
$x=2$

4. **(D)** $4(b-1)-3=17$
$4b-4-3=17$
$4b-7=17$
$4b=24$
$b=6$

5. **(E)** $12\left(\dfrac{2n}{3}-\dfrac{5n}{12}=\dfrac{5}{4}\right)$
$8n-5n=15$
$3n=15$
$n=5$

6. **(A)** $5+3(a+2)=14$
$5+3a+6=14$
$3a+11=14$
$3a=3$
$a=1$

## Practice Problems 15.2

1. **(B)** $\dfrac{3a+1}{4}=\dfrac{5}{2}$
$2(3a+1)=20$
$6a+2=20$
$6a=18$
$a=3$

2. **(C)** $\dfrac{m}{4}=\dfrac{5}{2}$
$2m=20$
$m=10$

3. **(D)** $\dfrac{x+1}{8}=\dfrac{2x+1}{4}$
$4(x+1)=8(2x+1)$
$4x+4=16x+8$
$12x=-4$
$x=-\frac{4}{12}=-\frac{1}{3}$

4. **(A)** $\dfrac{3l-1}{5l+2}=\dfrac{2}{7}$
$7(3l-1)=2(5l+2)$
$21l-7=10l+4$
$11l=11$
$l=1$

## Practice Problems 15.3

1. **(D)** $\dfrac{cx+1}{a}=\dfrac{b}{3}$
$3(cx+1)=ab$
$3cx+3=ab$
$\dfrac{ab}{b}=\dfrac{3cx+3}{b}$
$a=\dfrac{3cx+3}{b}$

2. **(A)** $3(x+2)=a(b+x)$
$3x+6=ab+ax$
$3x-ax=ab-6$
$\dfrac{x(3-a)}{3-a}=\dfrac{ab-6}{3-a}$
$x=\dfrac{ab-6}{3-a}$

3. **(C)**
$\dfrac{a}{1}=\dfrac{t+b}{t-c}$
$a(t-c)=1(t+b)$
$at-ac=t+b$
$at-t=b+ac$
$\dfrac{t(a-1)}{a-1}=\dfrac{b+ac}{a-1}$
$t=\dfrac{b+ac}{a-1}$

4. **(C)** $abn\left(\dfrac{1}{a}=\dfrac{1}{n}+\dfrac{1}{b}\right)$
$bn=ab+an$
$bn-an=ab$
$\dfrac{n(b-a)}{b-a}=\dfrac{ab}{b-a}$
$n=\dfrac{ab}{b-a}$

# Practice Problems 15.4

**1. (A)**
$$y - x = -1$$
$$2y + x = 4$$

Solve the first equation for $x$ and substitute into the second equation:

$$y - x = -1$$
$$-x = -1 - y$$
$$x = 1 + y$$

Substitute $x = 1 + y$ into the second equation.

$$2y + x = 4$$
$$2y + (1 + y) = 4$$
$$2y + 1 + y = 4$$
$$3y = 3$$
$$y = 1$$

Replace $y$ with 1 in one of the original equations:

$$2y + x = 4$$
$$2(1) + x = 4$$
$$2 + x = 4$$
$$x = 2$$
$$x = 2, \quad y = 1$$
$$(2, 1)$$

---

**2. (C)** Add the equations to eliminate $y$:

$$\begin{array}{r} 3x + 4y = 7 \\ x - 4y = 1 \\ \hline 4x \phantom{+4y} = 8 \\ x \phantom{+4y} = 2 \end{array}$$

Replace $x = 2$ into one of the original equations:

$$x - 4y = 1$$
$$2 - 4y = 1$$
$$-4y = -1$$
$$y = \tfrac{1}{4}$$
$$x = 2 \quad y = \tfrac{1}{4}$$
$$(2, \tfrac{1}{4})$$

**3. (A)** $3x - 2y = -1$
$\phantom{3. (A)}$ $2x + 3y = 8$

$$x = \frac{\begin{vmatrix} -1 & -2 \\ 8 & 3 \end{vmatrix}}{\begin{vmatrix} 3 & -2 \\ 2 & 3 \end{vmatrix}} = \frac{-1(3) - (-2)(8)}{9 - (-2)(2)} = \frac{13}{13} = 1$$

$$y = \frac{\begin{vmatrix} 3 & -1 \\ 2 & 8 \end{vmatrix}}{\begin{vmatrix} 3 & -2 \\ 2 & 3 \end{vmatrix}} = \frac{3(8) - (-1)(2)}{3(3) - (-2)(2)} = \frac{26}{13} = 2$$

$$x = 1 \quad y = 2$$
$$(1, 2)$$

**4. (D)** $2x - y = 6$
$\phantom{4. (D)}$ $x + y = 9$

Solve the second equation for $y$ and substitute into the first equation:

$$x + y = 9$$
$$y = 9 - x$$

Substitute $9 - x$ for $y$ in the first equation.

$$2x - y = 6$$
$$2x - (9 - x) = 6$$
$$2x - 9 + x = 6$$
$$3x = 15$$
$$x = 5$$

Replace $x = 5$ into one of the original equations:

$$x + y = 9$$
$$5 + y = 9$$
$$y = 4$$
$$x = 5 \quad y = 4$$
$$(5, 4)$$

**5. (D)** $y - 4 = x$
$\phantom{5. (D)}$ $5x - 20 = y$

Rearrange the equations:

$$x - y = -4$$
$$5x - y = 20$$

Multiply the first equation by $-1$ to eliminate $y$:

$$-1(x - y = -4) \Rightarrow -x + y = 4$$

$$\begin{array}{r} -x + y = \phantom{2}4 \\ 5x - y = 20 \\ \hline 4x \phantom{- y} = 24 \\ x \phantom{- y} = 6 \end{array}$$

Substitute $x = 6$ and solve for $y$:

$$y - 4 = x$$
$$y - 4 = 6$$
$$y = 10$$
$$x = 6 \quad y = 10$$
$$(6, 10)$$

**6. (A)** $4x - y = 10$
$\qquad x + y = 5$

$$x = \dfrac{\begin{vmatrix} 10 & -1 \\ 5 & 1 \end{vmatrix}}{\begin{vmatrix} 4 & -1 \\ 1 & 1 \end{vmatrix}} = \dfrac{10(1) - (5)(-1)}{4(1) - (-1)(1)} = \dfrac{15}{5} = 3$$

$$y = \dfrac{\begin{vmatrix} 4 & 10 \\ 1 & 5 \end{vmatrix}}{\begin{vmatrix} 4 & -1 \\ 1 & 1 \end{vmatrix}} = \dfrac{4(5) - (10)(1)}{4(1) - (-1)(1)} = \dfrac{10}{5} = 2$$

$$x = 3 \quad y = 2$$
$$(3, 2)$$

**7. (D)** Put into intercept form:

$$x + y = 3$$
$$2x - y = 6$$

$$\dfrac{x}{3} + \dfrac{y}{3} = 1 \qquad \begin{array}{l} x \text{ intercept} = 3 \\ y \text{ intercept} = 3 \end{array}$$

$$\dfrac{2x}{6} - \dfrac{y}{6} = \dfrac{6}{6}$$

$\downarrow$

$$\dfrac{x}{3} - \dfrac{y}{6} = 1 \qquad \begin{array}{l} x \text{ intercept} = 3 \\ y \text{ intercept} = -6 \end{array}$$

**8. (C)** $x + y = 12$
$\qquad y = 3x$

Substitute the second equation into the first:

$$x + y = 12$$
$$x + (3x) = 12$$
$$4x = 12$$
$$x = 3$$

Replace $x$ with 3, solve for $y$:

$$y = 3x$$
$$y = 3(3) = 9$$
$$x = 3 \quad y = 9$$
$$(3, 9)$$

## Practice Problems 15.5

**1. (A)**

$$2x^2 + 3x - 2 = 0$$
$$(2x - 1)(x + 2) = 0$$
$$2x - 1 = 0 \ \bigg| \ x + 2 = 0$$
$$x = \tfrac{1}{2} \ \bigg| \qquad x = -2$$

**2. (C)**

$$x^2 - 4x - 7 = 0$$
$$x^2 - 4x \qquad = 7$$
$$x^2 - 4x + (-2)^2 = 7 + (-2)^2$$
$$(x - 2)^2 = 11$$
$$\sqrt{(x - 2)^2} = \sqrt{11}$$
$$x - 2 = \pm\sqrt{11}$$
$$x = 2 \pm \sqrt{11}$$

**3. (B)**

$$f(x) = x^2 + 4x + 9 \quad a = 1, \ b = 4, \ c = 9$$

$$x = \dfrac{-b \pm \sqrt{b^2 - 4ac}}{2a}$$

$$= \dfrac{-4 \pm \sqrt{4^2 - 4(1)(9)}}{2(1)}$$

$$= \dfrac{-4 \pm \sqrt{16 - 36}}{2}$$

$$= \dfrac{-4 \pm \sqrt{-20}}{2} = \dfrac{-4 \pm 2i\sqrt{5}}{2}$$

$$= -2 \pm i\sqrt{5}$$

4. **(A)** axis of symmetry $x = \dfrac{-(-4)}{2} = 2$

| $x$ | $x^2 - 4x + 4$ | $y$ |
|---|---|---|
| $-1$ | $(-1)^2 - 4(-1) + 4$ | 9 |
| 0 | $0^2 - 4(0) + 4$ | 4 |
| 1 | $1^2 - 4(1) + 4$ | 1 |
| **2** | $2^2 - 4(2) + 4$ | 0 |
| 3 | $3^2 - 4(3) + 4$ | 1 |
| 4 | $4^2 - 4(4) + 4$ | 4 |
| 5 | $5^2 - 4(5) + 4$ | 9 |

root is x = 2

root is $x = 2$. When curve is tangent to the axis there are really two roots of $x = 2$.

∴ There is a double root at $x = 2$.

# Practice Problems 15.6

1. **(C)** $y = 5 + 2x$
$x^2 + y^2 = 25$

Substitute $y = 2x + 5$ into the quadratic:

$$x^2 + (2x + 5)^2 = 25$$
$$x^2 + 4x^2 + 20x + 25 - 25 = 0$$

Divide by 5:

$$5x^2 + 20x = 0$$
$$x^2 + 4x = 0$$
$$x(x + 4) = 0$$

| $x = 0$ | $x + 4 = 0$ |
|---|---|
|  | $x = -4$ |
| $y = 5 + 2x$ | $y = 5 + 2x$ |
| $y = 5 + 2(0)$ | $y = 5 + 2(-4)$ |
| $y = 5$ | $y = -3$ |
| $(0, 5)$ | $(-4, -3)$ |

2. **(E)** $x^2 + y^2 - 4x - 2y + 1 = 0$
$y + x = 1$

Solve the linear equation for $x$ and substitute:

$$y + x = 1$$
$$x = 1 - y$$
$$x^2 + y^2 - 4x - 2y + 1 = 0$$
$$(1 - y)^2 + y^2 - 4(1 - y) - 2y + 1 = 0$$
$$1 - 2y + y^2 + y^2 - 4 + 4y - 2y + 1 = 0$$

combine terms

$$2y^2 - 2 = 0$$

Divide by 2

$$y^2 - 1 = 0$$
$$(y + 1)\quad(y - 1) = 0$$

| $y + 1 = 0$ | $y - 1 = 0$ |
|---|---|
| $y = -1$ | $y = 1$ |
| $y + x = 1$ | $y + x = 1$ |
| $-1 + x = 1$ | $1 + x = 1$ |
| $x = 2$ | $x = 0$ |
| $(2, -1)$ | $(0, 1)$ |

3. **(B)** $y = 2x^2 - 5x + 5$

$y - x = 5$

Solve the linear equation for $y$ and substitute into the quadratic:

$$y - x = 5 \Rightarrow y = x + 5$$

$$y = 2x^2 - 5x + 5$$

$$x + 5 = 2x^2 - 5x + 5$$

combine like terms

$$2x^2 - 6x = 0$$

Divide by 2:

$$x^2 - 3x = 0$$

$$x \quad (x - 3) = 0$$

| | |
|---|---|
| $x = 0$ | $x - 3 = 0$ |
| | $x = 3$ |
| $y - x = 5$ | $y - x = 5$ |
| $y - 0 = 5$ | |
| $y = 5$ | $y = 8$ |
| $(0, 5)$ | $(3, 8)$ |

4. **(A)** $y = -x^2 + 4x - 3$

$y + 1 = x$

Solve the linear equation for $y$ and substitute:

$$y + 1 = x \Rightarrow y = x - 1$$

$$y = -x^2 + 4x - 3$$

$$x - 1 = -x^2 + 4x - 3$$

combine like terms

$$-x^2 + 3x - 2 = 0$$

Multiply by $-1$: $\quad x^2 - 3x + 2 = 0$

$$(x - 2) \quad (x - 1) = 0$$

| | |
|---|---|
| $x - 2 = 0$ | $x - 1 = 0$ |
| $x = 2$ | $x = 1$ |
| $y + 1 = x$ | $y + 1 = x$ |
| $y + 1 = 2$ | $y + 1 = 1$ |
| $y = 1$ | $y = 0$ |
| $(2, 1)$ | $(1, 0)$ |

## Practice Problems 15.7

1. **(E)** $4x^2 + y^2 = 20$

$x^2 + 4y^2 = 20$

Multiply the second equation by $-4$ and add it to the first equation to eliminate the $x^2$:

$$-4(x^2 + 4y^2 = 20) \Rightarrow -4x - 16y = -80$$

$$-4x^2 - 16y^2 = -80$$

$$4x^2 + y^2 = 20$$

$$\overline{\phantom{-4x^2 - 16y^2 = -80}}$$

$$-15y^2 = -60$$

$$y^2 = 4$$

$$y = \pm 2$$

For each value of $y$ determine the corresponding values of $x$:

| $y = 2$ | $y = -2$ |
|---|---|
| $x^2 + 4y^2 = 20$ | $x^2 + 4y^2 = 20$ |
| $x^2 + 4(2)^2 = 20$ | $x^2 + 4(-2)^2 = 20$ |
| $x^2 = 4$ | $x^2 = 4$ |
| $x = \pm 2$ | $x = \pm 2$ |
| $(2, 2)$ | $(2, -2)$ |
| $(-2, 2)$ | $(-2, -2)$ |

4 solutions

2. **(C)** Add the equations to eliminate the $y^2$.

$$y^2 + 2x = -1$$

$$-y^2 + x^2 = 16$$

$$\overline{\phantom{y^2 + 2x = -1}}$$

$$x^2 + 2x - 15 = 0$$

Solve the quadratic for $x$.

$$(x + 5) \quad (x - 3) = 0$$

| | |
|---|---|
| $x + 5 = 0$ | $x - 3 = 0$ |
| $x = -5$ | $x = 3$ |

For the corresponding $y$ values

| $y^2 - 2x = -1$ | $y^2 - 2x = -1$ |
|---|---|
| $y^2 - 2(-5) = -1$ | $y^2 - 2(3) = -1$ |
| $y^2 = -11$ | $y^2 = 5$ |
| $y = \pm\sqrt{11}i$ | $y = \pm\sqrt{5}$ |
| $(-5, \sqrt{11}i)$ | $(3, \sqrt{5})$ |
| $(-5, -\sqrt{11}i)$ | $(3, -\sqrt{5})$ |

2 real solutions

3. **(A)**

$$x^2 + y = 5$$
$$x^2 + y^2 = 1$$

Multiply the second equation by $-1$ and add to the first equation to eliminate $x$.

$$-1(x^2 + y^2 = 1) \Rightarrow -x^2 - y^2 = -1$$
$$\begin{aligned} -x^2 - y^2 &= -1 \\ \underline{x^2 + y = 5} \\ -y^2 + y &= 4 \end{aligned}$$

Multiply by $-1$: $-y^2 + y - 4 = 0$
$$y^2 - y + 4 = 0$$

Use the quadratic equation.

$$y = \frac{-b \pm \sqrt{b^2 - 4ac}}{2a} \qquad \begin{aligned} a &= 1 \\ b &= -1 \\ c &= 4 \end{aligned}$$

$$= \frac{1 \pm \sqrt{(-1)^2 - 4(1)(4)}}{2(1)}$$

$$= \frac{1 \pm \sqrt{1 - 16}}{2} = \frac{1 \pm \sqrt{-15}}{2}$$

All values of $y$ are imaginary. There are no real solutions.

4. **(A)** Solved graphically.

$xy = 12$ is a hyperbola

| $x$ | $y$ |
|-----|-----|
| 1 | 12 |
| 12 | 1 |
| $-1$ | $-12$ |
| $-12$ | $-1$ |
| 2 | 6 |
| 6 | 2 |
| $-2$ | $-6$ |
| $-6$ | $-2$ |
| 4 | 3 |
| $-4$ | $-3$ |
| $-3$ | $-4$ |
| $-3$ | 4 |

$x^2 + y^2 = 16$ is a circle with center at $(0,0)$ and radius $= 4$.

No real solutions.

# Practice Problems 15.8

1. **(B)** $\sqrt{x^2 + 3} = x + 1$

Square both sides:

$$(\sqrt{x^2 + 3})^2 = (x + 1)^2$$
$$x^2 + 3 = x^2 + 2x + 1$$
$$2x - 2 = 0$$
$$x = 1$$

*Check*

$$\sqrt{x^2 + 3} = x + 1$$
$$\sqrt{1^2 + 3} \stackrel{?}{=} 1 + 1$$
$$\sqrt{4} \stackrel{?}{=} 2$$
$$2 = 2 \checkmark$$

1 solution   $x = 1$

**2. (C)** Get the $\sqrt{\phantom{x}}$ by itself and square the equation

$$\sqrt{x+18}+2=x$$
$$\sqrt{x+18}=x-2$$
$$(\sqrt{x+18})^2=(x-2)^2$$
$$x+18=x^2-4x+4$$
$$x^2-5x-14=0$$
$$(x-7)\quad(x+2)=0$$
$$x-7=0 \mid x+2=0$$
$$x=7 \mid \quad x=-2$$

*Check*

| $x=7$ | $x=-2$ |
|---|---|
| $\sqrt{x+18}+2=x$ | $\sqrt{x+18}+2=x$ |
| $\sqrt{7+18}+2\overset{?}{=}7$ | $\sqrt{-2+18}+2\overset{?}{=}-2$ |
| $\sqrt{25}+2\overset{?}{=}7$ | $\sqrt{16}+2\overset{?}{=}-2$ |
| $5+2\overset{?}{=}7$ | $4+2\overset{?}{=}-2$ |
| $7=7$ | $6\neq-2$ |
| $\sqrt{}$ | X |

1 solution
$$x=7$$

**3. (E)**

$$\sqrt{x+1}+2=0$$

Get the $\sqrt{\phantom{x}}$ by itself and square:

$$(\sqrt{x+1})^2=(-2)^2$$
$$x+1=4$$
$$x=3$$

*Check*

$$\sqrt{x+1}+2=0$$
$$\sqrt{3+1}+2\overset{?}{=}0$$
$$\sqrt{4}+2\overset{?}{=}0$$
$$2+2\overset{?}{=}0$$
$$4\neq0$$

No solution

Solution set $\varnothing$ or { }

**4. (A)**

$$\sqrt{2x}-\sqrt{x+1}=0$$

Get one $\sqrt{\phantom{x}}$ by itself and square:

$$(\sqrt{2x})^2=(\sqrt{x+1})^2$$
$$2x=x+1$$
$$x=1$$

*Check*

$$\sqrt{2x}-\sqrt{x+1}=0$$
$$\sqrt{2(1)}-\sqrt{1+1}\overset{?}{=}0$$
$$\sqrt{2}-\sqrt{2}\overset{?}{=}0$$
$$0=0\;\sqrt{}$$

1 solution
$$x=1$$

# Practice Problems 15.9

**1. (A)**

$$|x-1|=3$$

$$x-1=-3 \qquad x-1=3$$
$$x=-2 \qquad\quad x=4$$

*Check*

| $|x-1|=3$ | $|x-1|=3$ |
|---|---|
| $|-2-1|\overset{?}{=}3$ | $|4-1|\overset{?}{=}3$ |
| $|-3|\overset{?}{=}3$ | $|3|\overset{?}{=}3$ |
| $3=3\;\sqrt{}$ | $3=3\;\sqrt{}$ |

2 solutions
$$\{-2,\,4\}$$

**2. (B)**

$$|x+7|=2x-1$$

$$x+7=-(2x-1) \qquad x+7=2x-1$$
$$x+7=-2x+1 \qquad\qquad 8=x$$
$$3x=-6$$
$$x=-2$$

*Check*

| $|x+7|=2x-1$ | $|x+7|=2x-1$ |
|---|---|
| $|-2+7|\overset{?}{=}2(-2)-1$ | D$8+7|\overset{?}{=}2(8)-1$ |
| $|5|\overset{?}{=}-5$ | $|15|\overset{?}{=}16-1$ |
| $5\neq-5$ | $15=15\;\sqrt{}$ |

1 solution
$$x=8$$
$$\{8\}$$

**3. (E)**

$$|x^2 - 5x| = 6$$
$$\downarrow$$
$$x^2 - 5x = -6$$
$$x^2 - 5x + 6 = 0$$
$$(x-3)(x-2) = 0$$

$$x - 3 = 0 \mid x - 2 = 0$$
$$x = 3 \mid \quad x = 2$$

*Check*

| $\|x^2 - 5x\| = 6$ | $\|x^2 - 5x\| = 6$ |
|---|---|
| $\|3^2 - 5(3)\| \overset{?}{=} 6$ | $\|2^2 - 5(2)\| \overset{?}{=} 6$ |
| $\|9 - 15\| \overset{?}{=} 6$ | $\|4 - 10\| \overset{?}{=} 6$ |
| $\|-6\| \overset{?}{=} 6$ | $\|-6\| \overset{?}{=} 6$ |
| $6 = 6 \checkmark$ | $6 = 6 \checkmark$ |

$$|x^2 - 5x| = 6$$
$$\downarrow$$
$$x^2 - 5x = 6$$
$$x^2 - 5x - 6 = 0$$
$$(x-6)(x+1) = 0$$

$$x - 6 = 0 \mid x + 1 = 0$$
$$x = 6 \mid \quad x = -1$$

*Check*

| $\|x^2 - 5x\| = 6$ | $\|x^2 - 5x\| = 6$ |
|---|---|
| $\|6^2 - 5(6)\| \overset{?}{=} 6$ | $\|(-1)^2 - 5(-1)\| \overset{?}{=} 6$ |
| $\|36 - 30\| \overset{?}{=} 6$ | $\|1 + 5\| \overset{?}{=} 6$ |
| $\|6\| \overset{?}{=} 6$ | $\|6\| \overset{?}{=} 6$ |
| $6 = 6 \checkmark$ | $6 = 6$ |

There are 4 solutions
$$\{-1, 2, 3, 6\}$$

**4. (A)**

$$|x - 7| = 2x - 10$$

$$x - 7 = -(2x - 10) \mid x - 7 = 2x - 10$$
$$x - 7 = -2x + 10 \mid \quad 3 = x$$
$$3x = 17 \mid$$
$$x = \tfrac{17}{3} \mid$$

*Check*

| $\|x - 7\| = 2x - 10$ | $\|x - 7\| = 2x - 10$ |
|---|---|
| $\|\tfrac{17}{3} - 7\| \overset{?}{=} 2(\tfrac{17}{3}) - 10$ | $\|3 - 7\| \overset{?}{=} 2(3) - 10$ |
| $\|-\tfrac{4}{3}\| \overset{?}{=} \tfrac{34}{3} - \tfrac{30}{3}$ | $\|-4\| \overset{?}{=} 6 - 10$ |
| $\tfrac{4}{3} = \tfrac{4}{3} \checkmark$ | $\|-4\| = -4$ |
| | $4 \neq -4$ |

1 solution
$$\left\{\tfrac{17}{3}\right\}$$

# Practice Problems 15.10

1. **(A)** Using Equations 1 with 2 and 1 with 3 eliminate $y$:

$$x+y+2z=1 \qquad \bigg| \qquad x+y+2z=1$$
$$2x+3y+4z=4 \qquad \bigg| \qquad 3x-y-z=2$$

Multiply the first equation by $-3$ and add it to second equation to eliminate $y$

$$-3x-3y-6z=-3$$
$$2x+3y+4z=\phantom{-}4$$
$$\overline{\boxed{-x\phantom{-3y}-2z=\phantom{-}1}}$$

Add the equations to eliminate $y$

$$x+y+2z=1$$
$$3x-y-\phantom{2}z=2$$
$$\overline{\boxed{4x\phantom{-y}+\phantom{2}z=3}}$$

$$-x-2z=1$$
$$4x+\phantom{2}z=3$$

Multiply second equation by 2 and add it to first equation to eliminate $z$

$$-x-2z=1$$
$$8x+2z=6$$
$$\overline{7x=7}$$
$$\boxed{x=1}$$

Substitute $x=1$ into one of the two variable equations

$$4x+z=3$$
$$4(1)+z=3$$
$$4+z=3$$
$$\boxed{z=-1}$$

Substitute $x=1$ and $z=-1$ into one of the 3-variable equations

$$3x-y-z=2$$
$$3(1)-y-(-1)=2$$
$$3-y+1=2$$
$$\boxed{y=2}$$

2. **(E)** Using equations 1 with 2 and 2 with 3 eliminate $x$:

$$2x-3y+2z=26 \qquad \bigg| \qquad x-2y-3z=-5$$
$$x-2y-3z=-5 \qquad \bigg| \qquad x+\phantom{2}y-\phantom{2}z=-7$$

Multiply second equation by $-2$ and add to first equation to eliminate $x$

$$2x-3y+2z=26$$
$$-2x+4y+6z=10$$
$$\overline{\boxed{y+8z=36}}$$

Multiply first equation by $-1$ and add to second equation to eliminate $x$

$$-x+2y+3z=\phantom{-}5$$
$$x+\phantom{2}y-\phantom{2}z=-7$$
$$\overline{\boxed{3y+2z=-2}}$$

$$y+8z=36$$
$$3y+2z=-2$$

Multiply first equation by $-3$ and add to second equation to eliminate $y$

$$-3y-24z=-108$$
$$3y+\phantom{2}2z=-2$$
$$\overline{-22z=-110}$$
$$\boxed{z=5}$$

Substitute $z=5$ into one of the 2-variable equations

$$y+8z=36$$
$$y+8(5)=36$$
$$y+40=36$$
$$\boxed{y=-4}$$

Substitute $z=5$ and $y=-4$ into one of the 3-variable equations

$$x+y-z=-7$$
$$x+(-4)-(5)=-7$$
$$x-4-5=-7$$
$$x-9=-7$$
$$\boxed{x=2}$$

**3. (B)**

$$x = \frac{\begin{vmatrix} 10 & 2 & -1 \\ 4 & 3 & 3 \\ 3 & 0 & 1 \end{vmatrix}}{\begin{vmatrix} 2 & 2 & -1 \\ 1 & 3 & 3 \\ 5 & 0 & 1 \end{vmatrix}} =$$

$$= \frac{[30+18+0]-[-9+8+0]}{[6+30+0]-[-15+2+0]} = \frac{49}{49} = 1$$

$$y = \frac{\begin{vmatrix} 2 & 10 & -1 \\ 1 & 4 & 3 \\ 5 & 3 & 1 \end{vmatrix}}{\begin{vmatrix} 2 & 2 & -1 \\ 1 & 3 & 3 \\ 5 & 0 & 1 \end{vmatrix}}$$

$$= \frac{[8+150-3]-[-20+10+18]}{49} = \frac{147}{49} = 3$$

$$z = \frac{\begin{vmatrix} 2 & 2 & 10 \\ 1 & 3 & 4 \\ 5 & 0 & 3 \end{vmatrix}}{\begin{vmatrix} 2 & 2 & -1 \\ 1 & 3 & 3 \\ 5 & 0 & 1 \end{vmatrix}}$$

$$= \frac{[18+40+0]-[150+6+0]}{49} = \frac{-98}{49} = -2$$

**4. (B)**

$$x = \frac{\begin{vmatrix} -6 & 1 & 2 \\ 11 & 0 & 1 \\ 16 & 2 & 0 \end{vmatrix}}{\begin{vmatrix} 0 & 1 & 2 \\ 2 & 0 & 1 \\ 1 & 2 & 0 \end{vmatrix}}$$

$$= \frac{[0+16+44]-[0+0-12]}{[0+1+8]-[0+0+0]} = \frac{72}{9} = 8$$

$$y = \frac{\begin{vmatrix} 0 & -6 & 2 \\ 2 & 11 & 1 \\ 1 & 16 & 0 \end{vmatrix}}{\begin{vmatrix} 0 & 1 & 2 \\ 2 & 0 & 1 \\ 1 & 2 & 0 \end{vmatrix}}$$

$$= \frac{[0-6+64]-[22+0+0]}{9} = \frac{36}{9} = 4$$

$$z = \frac{\begin{vmatrix} 0 & 1 & -6 \\ 2 & 0 & 11 \\ 1 & 2 & 16 \end{vmatrix}}{\begin{vmatrix} 0 & 1 & 2 \\ 2 & 0 & 1 \\ 1 & 2 & 0 \end{vmatrix}}$$

$$= \frac{[0+11-24]-[0+32+0]}{9} = \frac{-45}{9} = -5$$

# Practice Problems 15.11

**1. (B)** $3x+1>7$

$3x>6$

$x>2$

hollow, excludes $x=2$

**2. (A)** $1-3x \leqslant -5$

$-3x \leqslant -6$

$x \geqslant 2$

The inequality reverses when division is by $-3$.

solid, $x=2$ is included

**3. (E)** Graph the boundary: $x-y=0$

$x=y$

The line is dashed as $x-y<0$ does not include the $=$.

Test point $(6, 0)$:

$x-y<0$

$6-0<0$

False!

Shade the side of the line *not* including the test point.

4. **(B)** Graph the boundary $2x + y = -6$

| x | y |
|----|----|
| -3 | 0 |
| -4 | 2 |
| -2 | -2 |

The line is solid as $2x + y \leqslant -6$ (includes the $=$)

Test point $(0, 0)$:

$$2x + y \leqslant -6$$
$$2(0) + 0 \leqslant -6$$
$$0 \leqslant -6 \quad \text{False!}$$

Shade the side of the line *not* including the test point.

# Practice Problems 15.12

1. **(D)**

$$y > x + 1$$
$$x \leqslant -1$$

Graph the boundary $y = x + 1$ as a dashed line:

| x | y |
|---|---|
| 0 | 1 |
| 1 | 2 |
| 2 | 3 |

Test point $(0, 0)$:

$$y > x + 1$$
$$0 > 0 + 1$$
$$0 > 1$$

False

Shade the side of the boundary *not* including the test point.

Graph the boundary $x = -1$ as a solid line:

Test point $(0, 0)$:

$$x \leqslant -1$$
$$0 \leqslant -1$$

False

Shade the side of the boundary *not* including the test point.

Label the overlap shaded areas S.

2. **(B)** $y < x - 2$

$\quad\quad y \geqslant -2x + 4$

Graph the boundary $y = x - 2$ as a dashed line:

| $x$ | $y$ |
|---|---|
| 0 | $-2$ |
| 1 | $-1$ |
| 2 | 0 |

Graph the boundary $y = -2x + 4$ as a solid line.

| $x$ | $y$ |
|---|---|
| 0 | 4 |
| 1 | 2 |
| 2 | 0 |

Test point $(0, 0)$:

$$y < x - 2$$
$$0 < 0 - 2$$
$$0 < -2$$
$$\text{False}$$

Shade the side of the boundary *not* including the test point.

Test point $(0, 0)$

$$y \geqslant -2x + 4$$
$$0 \geqslant -2(0) + 4$$
$$0 \geqslant 4$$
$$\text{False}$$

Shade the side of the boundary *not* including the test point.

Label the overlap shaded area **N**.

## Practice Problems 15.13

1. **(E)** To find the critical values set $f(x) = 0$:

$$x^2 - 5x + 6 = 0$$
$$(x - 2)(x - 3) = 0$$
$$x = 2 \;\big|\; x = 3$$

The critical values are excluded.
Test the interval $x < 2$, e.g., $x = 0$:
$x^2 - 5x + 6 < 0 \quad 0^2 - 5(0) + 6 < 0$  False

Do *not* darken the number line in this interval.

Test the interval $2 < x < 3$, e.g., $x = \frac{5}{2}$:
$(\frac{5}{2})^2 - 5(\frac{5}{2}) + 6 < 0$  True

Darken the number line in this interval.

Test the interval $x > 3$, e.g., $x = 10$:
$10^2 - 5(10) + 6 < 0$  False

Do *not* darken the number line in this interval.

2. **(B)** To find the critical values set $f(x) = 0$:

$$(x + 3) \quad (x - 1) = 0$$
$$x = -3 \;\big|\; x = 1$$

The critical values are included.
Test the interval $x < -3$, e.g., $x = -10$:
$(x + 3)(x - 1) \leqslant 0$
$(-10 + 3)(-10 - 1) \leqslant 0$  False

Do *not* darken the number line in this interval.

Test the interval $-3 < x < 1$, e.g., $x = 0$:
$(0 + 3)(0 - 1) \leqslant 0$  True

Darken the number line in this interval.

Test the interval $x > 1$, e.g., $x = 10$:
$(10 + 3)(10 - 1) \leqslant 0$  False

Do *not* darken the number line in this interval.

3. **(A)** First sketch $\dfrac{x^2}{9}+\dfrac{y^2}{16}=1$.

This is an ellipse. The boundary is solid as the original equation has $a \geqslant 1$.

Test point (0, 0):
$$\frac{x^2}{9}+\frac{y^2}{16} \geqslant 1$$
$$\frac{0^2}{9}+\frac{0^2}{16} \geqslant 1$$
$$0 \geqslant 1$$

False

**Test Point**

The test point is *not* in the required space. Shade in the region outside the ellipse.

4. **(D)** First sketch $x^2 + y^2 = 16$.

This is a circle. The boundary is dashed as the original equation does not have a $\geqslant$.

center (0, 0)    radius = 4

Test point (0, 0):
$$x^2 + y^2 > 16$$
$$0^2 + 0^2 > 16$$
$$0 > 16$$

False

**Test Point**

The test point is *not* in the required space. Shade the region outside the circle.

# Practice Problems 15.14

1. **(E)** Graph the boundary $x = 1$. It is solid. It is a vertical line through $x = 1$.

Test point (0, 0):
$$x \leqslant 1$$
$$0 \leqslant 1 \qquad \text{True}$$

**Test Point**

The test point is in the included region.

Shade the side of the boundary including the test point.

Graph the boundary $y = x^2 + 3x$. It is dashed. It is a parabola with roots $x = 0$, $x = -3$.

| $x$ | $y$ |
|---|---|
| 0 | 0 |
| 1 | 4 |
| 2 | 10 |
| −1 | −2 |
| −2 | −2 |
| −3 | 0 |

**Test Point**

Test point (8, 0):
$$y < x^2 + 3x$$
$$0 < 8^2 + 3(8) \quad \text{True}$$

The test point is in the required space.

Shade the space outside the parabola.
Label the overlap region **S**.

2. **(E)** Graph the elliptical boundary $\dfrac{x^2}{9}+\dfrac{y^2}{4}=1$. It is dashed.

Test point (0, 0):

$\dfrac{x^2}{9}+\dfrac{y^2}{4}<1$

$0+0<1$   True

The test point is in the required region.

Shade the inside of the ellipse

Graph the hyperbolic boundary $\dfrac{x^2}{9}-\dfrac{y^2}{4}=1$. It is solid.

Test point (0, 0):

$\dfrac{x^2}{9}-\dfrac{y^2}{4}\geqslant1$

$0-0\geqslant1$   False

The test point is not in the required region.

Shade as shown.

There is no solution. There is no area or point of overlap.

# Practice Problems 16.1

**1. (E)**

$(2y-5)(y-4)=2y^2-8y-5y+20$
$\qquad\qquad\quad\ =2y^2-13y+20$

**2. (C)**

$6x^2-8x+6-(x^2-3x-6)$

$=6x^2-8x+6-x^2+3x+6$

$=\underbrace{6x^2-x^2}_{5x^2}-\underbrace{8x+3x}_{-5x}+\underbrace{6+6}_{+12}$

$\qquad 5x^2-5x+12$

**3. (B)**

$\dfrac{9x^6+3x^2}{3x^2}=\dfrac{9x^6}{3x^2}+\dfrac{3x^2}{3x^2}=3x^4+1$

☞ $\left(\dfrac{x^6}{x^2}\right)=x^{6-2}=x^4$

**4. (E)**

$(x-2)(x^2+4x-4)$

$=x(x^2+4x-4)-2(x^2+4x-4)$ or

$=x^3+4x^2-4x-2x^2-8x+8$

$=x^3+\underbrace{4x^2-2x^2}_{+2x^2}\ \underbrace{-4x-8x}_{-12x}\ +8$

$=x^3\quad+2x^2\qquad-12x\ +8$

$=x^3\ +2x^2\ -12x\ +8$

**5. (A)**

$(1+a^3b^3)(1-a^3b^3)$

$=1(1-a^3b^3)+a^3b^3(1-a^3b^3)$ or

$=1-a^3b^3+a^3b^3-a^6b^6$

$=1-a^6b^6$

**6. (B)**

$x^3+3x^2-6x\underbrace{-3(2x+3)}+\underbrace{2x(-3x+1)}$

$=x^3+3x^2-6x\ \underbrace{-6x-9}\ \underbrace{-6x^2+2x}$

$=x^3\ \underbrace{+3x^2-6x^2}\ \underbrace{-6x-6x+2x}\ -9$

$=x^3\qquad -3x^2\qquad\quad -10x\qquad -9$

$\qquad x^3-3x^2-10x-9$

**7. (C)**

$\dfrac{x^3+2x^2-4x-8}{x+2}$

| $a$ | $x^3$ | $x^2$ | $x$ | # |
|---|---|---|---|---|
| ↓ | ↓ | ↓ | ↓ | ↓ |
| $-2\rfloor$ | 1 | $+2$ | $-4$ | $-8$ |
| | | $-2$ | $0$ | $+8$ |
| | 1 | 0 | $-4$ | $\boxed{0}$ |
| | ↑ | ↑ | ↑ | |
| | $x^2$ | $x$ | # | |

$x^2+0x-4=x^2-4$

8. **(A)**

$$\frac{x^2 + 9x - 10}{x + 10}$$

$$
\begin{array}{c}
\quad x^2 \quad x \quad \# \\
\quad \downarrow \quad \downarrow \quad \downarrow \\
-10\big|\quad 1 \quad +9 \quad -10 \\
\quad\quad\quad\quad -10 \quad +10 \\
\hline
\quad 1 \quad -1 \quad \underline{|0} \\
\quad \uparrow \quad \uparrow \\
\quad x \quad \#
\end{array}
$$

$$x - 1$$

# Practice Problems 16.2

1. **(C)**

$$f(x) = \frac{g(x)}{h(x)} = \frac{x^3 + 2x^2 + 4x - 3}{x}$$

$h(x) = 0$ for vertical asymptotes
$\therefore x = 0$ is a vertical asymptote

2. **(D)**

$$f(x) = \frac{x^2}{x^2 + 4}$$

$$\lim_{x \to \infty} \frac{x^2}{x^2 + 4} = \lim_{x \to \infty} \frac{\frac{x^2}{x^2}}{\frac{x^2}{x^2} + \frac{4}{x^2}} = \lim_{x \to \infty} \frac{1}{1 + \frac{4}{x^2}} = 1$$

There is a horizontal asymptote at $y = 1$.

3. **(D)**

$$f(x) = \frac{g(x)}{h(x)} = \frac{x^2}{x^2 - 4}$$

$h(x) = x^2 - 4 = 0 \Rightarrow x = \pm 2$

There are two vertical asymptotes.

$$x = 2 \quad\quad x = -2$$

4. **(C)** Rearrange the equation to obtain $f(x)$:

$$x^2 - yx + y + 2 = 0$$

$$x^2 + 2 - y(x - 1) = 0$$

$$y(x - 1) = x^2 + 2$$

$$y = \frac{x^2 + 2}{x - 1}$$

$$y = f(x) = \frac{x^2 + 2}{x - 1}$$

Since degree $g(x) >$ degree $h(x)$ divide using synthetic division:

$$\frac{x^2 + 0x + 2}{x - 1}$$

$$
\begin{array}{c}
1\big|\quad 1 \quad 0 \quad 2 \\
\quad\quad\quad 1 \quad 1 \\
\hline
\quad 1 \quad 1 \quad \underline{|3|}
\end{array}
$$

$$\frac{x^2 + 2}{x - 1} = x + 1 + \frac{3}{x - 1}$$

Slant asymptote at $y = x + 1$.

# Practice Problems 16.3

**1. (B)** $P(x) = x^3 - 3x^2 + 3x - 1$

Factor out the root $+1$ using synthetic division:

Root
↓        original equation coefficient

$$\begin{array}{r|rrrr} 1 & 1 & -3 & +3 & -1 \\ & & 1 & -2 & 1 \\ \hline & 1 & -2 & 1 & \boxed{0} \end{array}$$

depressed    remainder
equation
coefficients

$$P(x) = x^3 - 3x^2 + 3x - 1$$
$$= (x-1)(x^2 - 2x - 1)$$
$$\underbrace{(x-1)(x-1)}$$
$$= (x-1)(x-1)(x-1)$$

There is a multiplicity of 3 for the root $x = +1$.

**2. (D)** $P(x) = 2x^3 + 3x^2 - 11x - 6$

$$\frac{p}{q} = \frac{\pm 6, \pm 3, \pm 2, \pm 1}{\pm 2, \pm 1}$$

By Descartes' Rule: 1 positive root
$P(-x) = -2x^3 + 3x^2 + 11x - 6$:
Either 2 negative and 0 complex or 0 negative and 2 complex

Try to find the positive root:
$$P(x) = 2x^3 + 3x^2 - 11x - 6$$
$$P(1) = 2 + 3 - 11 - 6 \neq 0$$
$$P(2) = 16 + 12 - 22 - 6 = 0 \checkmark$$

$x = 2$ is a root, $x - 2$ is a factor

$$\begin{array}{r|rrrr} 2 & 2 & 3 & -11 & -6 \\ & & 4 & 14 & 6 \\ \hline & 2 & 7 & 3 & \boxed{0} \end{array}$$

$$P(x) = 2x^2 + 3x^2 - 11x - 6$$
$$= (x-2)\underbrace{(2x^2 + 7x + 3)}$$
$$\qquad\qquad (2x+1)(x+3)$$

$$= (x-2)(2x+1)(x+3) = 0$$

$$x = 2 \bigg| x = -\tfrac{1}{2} \bigg| x = -3$$

There is one fractional root.

**3. (E)**

$$\frac{P(x)}{x-a} = Q(x) + R$$

$R = P(a)$ by the remainder theorem
$R = P(1) = x^{98} + 3 = 1^{98} + 3 = 4$

**4. (B)**

$$\text{Sum} = \frac{-a_{n-1}}{a_n} = \frac{-(-2)}{4} = \frac{1}{2}$$

**5. (C)** $P(x) = x^3 - 3x^2 + 4$

$$\frac{p}{q} = \frac{\pm 4, \pm 2, \pm 1}{\pm 1}$$

$P(x)$ has 2 sign changes
$P(-x) = -x^3 + 3x^2 + 4$ has one sign change

Possibilities

|        | 2 | 0 |
|--------|---|---|
| +      | 2 | 0 |
| −      | 1 | 1 |
| Complex| 0 | 2 |

Try for the negative root:

$$\begin{array}{r|rrrr} 1 & 1 & -3 & 0 & 4 \\ & & -1 & 4 & -4 \\ \hline & 1 & -4 & 4 & \boxed{0} \end{array}$$

$$P(x) = x^3 - 3x^2 + 4$$
$$= (x+1)\underbrace{(x^2 - 4x + 4)}$$
$$\qquad\qquad (x-2)(x-2)$$
$$= (x+1)\ (x-2)\ (x-2)$$
$$x = -1 \big| \quad x = 2 \big| x = 2$$

Note: The root $x = 2$ has multiplity 2
$x = -1$ is a root, $x + 1$ is a factor

**6. (C)** $P(x) = x^4 - 16$

By Descartes' Rule $P(x)$ has 1 sign change.
∴ 1 positive root.
$P(-x) = x^4 - 16$ has 1 sign change.
∴ 1 negative root.

Since there is only 1 positive real root and only 1 negative real root the other two roots must be complex.

7. **(D)** $P(x) = 4x^3 + 6x^2 + 8x + 9$

$$\frac{p}{q} = \frac{\pm 9, \pm 3, \pm 1}{\pm 4, \pm 2, \pm 1}$$

None of the $\frac{p}{q}$ values $= 2$.

8. **(B)** Since the coefficients are real, if $3 + 2i$ is a root, $3 - 2i$ is a root; if $1 - 3i$ is a root, $1 + 3i$ is a root. There is a minimum of four roots.

Hence a 4th degree equation is the minimum possible.

# Practice Problems 16.4

1. Let $x = 0$, $y = \dfrac{0}{-16} = 0$

(0, 0) is the $y$-intercept.

---

Let $y = 0$, $0 = \dfrac{x^2}{x^2 - 16}$

$x^2 = 0$   $x = 0$; the only root is at the origin.

---

Asymptotes

$$P(x) = \frac{g(x)}{h(x)} = \frac{x^2}{x^2 - 16}$$

$h(x) = x^2 - 16 = 0 \quad \begin{array}{l} x = +4 \\ x = -4 \end{array}$

Two vertical asymptotes

$$\lim_{x \to \infty} \frac{x^2}{x^2 - 16} = \lim_{x \to \infty} \frac{\dfrac{x^2}{x^2}}{\dfrac{x^2}{x^2} - \dfrac{16}{x^2}}$$

$$= \lim_{x \to \infty} \frac{1}{1 - 0} = 1$$

A horizontal asymptote at $y = 1$.

Test the intervals:

$-\infty$ to $-4$   $P(x) = +$
$-4$ to $0$   $P(x) = -$
$0$ to $4$   $P(x) = -$
$4$ to $\infty$   $P(x) = +$

These results are reflected by the shaded areas on the number line above.

2. Let $x = 0$.

$$y = 0^3 - 4(0)^2 - 3(0) + 18 = 18$$

The $y$ intercept is (0, 18).

---

Let $y = 0$.

$$0 = x^3 - 4x^2 - 3x + 18$$

$$\frac{p}{q} = \frac{\pm 18, \pm 9, \pm 6, \pm 3, \pm 2, \pm 1}{\pm 1}$$

By Descarte's Law of Signs: Either 2 positive or 0 positive roots

$P(-x) = -x^3 - 4x^2 + 3x + 18$: 1 sign change
$\therefore$ 1 negative root

Possibilities

| | | |
|---|---|---|
| + | 2 | 0 |
| − | 1 | 1 |
| Complex | 0 | 2 |

Look for the negative root:

$P(x) = x^3 - 4x^2 - 3x + 18$
$P(-1) = -1 - 4 + 3 + 18 \neq 0$
$P(-2) = -8 - 16 + 6 + 18 = 0 \checkmark$

$$
\underline{2 \rfloor} \quad \begin{array}{cccc} 1 & -4 & -3 & 18 \\ & -2 & +12 & -18 \\ \hline 1 & -6 & 9 & |0 \end{array}
$$

$P(x) = (x + 2)(x^2 - 6x + 9)$
$\qquad = (x + 2)(x - 3)(x - 3)$

A single root at $x = -2$.
The graph passes through the $x$ axis at $x = -2$ as a straight line.

A double root at $x = 3$.
The graph is tangent to the axis.

Test the intervals:

$$
\begin{array}{ll}
-\infty \text{ to } -2 & f(x) = - \\
-2 \text{ to } \quad 3 & f(x) = + \\
3 \text{ to } \quad \infty & f(x) = +
\end{array}
$$

The results are reflected by the shaded regions on the number line above.

# Cumulative Review 1-16

1. **(A)** $5^{3x-1} = 25$
$\qquad 5^{3x-1} = 5^2$
$\qquad 3x - 1 = 2$
$\qquad\qquad x = 1$

2. **(C)** Both the $x$ and $y$ variables are squared. They both have coefficients of the same sign, but different in value. The relation is an ellipse.

3. **(B)** $6x^2 - 7x - 3$
$\qquad a = 6, c = -3,$ middle term $= 7x$

1. List all the positive factors of $a$

$\qquad$ 3, 2
$\qquad$ 6, 1

2. Use the first pair

$\qquad (3x \quad )(2x \quad )$

3. Pairs of factors of $c$

$\qquad -3, \quad 1$
$\qquad \phantom{-}3, \; -1$

4. Insert and test (be sure to use each pair reversed).

$$(3x - 3)(2x + 1) \quad \overset{-6x}{\underset{+3x}{}} \quad -6x + 3x = -3x \neq -7x$$

$$(3x + 1)(2x - 3) \quad \overset{2x}{\underset{-9x}{}} \quad 2x + (-9x) = -7x = -7x \checkmark$$

The factors are: $(3x + 1)(2x - 3)$

4. **(B)**

$(x) + (2x - 20) + (3x - 10) = 180$
$\qquad\qquad\qquad 6x - 30 = 180$
$\qquad\qquad\qquad\qquad 6x = 210$
$\qquad\qquad\qquad\qquad\; x = 35$
$\qquad\qquad\qquad 2x - 20 = 50$
$\qquad\qquad\qquad 3x - 10 = 95$

The triangle is obtuse.

**5. (D)**

$$4x - 3y = 8$$
$$(2x + y = -1) - 2 \Rightarrow -4x - 2y = 2$$

$$4x - 3y = 8$$
$$\underline{-4x - 2y = 2}$$
$$-5y = 10$$
$$y = -2$$

$$\therefore 4x - 3y = 8$$
$$4x - 3(-2) = 8$$
$$4x = 2$$
$$x = \tfrac{1}{2}$$

**6. (C)**

$$\text{Sum} = -\frac{b}{a} = 7$$

**7. (A)** $y^{2n} - 16$ is the difference of two squares.

$$x^2 - y^2 = (x + y)(x - y)$$
$$y^{2n} - 16 = (y^n + 4)(y^n - 4)$$

**8. (D)**

$$y = x^2 + 3x + k$$
$$0 = (2)^2 + 3(2) + k$$
$$0 = 4 + 6 + k$$
$$k = -10$$

**9. (A)**

$$m = \frac{\Delta y}{\Delta x} = \frac{6 - (-2)}{1 - (-1)} = \frac{8}{2} = 4$$

Using point (1, 6)

$$(y - y_1) = m(x - x_1)$$
$$y - 6 = 4(x - 1)$$

**10. (D)**

$$6.5 \times 10^4 = 6.50000 = 65000$$

**11. (B)**

$$\sqrt{-81} = 9i \quad \sqrt{-25} = 5i \quad \sqrt{-49} = 7i$$
$$\sqrt{-81} + \sqrt{-25} + 3\sqrt{-49} = 9i + 5i + 3(7i) = 35i$$

**12. (C)**

$$2 + 3\sqrt{x} = 8$$
$$3\sqrt{x} = 6$$
$$\sqrt{x} = 2$$
$$(\sqrt{x})^2 = (2)^2$$
$$x = 4$$

*Check*

$$2 + 3\sqrt{x} = 8$$
$$2 + 3\sqrt{4} \overset{?}{=} 8$$
$$2 + 3(2) \overset{?}{=} 8$$
$$8 = 8 \checkmark$$

**13. (A)** $12x^2 - 5x - 3$

$a = 12$, $c = -3$, the middle term $= -5x$

1. List all the positive factors of $a$

   12, 1
   3, 4
   6, 2

2. Use the first pair

   $(12x \quad )(x \quad )$

3. List the factors of $c$

   $c = -3$, middle term $= -5x$

   $-3, 1$
   $3, -1$

4. Insert and try

4. $(12\underset{12x}{\underline{x - 3}})(x + 1)$    $-3x + 12x = 9x \neq -5x$

---

$(12\underset{-36x}{\underline{x + 1}})(x - 3)$    $x + (-36x) = -35x \neq -5x$

---

$(12\underset{-12x}{\underline{x + 3}})(x - 1)$    $3x + (-12x) = -9x \neq -5x$

---

$(12\underset{36x}{\underline{x - 1}})(x + 3)$    $36x + (-x) = 35x \neq -5x$

---

$(3\underset{+3x}{\underline{x - 3}})(4x + 1)$    $-12x + 3x = -9x \neq -5x$

---

$(3\underset{-9x}{\underline{x + 1}})(4x - 3)$    $4x + (-9x) = -9x = -5x \checkmark$

The factors are:

$(3x + 1)(4x - 3)$

**14. (D)**

$$y=\frac{x-3}{x+2}$$

$$x=\frac{y-3}{y+2} \Rightarrow xy+2x=y-3$$

$$xy-y=-2x-3$$

$$y(x-1)=-2x-3$$

$$y=\frac{-2x-3}{x-1}$$

$$y=\frac{2x+3}{1-x}$$

**15. (A)**

There are no points that share the individual loci. There are 0 points which satisfy the given conditions.

**16. (B)**

$$S=\frac{rl-a}{r-1}$$

$$Sr-S=rl-a$$

$$Sr-rl=S-a$$

$$r(S-l)=S-a$$

$$r=\frac{S-a}{S-l}$$

**17. (D)**

$$x=\frac{-b\pm\sqrt{b^2-4ac}}{2a}$$

$$x=\frac{-2\pm\sqrt{(2)^2-4(3)(-6)}}{2(3)}$$

$$x=\frac{-2\pm\sqrt{4+72}}{6}$$

$$x=\frac{-2\pm\sqrt{76}}{6}=\frac{-2\pm2\sqrt{19}}{6}$$

$$=\frac{-1\pm\sqrt{19}}{3}$$

**18. (C)** $f(x)=x^2+bx+c$

Product of the roots $=\dfrac{c}{a}=\dfrac{c}{1}=c$

$$=(1+i)(1-i)=1-i^2=2$$

$$c=2$$

**19. (E)**

$$P(x)=2x^3-3x^2-23x+12$$

$$\frac{p}{q}=\frac{\pm12,\ \pm6,\ \pm4,\ \pm3,\ \pm2,\ \pm1}{\pm2,\ \pm1}$$

$$P(1)=2-3-23+12\neq0$$

$$P(2)=16-12-46+12\neq0$$

$$P(3)=54-27-69+12\neq0$$

$$P(4)=128-48-92+12=0$$

$$
\begin{array}{r|rrrr}
4 & 2 & -3 & -23 & 12 \\
  &   & 8  & 20  & -12 \\
\hline
  & 2 & 5  & -3  & \boxed{0}
\end{array}
$$

$$P(x)=2x^3-3x^2-23x+12$$

$$(x-4)\underbrace{(2x^2+5x-3)}_{(2x-1)(x+3)}$$

$$=(x-4)(2x-1)(x+3)=0$$

$$x=4\ |\ x=\tfrac{1}{2}\ |\ x=-3$$

$$\{4,\ \tfrac{1}{2},\ -3\}$$

**20. (A)**

$$3+|x-1|=2x$$

$$|x-1|=2x-3$$

| $-(x-1)=2x-3$ | $x-1=2x-3$ |
|---|---|
| $-x+1=2x-3$ | $x=2$ |
| $3x=4$ | |
| $x=\frac{4}{3}$ | |

| *check* | *check* |
|---|---|
| $3+|x-1|=2x$ | $3+|x-1|=2x$ |
| $3+|\frac{4}{3}-1|=2(\frac{4}{3})$ | $3+|2-1|=2(2)$ |
| $3+|\frac{1}{3}|=\frac{8}{3}$ | $3+|1|=2(2)$ |
| $\frac{10}{3}\neq\frac{8}{3}$ | $3+1=2(2)$ |
| | $4=4\ \checkmark$ |

# Unit Five

# 17. Trigonometry

## 17.1 Basics and Beyond

| | | |
|---|---|---|
| $\sin \theta = \dfrac{\text{leg Opposite angle } \theta}{\text{Hypotenuse}} = \dfrac{\text{Opp}}{\text{Hyp}}$ | | Sine is abbreviated sin. |
| $\cos \theta = \dfrac{\text{leg Adjacent to angle } \theta}{\text{Hypotenuse}} = \dfrac{\text{Adj}}{\text{Hyp}}$ | | Cosine is abbreviated cos. |
| $\tan \theta = \dfrac{\text{leg Opposite angle } \theta}{\text{leg Adjacent to angle } \theta} = \dfrac{\text{Opp}}{\text{Adj}}$ | | Tangent is abbreviated tan. |

A mnemonic to remember these rules is: **SOH   CAH   TOA**

☞ The **opposite** leg is always directly across (opposite) from the angle.
The **adjacent** leg is next to (adjacent) to the angle. It is *never* the hypotenuse.

**Examples:**

1. Find sin $\phi$, cos $\phi$ and tan $\phi$.

**Solutions:**

1. $\sin \phi = \dfrac{\text{leg opposite } \angle \phi}{\text{hypotenuse}} = \dfrac{4}{5}$

   $\cos \phi = \dfrac{\text{leg adjacent to } \angle \phi}{\text{hypotenuse}} = \dfrac{3}{5}$

   $\tan \phi = \dfrac{\text{leg opposite } \angle \phi}{\text{leg adjacent to } \angle \phi} = \dfrac{4}{3}$

2. Find: sin $N$, cos $N$, tan $N$, sin $J$, cos $J$ and tan $J$.

2. First find the hypotenuse using the pythagorean theorem:

$$(\text{leg})^2 + (\text{leg})^2 = (\text{hypotenuse})^2$$
$$1^2 + 3^2 = x^2$$
$$x^2 = 10$$
$$x = \sqrt{10}$$

$$\sin N = \frac{1}{\sqrt{10}} \qquad \sin J = \frac{3}{\sqrt{10}}$$

$$\cos N = \frac{3}{\sqrt{10}} \qquad \cos J = \frac{1}{\sqrt{10}}$$

$$\tan N = \frac{1}{3} \qquad \tan J = \frac{3}{1}$$

☞ The fractions have *not* been rationalized in order to provide clarity of example.

# Reciprocal Trigonometric Functions

The **inverse or reciprocal trigonometric functions** are the reciprocals of the basic sine, cosine and tangent functions.

| | |
|---|---|
| $\cot \theta = \dfrac{1}{\tan \theta} = \dfrac{\text{leg adjacent}}{\text{leg opposite}}$ | Cotangent is abbreviated cot. |
| $\sec \theta = \dfrac{1}{\cos \theta} = \dfrac{\text{hypotenuse}}{\text{leg adjacent}}$ | Secant is abbreviated sec. |
| $\csc \theta = \dfrac{1}{\sin \theta} = \dfrac{\text{hypotenuse}}{\text{leg opposite}}$ | Cosecant is abbreviated csc. |

**Examples:**

3. Find sin $\alpha$, cos $\alpha$, tan $\alpha$, cot $\alpha$, sec $\alpha$, and csc $\alpha$.

**Solutions:**

3. $\sin \alpha = \dfrac{\text{opp}}{\text{hyp}} = \dfrac{5}{13}$  $\qquad$  $\csc \alpha = \dfrac{1}{\sin \alpha} = \dfrac{13}{5}$

$\cos \alpha = \dfrac{\text{adj}}{\text{hyp}} = \dfrac{12}{13}$  $\qquad$  $\sec \alpha = \dfrac{1}{\cos \alpha} = \dfrac{13}{12}$

$\tan \alpha = \dfrac{\text{opp}}{\text{adj}} = \dfrac{5}{12}$  $\qquad$  $\cot \alpha = \dfrac{1}{\tan \alpha} = \dfrac{12}{5}$

4. *a* is an acute angle such that $\sin a = \frac{8}{17}$. Find the value of the other five trigonometric functions.

4. First draw a right triangle such that

$$\sin a = \frac{\text{opp}}{\text{hyp}} = \frac{8}{17}.$$

Then find the third side of the triangle.

$$8^2 + x^2 = 17^2$$
$$x^2 = 225$$
$$x = 15$$

$$\sin a = \frac{\text{opp}}{\text{hyp}} = \frac{8}{17}$$

$$\csc a = \frac{1}{\sin a} = \frac{\text{hyp}}{\text{opp}} = \frac{17}{8}$$

$$\cos a = \frac{\text{adj}}{\text{hyp}} = \frac{15}{17}$$

$$\sec a = \frac{1}{\cos a} = \frac{\text{hyp}}{\text{adj}} = \frac{17}{15}$$

$$\tan a = \frac{\text{opp}}{\text{adj}} = \frac{8}{15}$$

$$\cot a = \frac{1}{\tan a} = \frac{\text{adj}}{\text{opp}} = \frac{15}{8}$$

# Complementary Functions (Cofunctions)

**Complementary trigonometric functions** are those trigonometric functions which are equal when their angles are complementary (add to 90°)

Cosine is the complementary function of sine.              sin and cos are cofunctions.
Cotangent is the complementary function of tangent.        tan and cot are cofunctions.
Cosecant is the complementary function of secant.          sec and csc are cofunctions.

$$\sin\theta = \cos(90-\theta) \quad \Rightarrow \quad \cos\theta = \sin(90-\theta)$$
$$\tan\theta = \cot(90-\theta) \quad \Rightarrow \quad \cot\theta = \tan(90-\theta)$$
$$\sec\theta = \csc(90-\theta) \quad \Rightarrow \quad \csc\theta = \sec(90-\theta)$$

Example:   $\sin 60° = \cos 30°$
cofunctions
angles add to 90°

**Examples:**

5. Express csc 3° in terms of its cofunction.

6. If $\sin(3x-26)° = \cos(5x-60)°$, find $x$.

7. If $\sin(3x-26)° = \dfrac{1}{\csc(5x-60)°}$, find $x$.

**Solutions:**

5.   cofunction
$\csc 3° = \sec 87°$
adds to 90

6. If $\sin x = \cos y$, then $x + y = 90°$.

   If $\sin a = \cos b$, this is a **cofunction** relationship.

$$(3x-26)+(5x-60)=90$$
$$8x-86=90$$
$$8x=176$$
$$x=22$$

7. If $\sin x = \dfrac{1}{\csc y}$, this is an **inverse** relationship, not a cofunction relationship.

   If $\sin x = \dfrac{1}{\csc y}$, then $x = y$.

$$(3x-26)=(5x-60)$$
$$34=2x$$
$$x=17$$

$$\tan\theta = \frac{\sin\theta}{\cos\theta} \qquad \cot\theta = \frac{\cos\theta}{\sin\theta}$$

# Pythagorean Identities

$$\sin^2 \theta + \cos^2 \theta = 1 \qquad \tan^2 \theta + 1 = \sec^2 \theta \qquad \cot^2 \theta + 1 = \csc^2 \theta$$

| 30-60°-90° Triangle | 45°-45°-90° Triangle | Unit Circle (radius = 1) |
|---|---|---|
| Used to determine the value of all six trig. functions for 30° and 60° angles | Used to determine the value of all six trig. functions for 45° angle | Used to determine the value of $(\cos \theta, \sin \theta)$ for 0°, 90°, 180°, 270° angles. Remember $\tan \theta$ is $\dfrac{\sin \theta}{\cos \theta}$ |
| $y = \sin \theta$ | $y = \cos \theta$ | $y = \tan \theta$ |
| Used to determine the value of sin for 0, 90°, 180°, 270° angles | Used to determine the value of cos for 0°, 90°, 180°, 270° angles | Used to estimate the value of $\tan \theta$ |

| $\theta =$ | $\theta = 0°$ | 30° | 45° | 60° | 90° | 180° | 270° |
|---|---|---|---|---|---|---|---|
| **Trigonometric Function** | **From:** Unit circle | 30°-60°-90° | 45°-45°-90° | 30°-60°-90° | Unit circle | Unit circle | Unit circle |
| $\sin \theta$ | 0 | $\dfrac{1}{2}$ | $\dfrac{\sqrt{2}}{2}$ | $\dfrac{\sqrt{3}}{2}$ | 1 | 0 | $-1$ |
| $\cos \theta$ | 1 | $\dfrac{\sqrt{3}}{2}$ | $\dfrac{\sqrt{2}}{2}$ | $\dfrac{1}{2}$ | 0 | $-1$ | 0 |
| $\tan \theta = \dfrac{\sin \theta}{\cos \theta}$ | 0 | $\dfrac{\sqrt{3}}{3}$ | 1 | $\sqrt{3}$ | Undefined or $\pm\infty$ | 0 | Undefined or $\pm\infty$ |
| $\cot \theta = \dfrac{1}{\tan \theta}$ | Undefined or $\pm\infty$ | $\sqrt{3}$ | 1 | $\dfrac{\sqrt{3}}{3}$ | 0 | Undefined or $\pm\infty$ | 0 |
| $\sec \theta = \dfrac{1}{\cos \theta}$ | 1 | $\dfrac{2\sqrt{3}}{3}$ | $\sqrt{2}$ | 2 | Undefined or $\pm\infty$ | $-1$ | Undefined or $\pm\infty$ |
| $\csc \theta = \dfrac{1}{\sin \theta}$ | Undefined or $\pm\infty$ | 2 | $\sqrt{2}$ | $\dfrac{2\sqrt{3}}{3}$ | 1 | Undefined or $\pm\infty$ | $-1$ |

**Examples:**

**8.** Express $\tan x$ in terms of $\sec x$.

**9.** Evaluate: $\dfrac{(\sin 30°)(\sec 45°)}{\tan 0° + \csc 30°}$

**Solutions:**

**8.**
$$\tan^2 x + 1 = \sec^2 x$$
$$\tan^2 x = \sec^2 x - 1$$
$$\sqrt{\tan^2 x} = \pm\sqrt{\sec^2 x - 1}$$
$$\tan x = \pm\sqrt{\sec^2 x - 1}$$

**9.**

$$\sin 30° = \frac{\text{opp}}{\text{hyp}} = \frac{1}{2}$$

$$\sec 45° = \frac{1}{\cos 45°} = \frac{1}{\dfrac{1}{\sqrt{2}}} = \sqrt{2}$$

$$\tan 0° = \frac{\sin 0°}{\cos 0°} = \frac{0}{1} = 0$$

$$\csc 30° = \frac{1}{\sin 30°} = \frac{2}{1}$$

$$\frac{(\sin 30°)(\sec 45°)}{\tan 0° + \csc 30°} = \frac{\left(\dfrac{1}{2}\right)\left(\dfrac{\sqrt{2}}{1}\right)}{0 + 2} = \frac{\dfrac{\sqrt{2}}{2}}{2} = \frac{\sqrt{2}}{4}$$

10. Show: $\dfrac{\cos\theta + \cot\theta}{\cos\theta\,\cot\theta} = \tan\theta + \sec\theta$

10. Express all the trigonometric expressions in terms of $\sin\theta$ and $\cos\theta$ only.

$$\frac{\cos\theta + \cot\theta}{\cos\theta\,\cot\theta} \overset{?}{=} \tan\theta + \sec\theta$$

$$\frac{\left(\dfrac{\sin\theta}{\sin\theta}\right)\dfrac{\cos\theta}{1} + \dfrac{\cos\theta}{\sin\theta}}{\cos\theta \cdot \dfrac{\cos\theta}{\sin\theta}} \quad \Bigg| \quad \frac{\sin\theta}{\cos\theta} + \frac{1}{\cos\theta}$$

$$\frac{\dfrac{\sin\theta\cos\theta + \cos\theta}{\sin\theta}}{\dfrac{\cos^2\theta}{\sin\theta}} \quad \Bigg| \quad = \frac{\sin\theta + 1}{\cos\theta}$$

$$\frac{(\sin\theta\cos\theta + \cos\theta)}{\sin\theta} \cdot \frac{\sin\theta}{\cos^2\theta}$$

$$\frac{\cos\theta(\sin\theta + 1)}{\cos^2\theta}$$

$$\frac{\sin\theta + 1}{\cos\theta} \quad \Bigg| \quad \frac{\sin\theta + 1}{\cos\theta}$$

Q.E.D.

# Quadrants

Quadrants are the divisions of the coordinate plane into four equal sections, usually denoted by roman numerals. Each quadrant is 90° in measure.

The six trigonometric functions have different signs (+ or −) in the different quadrants. The diagram below indicates which functions are positive (+) in which of the quadrants.

Many students remember the sequence using the sentence:
**ALL** **S**tudents **T**ake **C**alculus indicating in sequence I, II, III, IV ... **All** sin tan cos.
The inverse functions follow the sign of their reciprocal trigonometric functions.

**Examples:**

11. If $\sin \theta > 0$ and $\tan \theta < 0$, in which quadrant is $\theta$?

**Solutions:**

11. sin is positive in I and II , tan is negative in II and IV.

| Sin (csc) | A |
|---|---|
| Tan (cot) | Cos (sec) |

$\theta$ is in the second quadrant.

12. If $\csc \theta$ is negative, is $\sin \theta$ positive or negative?

12. $\csc \theta$ and $\sin \theta$ are reciprocal functions.

$$\left( \sin \theta = \frac{1}{\csc \theta} \right) \quad \therefore \sin \theta < 0$$

13. If $\theta = 291°$. What are the signs of the six trigonometric functions?

13. $291°$ is in the fourth quadrant.

$\sin 291° < 0$

$\cos 291° > 0$ (cos is positive in the fourth quadrant.)

$\tan 291° < 0$

$\cot 291° < 0$

$\sec 291° > 0$ (sec is positive in the fourth quadrant.)

$\csc 291° < 0$

# Evaluating Trigonometric Functions

There are **four categories** of angles to be analyzed:

- Those angles **less than 0°**, e.g., $-30°$, $-240°$
- Those angles **between 0° and 90°**, e.g., $45°$, $38°$
- Those angles **between 90° and 360°**, e.g., $225°$, $300°$
- Those angles **greater than 360°**, e.g., $420°$, $2000°$

I. **Angles between 0° and 90°.** These are first quadrant angles. For 30°, 45° and 60° use the 30°-60°-90° and 45°-45°-90° triangles. For other angles such as 28°, 86° use the trigonometric tables.

**Examples:**

14. Evaluate cos 60°.

15. Evaluate tan 17°.

**Solutions:**

14. $\cos 60° \doteq \dfrac{adj}{hyp} = \dfrac{1}{2}$

15. Refer to trigonometric tables:

| Angle | sin | cos | **tan** |
|-------|------|------|------|
| 13° | .2250 | .9744 | .2309 |
| 14° | .2419 | .9703 | .2493 |
| 15° | .2588 | .9659 | .2679 |
| 16° | .2756 | .9613 | .2867 |
| **17°** | .2924 | .9563 | **.3057** |
| 18° | .3090 | .9511 | .3249 |
| 19° | .3256 | .9455 | .3443 |

tan 17° = .3057

To change from angles between 0° and 45° to angles between 45° and 90°, and vice versa:

Use cofunctions.

**Examples:**

16. Express csc 10° as a trigonometric function with an angle between 45° and 90°.

17. Express sin 79° as a trigonometric function of a positive acute angle less than 45°.

**Solutions:**

16. Using cofunctions: csc 10° = sec 80°
Ans: sec 80°.

17. Using cofunctions: sin 79° = cos 11°
Ans: cos 11°.

---

II. **Angles between 90° and 360°.** These are second, third and fourth quadrant angles. The procedure is to convert these trigonometric functions into their equivalent in the first quadrant and utilize the procedures for 0° to 90° angles.

---

Procedure:

1. Draw a picture of the angle. Label which quadrants have positive trig. values. (Use: **ALL** Students Take Calculus.)

2. The intent is to convert the given trig. function and angle into three (3) items.

| + or − | Trig. function | Reference angle |
|:---:|:---:|:---:|
| ↑ | ↑ | ↑ |
| The first item is + or − This is determined by comparing the trigonometric function involved with the quadrant rules associated with the angle location. | The second item is a repeat of the given trig. function. | The third item is the reference angle. It is the angle from the given angle to the nearest horizontal axis. It is always a positive value. Label the angle in the diagram. |

3. Evaluate the first quadrant angle.

Evaluate sin 240°

sin 240°

⇓

| — | Sin | 60° |
|---|---|---|
| ↑ | ↑ | ↑ |
| The trig. function involved is sin. The angle is in the third quadrant. ∴ sin is negative. | A repeat of the given trig. function. | The angle to the nearest horizontal axis is 240° − 180° = 60° |

⇓

$$\sin 240° = -\sin 60°$$

$$\sin 60° = \frac{\text{opp}}{\text{hyp}} = \frac{\sqrt{3}}{2}$$

$$\therefore \sin 240° = -\sin 60° = -\left(\frac{\sqrt{3}}{2}\right) = \frac{-\sqrt{3}}{2}$$

- − sin 60°: the angle is 60°. It is positive and acute. The " − " relates to the value not the angle.
- sin (− 60°): the angle is − 60°. It is negative and acute.

---

III. **Angles greater than 360°.** These are angles measuring more than one complete revolution. Subtract multiples of 360° until the angle is in the first, second, third or fourth quadrant. Then follow the rules for that quadrant.

**Examples:**

18. Evaluate cos 390°.

**Solutions:**

18.

$$
\begin{array}{r}
390° \\
-360° \\
\hline
30°
\end{array}
$$

$$\cos 390° = \cos 30° = \frac{\sqrt{3}}{2}$$

19. Evaluate cos 855°.

19.

$$
\begin{array}{r}
855° \\
-360° \\
\hline
495° \\
-360° \\
\hline
135°
\end{array}
$$

$$\cos 855° = \cos 135°$$

$$\underbrace{-}_{\substack{\uparrow \\ \cos \text{ is} \\ \text{negative} \\ \text{in the second} \\ \text{quadrant}}} \quad \underbrace{\cos}_{\text{Trig. funct.}} \quad \underbrace{45°}_{\substack{\uparrow \\ 180° - 135° = 45° \\ \text{The angle to the} \\ \text{nearest} \\ \text{horizontal axis}}}$$

+ or −        cos        45°
              Trig. funct.    Ref. angle

$$\cos 855° = \cos 135° = -\cos 45°$$

$$= -\left(\frac{\sqrt{2}}{2}\right) = -\frac{\sqrt{2}}{2}$$

> IV. **Angles less than 0°.** These are negative angles. Add multiples of 360° until the angle is in the first, second, third or fourth equadrant. Then follow the rules for that quadrant.

**Examples:**

20. Evaluate $\sin(-330°)$.

21. Evaluate $\tan(-120°)$.

**Solutions:**

20.

$$\begin{array}{r} -330° \\ +360° \\ \hline +30° \end{array}$$

$\sin(-330°) = \sin 30° = \frac{1}{2}$

21.

$$\begin{array}{r} -120° \\ +360° \\ \hline 240° \end{array}$$

$\tan(-120°) = \tan 240°$

$$\underset{\substack{\uparrow \\ \text{tangent is} \\ \text{positive in the} \\ \text{third} \\ \text{quadrant}}}{+} \quad \underset{\text{Trig. function}}{\text{tan}} \quad \underset{\substack{\uparrow \\ 240° - 180° = 60° \\ \text{(the angle to the} \\ \text{nearest} \\ \text{horizontal axis)}}}{60°}$$

# Solving for an Angle

## Linear

### Procedure:

1. Rearrange the equation so that the trigonometric function equals a number.
2. The sign $(+/-)$ of the value indicates the possible quadrants. The numerical part of the value indicates the values of the reference angle.
3. Draw a picture: indicate all possible quadrants. Include the reference angle in each possible quadrant.
4. Determine the value of the angle.

## Quadratic

1. Solve the quadratic for the trigonometric function. Use factoring or the quadratic formula as required.
2. Determine the angles for each linear term.

**Example:** Solve $2 \sin \theta + 1 = 0$ for all values of $\theta$, $0 \leqslant \theta < 360$.

$$2 \sin \theta = -1$$
$$\sin \theta = -\tfrac{1}{2}$$

sin is "$-$" in the third and fourth quadrants.

Ref. $\angle = 30°$ ($\sin 30° = \tfrac{1}{2}$)

$$\theta = 210°, 330°$$

---

Solve $\tan\theta - 2\tan\theta\sin\theta = 0$ for all values of $\theta$, $0 \leqslant \theta < 360$.

$$\tan \theta(1 - 2 \sin \theta) = 0$$

| | |
|---|---|
| $\tan \theta = 0$ | $1 - 2 \sin \theta = 0$ |
| $\tan \theta = \dfrac{\sin \theta}{\cos \theta}$ | $\sin \theta = +\dfrac{1}{2}$ |
| $\tan \theta$ is 0 when $\sin \theta$ is 0. | sin is "$+$" in the first and second quadrants. Ref. $\angle = 30°$ |

| | |
|---|---|
| $\theta = 0°, 180°$ | $\theta = 30°, 150°$ |

$$\theta = 0°, 30°, 150°, 180°$$

$\theta$ only has values if:

Example: Solve csc $\theta = \frac{1}{2}$, for $\theta$. There are no solutions, csc $\theta$ is never $\frac{1}{2}$.
Solve sin $\theta = 2$, for $\theta$. There are no solutions, sin $\theta$ is never 2.

**Examples:**

22. Solve $4 \csc^2 \theta - 1 = 0$ for all values $0 \leqslant \theta < 360$.

**Solutions:**

22.

$$4 \csc^2 \theta - 1 = 0$$
$$\sqrt{\csc^2 \theta} = \pm \sqrt{\tfrac{1}{4}}$$
$$\csc \theta = \pm \tfrac{1}{2}$$

| | |
|---|---|
| csc $\theta = \frac{1}{2}$ | csc $\theta = -\frac{1}{2}$ |
| No solutions | No solutions |
| csc $\theta$ is never $\frac{1}{2}$. | csc $\theta$ is never $-\frac{1}{2}$. |

Answer: { } or $\varnothing$ (null set).

# Radians

Radians are a measure of angle size. The relationship between degrees and radians is:

$$2\pi \text{ radians} = 360°$$

| | From | To | Multiply By |
|---|---|---|---|
| **To Convert:** | Degree measure | Radian measure | $\frac{\pi}{180}$ |
| **To Convert:** | Radian measure | Degree measure | $\frac{180}{\pi}$ |

**Examples:**

23. Convert 210° to radians.

**Solutions:**

23.

$$210 \cdot \frac{\pi}{180} = \frac{210\pi}{180} = \frac{7\pi}{6} \text{ radians}$$

24. Convert $\dfrac{3\pi}{2}$ radians to degrees.

24.
$$\frac{3\pi}{2}\cdot\frac{180}{\pi}=\frac{3\pi(180)}{2\pi}=270°$$

25. Evaluate $\csc\dfrac{5\pi}{4}$.

25.
$$\frac{5\pi}{4}\cdot\frac{180}{\pi}=225°$$

$$\csc\frac{5\pi}{4}=\csc 225°$$
$$=-\csc 45°$$
$$=-\sqrt{2}$$

$$\sin 45°=\frac{1}{\sqrt{2}}$$
$$\csc 45°=\frac{\sqrt{2}}{1}$$

# Practice Problems 17.1

1. $\cos 52°$ is equivalent to

    A) $\sec 52°$  B) $\sin 52°$
    C) $\sec 38°$  D) $\sin 38°$
    E) $\csc 38°$

2. Express $\sin(-140°)$ as a function of a positive acute angle.

    A) $-\sin 40°$  B) $\sin 40°$
    C) $\sin 50°$  D) $-\sin 50°$
    E) $\cos 50°$

3. If $\sec A=\tfrac{5}{4}$, find the value of $\cos A$.

    A) $\dfrac{3}{5}$  B) $\dfrac{4}{5}$  C) $\dfrac{-3}{4}$

    D) $\dfrac{-3}{5}$  E) $\dfrac{-4}{5}$

4. What is the smallest positive value of $\theta$ which satisfies the equation $2\cos^2\theta-\cos\theta=0$?

    A) $30°$  B) $60°$  C) $90°$
    D) $150°$  E) $180°$

5. For all values of $A$ for which the expression is defined, which is an equivalent expression to $4+\cos^2 A$?

    A) $5-\sec^2 A$  B) $5-\sin^2 A$
    C) $\dfrac{5}{\sec^2 A}$  D) $5+\sin^2 A$  E) $5$

6. Which values of $x$ in the interval $0°\leqslant x<360°$ satisfy the equation $2\cos x-1=0$?

    A) $30°$ and $330°$  B) $60°$ and $300°$
    C) $30°$ and $150°$  D) $60°$ and $120°$
    E) $60°$ and $240°$

7. Express $\dfrac{7\pi}{6}$ radians in degrees.

    A) $120°$  B) $150°$  C) $210°$
    D) $240°$  E) $420°$

8. Which value of $x$ satisfies the equation $6\tan x-2=4$?

    A) $\dfrac{\pi}{6}$  B) $\dfrac{\pi}{4}$  C) $\dfrac{\pi}{3}$  D) $\dfrac{\pi}{2}$  E) $\pi$

9. Which value of $x$ satisfies the equation
   $\sin(2x+30)° = \cos(3x+20)°$?

   A) 8   B) 2   C) 10   D) 28   E) 30

10. A value of $x$ for which the expression $\dfrac{1}{1-\cos x}$ is undefined is

    A) 0°   B) 30°   C) 45°   D) 60°   E) 90°

Solutions on page 324

## 17.2 Graph of Trigonometric Functions

**Amplitude:** One-half the range of the periodic function $= \frac{1}{2}|\text{maximum value} - \text{minimum value}|$
**Frequency:** Number of times a periodic function repeats in an interval
**Period:** The interval for one complete cycle of the curve (wavelength)

| | Domain | Range | Amplitude | Interval (for the basic curve) | Frequency | Period | Shift to left | to rt. |
|---|---|---|---|---|---|---|---|---|
| $y = a\sin(bx+c)$ | All real | $-a \leqslant y \leqslant a$ | $\lvert a \rvert$ | 360° or $2\pi$ | $\lvert b \rvert$ | $\dfrac{360°}{\lvert b \rvert}$ or $\dfrac{2\pi}{\lvert b \rvert}$ | $\dfrac{c}{b}$ $\dfrac{c}{b}>0$ | $\dfrac{c}{b}<0$ |
| $y = a\cos(bx+c)$ | All real | $-a \leqslant y \leqslant a$ | $\lvert a \rvert$ | 360° or $2\pi$ | $\lvert b \rvert$ | $\dfrac{360°}{\lvert b \rvert}$ or $\dfrac{2\pi}{\lvert b \rvert}$ | $\dfrac{c}{b}$ $\dfrac{c}{b}>0$ | $\dfrac{c}{b}<0$ |
| $y = a\tan(bx+c)$ | All real except $[90(2n+1)]°$ $n$ = integers | All real | Has none | 180° or $\pi$ | $\lvert b \rvert$ | $\dfrac{180°}{\lvert b \rvert}$ or $\dfrac{\pi}{\lvert b \rvert}$ | $\dfrac{c}{b}$ $\dfrac{c}{b}>0$ | $\dfrac{c}{b}<0$ |
| $y = a\cot(bx+c)$ | All real except $(180n)°$ $n$ = integers | All real | Has none | 180° or $\pi$ | $\lvert b \rvert$ | $\dfrac{180°}{\lvert b \rvert}$ or $\dfrac{\pi}{\lvert b \rvert}$ | $\dfrac{c}{b}$ $\dfrac{c}{b}>0$ | $\dfrac{c}{b}<0$ |
| $y = a\sec(bx+c)$ | All real except $[90(2n+1)]°$ $n$ = integers | $\lvert y \rvert \geqslant a$ | Has none | 360° or $2\pi$ | $\lvert b \rvert$ | $\dfrac{360°}{\lvert b \rvert}$ or $\dfrac{2\pi}{\lvert b \rvert}$ | $\dfrac{c}{b}$ $\dfrac{c}{b}>0$ | $\dfrac{c}{b}<0$ |
| $y = a\csc(bx+c)$ | All real except $(180n)°$ $n$ = integers | $\lvert y \rvert \geqslant a$ | Has none | 360° or $2\pi$ | $\lvert b \rvert$ | $\dfrac{360°}{\lvert b \rvert}$ or $\dfrac{2\pi}{\lvert b \rvert}$ | $\dfrac{c}{b}$ $\dfrac{c}{b}>0$ | $\dfrac{c}{b}<0$ |

# Practice Problems 17.2

1. The accompanying diagram represents the graph in the interval $-\pi \leqslant x \leqslant \pi$ of which function?

A) $y = \sin 2x$  B) $y = \cos 2x$
C) $y = \cos x$  D) $y = 2 \sin x$
E) $y = 2 \cos x$

2. What is the amplitude of $y = -3 \cos (2x - \pi)$?

A) 3  B) 2  C) $\dfrac{\pi}{2}$  D) 6  E) $-3$

3. What is the amplitude of the graph shown below?

A) $2\pi$  B) 2  C) 1  D) $-2$  E) $4\pi$

4. What is the period of $f(x) = 3 \sin\left(2x + \dfrac{\pi}{7}\right)$?

A) 3  B) 2  C) $\dfrac{\pi}{7}$  D) $\dfrac{\pi}{4}$  E) $\pi$

5. What is the phase shift if $y = 3 \sin\left(2x - \dfrac{\pi}{3}\right)$?

A) $\dfrac{\pi}{3}$ to the right  B) $\dfrac{\pi}{3}$ to the left

C) $\dfrac{\pi}{6}$ to the right  D) $\dfrac{\pi}{6}$ to the left

E) 3 to the left

6. What is the amplitude of $5 \tan(2x - 180)°$?

A) 5  B) $-5$  C) 2  D) 180  E) has none

7. What are the only value(s) all six basic trigonometric functions can have?

A) 0 only  B) 1 only  C) $-1$ only
D) 1 and $-1$ only  E) 0 and 1 only

8. Find the period of $y = 4 \sec\left(\dfrac{3\pi}{4}x - \dfrac{\pi}{2}\right)$.

A) 4   B) $\dfrac{3\pi}{4}$   C) $\dfrac{-\pi}{2}$

D) $\dfrac{8}{3}$   E) $\dfrac{-2}{3}$

9. Which best expresses the equation of the graph shown?

A) $y = \sin \frac{1}{2}x$   B) $y = \cos \frac{1}{2}x$
C) $y = \frac{1}{2}\sin x$   D) $y = \frac{1}{2}\cos x$
E) $y = \frac{1}{2}$

10. As angle $x$ increases from 90° to 180°, the value of $\sin x$:

A) increases from $-1$ to 0
B) increases from 0 to 1
C) decreases from 0 to $-1$
D) decreases from 1 to 0
E) increases to 1 and then decreases

Solutions on page 325

# 17.3 Trigonometric Identities for Two Angles, Double Angles and Half-Angles

|  | sin | cos | tan |
|---|---|---|---|
| **Sum of Two Angles** | $\sin(x+y) =$ $\sin x \cos y + \sin y \cos x$ | $\cos(x+y) =$ $\cos x \cos y - \sin x \sin y$ | $\tan(x+y) =$ $\dfrac{\tan x + \tan y}{1 - \tan x \tan y}$ |
| **Difference of Two Angles** | $\sin(x-y) =$ $\sin x \cos y - \sin y \cos x$ | $\cos(x-y) =$ $\cos x \cos y + \sin x \sin y$ | $\tan(x-y) =$ $\dfrac{\tan x - \tan y}{1 + \tan x \tan y}$ |
| **Double-Angle** | $\sin 2x =$ $2\sin x \cos x$ | $\cos 2x =$ • $\cos^2 x - \sin^2 x$ • $2\cos^2 x - 1$ • $1 - 2\sin^2 x$ | $\tan 2x =$ $\dfrac{2\tan x}{1 - \tan^2 x}$ |
| **Half-Angle** | $\sin \frac{1}{2}x =$ $\pm\sqrt{\dfrac{1 - \cos x}{2}}$ | $\cos \frac{1}{2}x =$ $\pm\sqrt{\dfrac{1 + \cos x}{2}}$ | $\tan \frac{1}{2}x =$ • $\pm\sqrt{\dfrac{1 - \cos x}{1 + \cos x}}$ • $\dfrac{\sin x}{1 + \cos x}$ • $\dfrac{1 - \cos x}{\sin x}$ |

**Examples:**

1. Evaluate cos 15° without use of a cosine table. (Use half-angle rules.)

**Solutions:**

1. 15° is half of 30°.

$$\cos \tfrac{1}{2}x = \pm \sqrt{\frac{1+\cos x}{2}}$$

Let $x = 30°$.

$$\cos \tfrac{1}{2}(30°) = \pm \sqrt{\frac{1+\cos 30°}{2}}$$

$$\cos 30° = \frac{\sqrt{3}}{2}$$

$$\cos 15° = \pm \sqrt{\frac{1+(\sqrt{3}/2)}{2}}$$

$$= \pm \sqrt{\frac{2+\sqrt{3}}{2} \cdot \frac{1}{2}}$$

$$= \pm \frac{\sqrt{2+\sqrt{3}}}{2}$$

Since 15° is in the first quadrant cos 15° is " + ".

$$= \frac{1}{2}\sqrt{2+\sqrt{3}}$$

2. Evaluate cos 15° without use of a cosine table. (Use the difference of two angles.)

2. $15° = 45° - 30°$

$$\cos(x-y) = \cos x \cos y + \sin x \sin y$$

Let $x = 45, \quad y = 30$.

$$\cos(45-30) = \cos 45 \cos 30 + \sin 45 \sin 30$$

$$\cos 15 = \left(\frac{\sqrt{2}}{2}\right)\left(\frac{\sqrt{3}}{2}\right) + \left(\frac{\sqrt{2}}{2}\right)\left(\frac{1}{2}\right)$$

$$= \frac{\sqrt{6}}{4} + \frac{\sqrt{2}}{4} = \frac{\sqrt{6}+\sqrt{2}}{4}$$

☞ $\dfrac{\sqrt{6}+\sqrt{2}}{4}$ from example 2

$$= \tfrac{1}{2}\sqrt{2+\sqrt{3}} \text{ from example 1.}$$

3. If $\cos x = \frac{3}{5}$ and $x$ is a positive acute angle, find $\tan \frac{1}{2}x$.

3. Since $x$ is positive acute, $x$ is in the first quadrant.
$\frac{1}{2}x$ is also in the first quadrant.
tan is '+' in the first quadrant.

$$\tan \tfrac{1}{2}x = \sqrt{\frac{1-\cos x}{1+\cos x}} = \sqrt{\frac{1-\frac{3}{5}}{1+\frac{3}{5}}}$$

$$= \sqrt{\frac{\frac{2}{5}}{\frac{8}{5}}} = \sqrt{\tfrac{1}{4}} = \tfrac{1}{2}$$

## Practice Problems 17.3

1. If $\tan A = \frac{1}{3}$, find the value of $\tan 2A$.

A) $\frac{2}{3}$   B) $\frac{1}{9}$   C) $\frac{1}{2}$   D) $\frac{3}{4}$   E) $\frac{4}{3}$

2. If $\theta$ is a positive acute angle and $\cos\theta = \frac{7}{25}$, then $\sin \frac{1}{2}\theta$ is equal to

A) $\frac{4}{5}$   B) $-\frac{4}{5}$   C) $\frac{3}{5}$   D) $-\frac{3}{5}$   E) $\frac{3}{4}$

3. If $\cos \theta = \frac{3}{7}$ and $\tan \theta < 0$, find $\sin 2\theta$.

A) $\frac{8}{7}$   B) $\frac{6\sqrt{10}}{49}$   C) $\frac{-12\sqrt{10}}{58}$

D) $\frac{-7}{8}$   E) $\frac{-12\sqrt{10}}{49}$

4. If $\sin x = \frac{5}{6}$, what is the value of $\cos 2x$?

A) $\frac{-7}{18}$   B) $\frac{1}{6}$   C) $-\frac{2}{3}$

D) $-\frac{22}{3}$   E) $-\frac{1}{2}$

Solutions on page 326

## 17.4 Arc Functions and Relations

**The inverse of** $y = \sin x$ **is** $x = \sin y$. After swapping the $x$ and $y$ the next step in the inverse process is to solve for $y$.

$$\boxed{\text{If } x = \sin y \Rightarrow y = \text{arc sin } x \text{ or } y = \sin^{-1} x}$$

- The '$a$' in arc and the '$s$' in $\sin^{-1}$ are lower case letters.
- The equations $y = \text{arc sin } x$ or $y = \sin^{-1} x$ are read: $y$ represents the angles whose sine is $x$.

In this context, $\sin^{-1} x \neq \dfrac{1}{\sin x}$. This is a point of confusion.

**Example:**

For $0° \leqslant y < 360°$ solve $\sin y = \frac{1}{2}$

    or

For $0° \leqslant y < 360°$ solve $y = \text{arc sin } \frac{1}{2}$

    or

For $0° \leqslant y < 360°$ solve $y = \sin^{-1} \frac{1}{2}$

**Solution:**

$$\sin y = +\tfrac{1}{2}$$

Indicates solutions in the first and second quadrants      Ref. $\angle = 30°$

$$\boxed{y = 30°,\ 150°}$$

The relations $y = \text{arc sin } x$ and $y = \sin^{-1} x$ are not functions. By restricting the range of the $y$ values, an inverse trigonometric function can be defined:

$$\boxed{y = \text{Arc sin } x \quad \text{or} \quad y = \text{Sin}^{-1} x}$$

☞ • The '$A$' in Arc and the '$S$' in $\text{Sin}^{-1}$ are upper case letters.

The **principal value** of $y$ is the angle defined within the restricted range of $y$.

| | For Positive $x$ ($x \geqslant 0$) | | For Negative $x$ ($x < 0$) | |
|---|---|---|---|---|
| | Restricted Range (Principal Value) | Comment | Restricted Range (Principal Value) | Comment |
| $\theta = \text{Arc sin } x$ $= \text{Sin}^{-1} x$ | $0° \leqslant \theta \leqslant 90°$ $0 \leqslant \theta \leqslant \dfrac{\pi}{2}$ | $\theta$ represents an angle in the first quadrant | $-90° \leqslant \theta < 0°$ $-\dfrac{\pi}{2} \leqslant \theta < 0$ | $\theta$ represents a **negative** angle in the fourth quadrant |
| $\theta = \text{Arc cos } x$ $= \text{Cos}^{-1} x$ | $0° \leqslant \theta \leqslant 90°$ $0 \leqslant \theta \leqslant \dfrac{\pi}{2}$ | $\theta$ represents an angle in the first quadrant | $90° < \theta \leqslant 180°$ $\dfrac{\pi}{2} < \theta \leqslant \pi$ | $\theta$ represents a **positive** angle in the second quadrant |

| $\theta = $ Arc tan $x$ $= $ Tan$^{-1}\,x$ | $0° \leqslant \theta \leqslant 90°$ $0 \leqslant \theta \leqslant \dfrac{\pi}{2}$ | $\theta$ represents an angle in the first quadrant | $-90° \leqslant \theta < 0°$ $-\dfrac{\pi}{2} \leqslant \theta < 0$ | $\theta$ represents a **negative** angle in the fourth quadrant |
|---|---|---|---|---|

y = arc sin x

y = arc cos x

y = arc tan x

y = arc csc x

y = arc sec x

y = arc cot x

**Examples:**

1. Find $\theta$, if $\theta = \text{Arc cos } \frac{\sqrt{3}}{2}$.

2. Find $\theta$, if $\theta = \text{Arc sin } \left(-\frac{\sqrt{3}}{2}\right)$.

3. Evaluate $\sin\left(\text{Arc cos }\left(-\frac{3}{5}\right)\right)$.

**Solutions:**

1.

Answer: $\theta = 30°$

2.

$\theta$ is a negative angle in the fourth quadrant.

Answer: $\theta = -60°$

☞ Negative angles are measured clockwise.

3.

Let $\theta = \text{Arc cos }\left(-\frac{3}{5}\right)$

$\theta$ is a positive angle in the second quadrant.

$$\overbrace{\sin(\text{Arc cos }(-\tfrac{3}{5}))}^{\theta}$$

Draw a representation of $\cos\theta = \frac{3}{5}$.

Evaluate $\sin\theta$: $\sin\theta = \frac{4}{5}$ and since $\theta$ is in the second quadrant

$$\sin\theta = +\frac{4}{5}$$

# Practice Problems 17.4

1. Find $\theta$, if $\theta = \text{Arc tan}\left(-\dfrac{\sqrt{3}}{3}\right)$.

   A) $-30°$   B) $-150°$   C) $30°$
   D) $150°$   E) $330°$

2. Find the value of $\cos(\text{Arc tan } \tfrac{3}{4})$.

   A) $\tfrac{4}{3}$   B) $\tfrac{3}{4}$   C) $\tfrac{3}{5}$   D) $\tfrac{4}{5}$   E) $-\tfrac{1}{4}$

3. The value of $\text{Arc sin}\left(\dfrac{\sqrt{3}}{2}\right) + \text{Arc tan } 1$ is

   A) $120°$   B) $105°$   C) $90°$
   D) $\dfrac{2+\sqrt{3}}{2}$   E) $75°$

4. Evaluate $\sin(2 \text{ Arc cos } \tfrac{1}{3})$.

   A) $\tfrac{2}{3}$   B) $17°$   C) $6$
   D) $\dfrac{4\sqrt{2}}{9}$   E) $78°$

Solutions on page 327

## 17.5 Oblique Triangles

**Oblique triangles are non-right triangles.**

| LAW OF SINES |
|---|
| $\dfrac{a}{\sin A} = \dfrac{b}{\sin B} = \dfrac{c}{\sin C}$ |

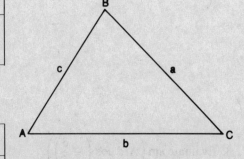

| LAW OF COSINES |
|---|
| $a^2 = b^2 + c^2 - 2bc \cos A$ <br> or <br> $b^2 = a^2 + c^2 - 2ac \cos B$ <br> or <br> $c^2 = a^2 + b^2 - 2ab \cos C$ |

| Area | Projection Equations |
|---|---|
| $\text{Area} = \tfrac{1}{2}ab \sin C$ <br> $= \tfrac{1}{2}ac \sin B$ <br> $= \tfrac{1}{2}bc \sin A$ | $a = b \cos C + c \cos B$ <br> $b = c \cos A + a \cos C$ <br> $c = a \cos B + b \cos A$ |

**Examples:**

1. In triangle $ABC$, $a=6$, $b=8$ and $\cos C=\frac{3}{8}$. What is the length of side $c$?

2. In $\triangle NJL$, if $j=10$, $l=16$ and $m\angle N=30$, find the area of the triangle.

3. In $\triangle ABC$, $a=6$, $b=3$ and $\sin B=\frac{1}{4}$. What is the value of $\sin A$?

**Solutions:**

1. Use the law of cosines.

$$c^2 = a^2 + b^2 - 2ab \cos C$$
$$= 6^2 + 8^2 - 2(6)(8)(\tfrac{3}{8})$$
$$= 36 + 64 - 36$$
$$= 64$$
$$c = \sqrt{64} = 8$$

2. $A = \frac{1}{2}jl \sin N$
$$= \tfrac{1}{2}(10)(16) \sin 30°$$
$$= \tfrac{1}{2}(10)(16)(\tfrac{1}{2})$$
$$= 40$$

3.

$$\frac{a}{\sin A} = \frac{b}{\sin B}$$

$$\frac{6}{\sin A} = \frac{3}{\frac{1}{4}}$$

$$\frac{6}{\sin A} = \frac{12}{1}$$

$$\sin A = \tfrac{1}{2}$$

# Practice Problems 17.5

1. In triangle $NJL$, $\sin N = 8$ and $n = 4$. The ratio $\dfrac{\sin L}{l}$ is

   A) $\frac{1}{3}$  B) $\frac{1}{2}$  C) $\frac{2}{1}$  D) $\frac{3}{1}$  E) .6

2. In $\triangle ABC$, $a=5$, $b=8$ and $c=9$. The value of $\cos A$ is

   A) $\frac{1}{12}$  B) $\frac{5}{6}$  C) $-\frac{5}{6}$  D) 25  E) $-\frac{1}{12}$

3. In $\triangle ABC$ if $c=20$, $m\angle A = 35$, $m\angle B = 68$, find $a$ to the nearest integer.
   ($\cos 35° = .8192$,   $\cos 68° = .3746$,
    $\sin 35° = .5736$,   $\sin 68° = .9272$,
    $\sin 77° = .9563$.)

   A) 11  B) 12  C) 13  D) 14  E) 15

4. In the figure below $\square ABCD$, $DB=8$, $AD=6$ and $m\angle ADB = 38$. Find $AB$ to the nearest integer. ($\cos 38° = .7880$.)

   A) 2  B) 3  C) 4  D) 5  E) 6

Solutions on page 328

## 17.6 Trigonometry Summary

| Basics | |
|---|---|
|  | $\sin \theta = \dfrac{\text{opp}}{\text{hyp}} = \dfrac{b}{c}$   $\csc \theta = \dfrac{\text{hyp}}{\text{opp}} = \dfrac{c}{b}$ <br><br> $\cos \theta = \dfrac{\text{adj}}{\text{hyp}} = \dfrac{a}{c}$   $\sec \theta = \dfrac{\text{hyp}}{\text{adj}} = \dfrac{c}{a}$ <br><br> $\tan \theta = \dfrac{\text{opp}}{\text{adj}} = \dfrac{b}{a}$   $\cot \theta = \dfrac{\text{adj}}{\text{opp}} = \dfrac{a}{b}$ <br><br> SOH   CAH   TOA |

**Inverse (Reciprocal)**

$$\sin \theta = \frac{1}{\csc \theta} \Rightarrow \csc \theta = \frac{1}{\sin \theta} \Rightarrow \sin \theta \cdot \csc \theta = 1$$

$$\cos \theta = \frac{1}{\sec \theta} \Rightarrow \sec \theta = \frac{1}{\cos \theta} \Rightarrow \cos \theta \cdot \sec \theta = 1$$

$$\tan \theta = \frac{1}{\cot \theta} \Rightarrow \cot \theta = \frac{1}{\tan \theta} \Rightarrow \tan \theta \cdot \cot \theta = 1$$

**Cofunctions**

$$\sin \theta = \cos(90 - \theta) \qquad \tan \theta = \cot(90 - \theta) \qquad \sec \theta = \csc(90 - \theta)$$

$$\cos \theta = \sin(90 - \theta) \qquad \cot \theta = \tan(90 - \theta) \qquad \csc \theta = \sec(90 - \theta)$$

**Basics**

$$\tan \theta = \frac{\sin \theta}{\cos \theta} \qquad \cot \theta = \frac{\cos \theta}{\sin \theta}$$

**Pythagorean Equalities**

$$\sin^2 \theta + \cos^2 \theta = 1 \qquad \tan^2 \theta + 1 = \sec^2 \theta \qquad \cot^2 \theta + 1 = \csc^2 \theta$$

| **sin** | **cos** | **tan** |
|---|---|---|
| **Sum of two angles:** | | |
| $\sin(x + y) = \sin x \cos y + \sin y \cos x$ | $\cos(x + y) = \cos x \cos y - \sin x \sin y$ | $\tan(x + y) = \dfrac{\tan x + \tan y}{1 - \tan x \tan y}$ |
| **Difference of two angles:** | | |
| $\sin(x - y) = \sin x \cos y - \sin y \cos x$ | $\cos(x - y) = \cos x \cos y + \sin x \sin y$ | $\tan(x - y) = \dfrac{\tan x - \tan y}{1 + \tan x \tan y}$ |

| **Double angle** | | |
|---|---|---|
| $\sin 2x = 2 \sin x \cos x$ | $\cos 2x = \begin{cases} \cos^2 x - \sin^2 x \\ 2\cos^2 x - 1 \\ 1 - 2\sin^2 x \end{cases}$ | $\tan 2x = \dfrac{2\tan x}{1 - \tan^2 x}$ |

| **Half angle** | | |
|---|---|---|
| $\sin \tfrac{1}{2}x = \pm \sqrt{\dfrac{1 - \cos x}{2}}$ | $\cos \tfrac{1}{2}x = \pm \sqrt{\dfrac{1 + \cos x}{2}}$ | $\tan \tfrac{1}{2}x = \begin{cases} \dfrac{\sin x}{1 + \cos x},\ \dfrac{1 - \cos x}{\sin x} \\ \pm \sqrt{\dfrac{1 - \cos x}{1 + \cos x}} \end{cases}$ |

| **Law of sines** | **Law of cosines** | **Area ($K$)** |
|---|---|---|
| $\dfrac{a}{\sin A} = \dfrac{b}{\sin B} = \dfrac{c}{\sin C}$ | $a^2 = b^2 + c^2 - 2bc \cos A$ | $K = \tfrac{1}{2}ab \sin C$ |
| | or | or |
| | $b^2 = a^2 + c^2 - 2ac \cos B$ | $\tfrac{1}{2}bc \sin A$ |
| | or | or |
| | $c^2 = a^2 + b^2 - 2ab \cos C$ | $\tfrac{1}{2}ac \sin B$ |

# 18. Polar Coordinates

## 18.1 Essentials

Polar coordinates use an ordered pair $(r,\theta)$ to represent a point.

**θ:** The smallest angle of rotation from the polar axis to $OP$
  $\theta > 0$ counterclockwise
  $\theta < 0$ clockwise

**r:** The distance from the pole to point $P$
  : The magnitude of $OP$
  $r > 0$ extends in the direction of $\theta$
  $r < 0$ reflects through the pole (extends 180° from $\theta$)

**Example:**

Plot $(r, \theta)$    A(2, 30°)
                 B(2, 170°)
                 C(−2, 60°)
                 D(2, 240°)
                 E(2, 390°)
                 F(3, −30°)

 Note: In polar coordinates $(r,\theta)$, $(r,\theta + (360n°)$ and $(-r,\theta + 180°)$ all represent the same point ($n$ is any integer).

**Solution:**

| Conversion Table | |
|---|---|
| $x = r \cos \theta$ | $r^2 = x^2 + y^2$ |
| $y = r \sin \theta$ | $\theta = \text{Arc tan } \dfrac{y}{x}$ |
| $x + yi = r(\cos \theta + i \sin \theta)$ | |

**Examples:**

1. Find the ordered pair of rectangular coordinates whose polar coordinates are $(4, 240°)$.

2. Convert the rectangular coordinates $(0, 2)$ into polar coordinates.

**Solutions:**

1.
$$x = r \cos \theta = 4 \cos 240°$$
$$= 4(-\tfrac{1}{2}) = -2$$
$$y = r \sin \theta = 4 \sin 240°$$
$$= 4\left(-\frac{\sqrt{3}}{2}\right) = -2\sqrt{3}$$

$$\cos 240° = -\cos 60° = -\tfrac{1}{2}$$
$$\sin 240° = -\sin 60° = -\frac{\sqrt{3}}{2}$$

$$(-2, -2\sqrt{3})$$

2. $r^2 = x^2 + y^2 = 0^2 + 2^2 = 4$
$$r = \pm 2$$

$$\theta = \text{Tan}^{-1} \frac{y}{x} = \text{Tan}^{-1} \frac{2}{0} = \text{Tan}^{-1} \infty$$

$\theta = 90°$ is the reference angle.
  From the graph

$$\text{if } \theta = 90 \quad r = 2$$

we could also refer to the point as

$$\theta = 270 \quad r = -2$$

Solution: $(2, 90°)$ or $(-2, 270°)$

# Practice Problems 18.1

1. Find the polar coordinates given by the rectangular coordinates $(-\sqrt{3}, 1)$.

   A) $(2, 30°)$   B) $(2, 60°)$   C) $(2, 120°)$
   D) $(2, 150°)$   E) $(-2, 315°)$

2. Transform the equation $r \sin \theta = -2$ into rectangular form.

   A) $x + y = -2$   B) $xy = 2$   C) $x = -2$
   D) $y = -2$   E) $\dfrac{x}{2} = 1$

3. Convert $1 + i\sqrt{3}$ to polar form.

   A) $2(\cos 60° + i \sin 60°)$
   B) $3(\cos 45° + i \sin 45°)$
   C) $(\cos 30° + i \sin 30°)$
   D) $4(\cos 60° + i \sin 60°)$
   E) $\sqrt{3}(\cos 30° + i \sin 30°)$

4. Convert $2\sqrt{6}(\cos 120° + i \sin 120°)$ into rectangular form.

   A) $-3\sqrt{2} - \sqrt{6}i$   B) $3\sqrt{2} + \sqrt{6}i$
   C) $\sqrt{6} + 3\sqrt{2}i$   D) $-\sqrt{6} - 3\sqrt{2}i$
   E) $-\sqrt{6} + 3\sqrt{2}i$

Solutions on page 328

# 19. DeMoivre's Theorem and Euler's Formula

## 19.1 Basics and Beyond

---

**De Moivre's Theorem:**

$$[r(\cos \theta + i \sin \theta)]^n = r^n(\cos n\theta + i \sin n\theta)$$

$$[r(\cos \theta + i \sin \theta)]^{1/n} = r^{1/n}\left(\cos \frac{\theta + 2\pi k}{n} + i \sin \frac{\theta + 2\pi k}{n}\right) \quad k = 0 \text{ to } n-1$$

$$= r^{1/n}\left(\cos \frac{\theta + 360k}{n} + i \sin \frac{\theta + 360k}{n}\right)$$

---

**Euler's Formula:**

$$e^{i\theta} = \cos \theta + i \sin \theta$$

---

**Combined DeMoivre and Euler:**

$$(e^{i\theta})^n = (\cos \theta + i \sin \theta)^n = \cos n\theta + i \sin n\theta$$

---

**Polar Multiplication:**

$$[r_1(\cos \theta_1 + i \sin \theta_2] \cdot [r_2(\cos \theta_2 + i \sin \theta_2] = r_1 r_2[\cos(\theta_1 + \theta_2) + i \sin(\theta_1 + \theta_2)]$$

---

**Polar Division:**

$$\frac{r_1(\cos \theta_1 + i \sin \theta_1)}{r_2(\cos \theta_2 + i \sin \theta_2)} = \frac{r_1}{r_2}(\cos(\theta_1 - \theta_2) + i \sin(\theta_1 - \theta_2))$$

---

**Examples:**

1. Evaluate $e^{i\pi}$.

**Solutions:**

1. $e^{i\pi} = \cos \pi + i \sin \pi$

   $= -1 + i(0) = -1$

---

317

2. Use DeMoivre's theorem to find $(1+i)^4$.

2. $1+i=x+yi \quad x=1, \quad y=1$

$$r=\sqrt{x^2+y^2}=\sqrt{1^2+1^2}=\sqrt{2}$$

$$\tan\theta=\frac{y}{x}=\frac{1}{1}=1 \quad \theta=\frac{\pi}{4}$$

$$1+i=\sqrt{2}\left(\cos\frac{\pi}{4}+i\sin\frac{\pi}{4}\right)$$

$$(1+i)^4=\left[\sqrt{2}\left(\cos\frac{\pi}{4}+i\sin\frac{\pi}{4}\right)\right]^4$$

$$=(\sqrt{2})^4\left(\cos 4\left(\frac{\pi}{4}\right)+i\sin 4\left(\frac{\pi}{4}\right)\right)$$

$$=4(\cos\pi+i\sin\pi)$$

$$=4(-1+i(0))=-4$$

3. Find the three cube roots of 1.

3. Rewrite the cube root of 1 as $1^{1/3}$

$1=x+yi \quad x=1, y=0$

$$r=\sqrt{x^2+y^2}=\sqrt{1^2+0^2}=1$$

$$\tan\theta=\frac{y}{x}=\frac{0}{1}=0. \quad \theta=0$$

$$1=1(\cos 0+i\sin 0)$$

$$1^{1/3}=[1(\cos(0+2\pi k)+i\sin(0+2\pi k)]^{1/3}$$

$$=1^{1/3}\left[\cos\left(\frac{0+2\pi k}{3}\right)+i\sin\left(\frac{0+2\pi k}{3}\right)\right]$$

$k=0$

$$=1^{1/3}(\cos 0+i\sin 0]$$

$$=1(1+i(0))=1$$

$k=1$

$$=1^{1/3}\left(\cos\frac{2\pi}{3}+i\sin\frac{2\pi}{3}\right)$$

$$=1\left(-\frac{1}{2}+\frac{i\sqrt{3}}{2}\right)=\frac{-1+i\sqrt{3}}{2}$$

$k=2$

$$=1^{1/3}\left(\cos\frac{4\pi}{3}+i\sin\frac{4\pi}{3}\right)$$

$$=1\left(-\frac{1}{2}-\frac{i\sqrt{3}}{2}\right)=\frac{-1-\sqrt{3}}{2}$$

$$\left\{1,\ \frac{-1+i\sqrt{3}}{2},\ -\frac{1-i\sqrt{3}}{2}\right\}$$

# Practice Problems 19.1

1. Evaluate $e^{i(\pi/2)}$.

   A) 1  B) $e$  C) $i$  D) $-1$  E) $-i$

2. Use DeMoivre's theorem to evaluate $(\sqrt{3}-i)^5$.

   A) $32(\sqrt{3}-i)$  B) $-16(\sqrt{3}+i)$
   C) $16(-\sqrt{3}+i)$  D) $32(-\sqrt{3}+i)$
   E) $32(-\sqrt{3}-i)$

3. Find the two square roots of $2+2i\sqrt{3}$.

   A) $\sqrt{3}+1i$ and $\sqrt{3}-1i$
   B) $\sqrt{3}+1i$ and $-\sqrt{3}-1i$
   C) $\sqrt{3}-1i$ and $-\sqrt{3}-1i$
   D) $-\sqrt{3}-1i$ and $-\sqrt{3}+1i$
   E) $-\sqrt{3}-\sqrt{2}i$ and $\sqrt{3}+\sqrt{2}i$

4. Multiply $2(\cos 30° + i \sin 30°)$ by $7(\cos 10° + i \sin 10°)$.

   A) $14(\cos 300° + i \sin 300°)$
   B) $9(\cos 40° + i \sin 40°)$
   C) $14(\cos 40° + i \sin 40°)$
   D) $9(\cos 300° + i \sin 300°)$
   E) $2^7(\cos 3° + i \sin 3°)$

Solutions on page 329

# 20. Parametric Equations

## 20.1 Fundamental Facts

When a point $(x, y)$ can be described in terms of a third variable $t$, the equations $x = f(t)$ and $y = g(t)$ are called **parametric equations** and the variable $t$ is called the **parameter**.

**Examples:**

1. Express the equation $x = 2\cos t$ and $y = 3\sin t$ in terms of the rectangular coordinates $x$ and $y$ by eliminating the parameter $t$.

2. Eliminate the parameter $t$ and express the equation in terms of the rectangular coordinates $x$ and $y$.

$$x = 2 + t \qquad y = 2 + 3t$$

**Solutions:**

1. $\dfrac{x}{2} = \cos t \qquad \dfrac{y}{3} = \sin t$

$$\sin^2 t + \cos^2 t = 1 = \left(\frac{y}{3}\right)^3 + \left(\frac{x}{2}\right)^2$$

$$\frac{x^2}{4} + \frac{y^2}{9} = 1$$

This is an ellipse.

2.

$$t = x - 2 \quad t = \frac{y - 2}{3}$$

$$\therefore x - 2 = \frac{y - 2}{3}$$

$$3x - 6 = y - 2$$

$$y = 3x - 4$$

This is a line.

# Practice Problems 20.1

1. Eliminate the parameter $t$.

$$x = t^2 - 2 \qquad y = t + 1$$

A) $y = x^2 - 2y + 1$  B) $x = y^2 - 2y - 1$
C) $x = y^2 - 2y + 1$  D) $x = y - 1$
E) $y = x^2 - 2x + 1$

2. Eliminate the parameter $\theta$ and identify the curve.

$$y = \tan\theta \qquad x = 2\cot\theta$$

A) line  B) circle  C) ellipse
D) parabola  E) hyperbola

3. Eliminate the parameter $t$ and identify the curve.

$$x = e^t \qquad y = e^{-t}$$

A) line  B) circle  C) ellipse
D) parabola  E) hyperbola

4. Eliminate the parameter $\phi$ and identify the curve.

$$y = 3\sin\phi \qquad x = 3\cos\phi$$

A) line  B) circle  C) ellipse
D) parabola  E) hyperbola

Solutions on page 330

# Cumulative Review Test

1. Factor completely $ax + ay + a^2 + a$.

    A) $a(x+y)(a+1)$    B) $a(x+y+a)$
    C) $a(x+y+1)$    D) $a(x+y+a+1)$
    E) $a(x+y)+(a+1)$

2. $\dfrac{20(\cos 15° + i \sin 15°)}{5(\cos 3° + i \sin 3°)} =$

    A) 4    B) $15(\cos 5° + i \sin 5°)$
    C) $4(\cos 12° + i \sin 12°)$
    D) $4(\cos 5° + i \sin 5°)$
    E) $15(\cos 12° + i \sin 12°)$

3. $\tan 810° =$

    A) 1    B) 0    C) $\dfrac{\sqrt{3}}{3}$

    D) Undefined    E) $\cot 90°$

4. Simplify $b^{-3} \cdot b^{2/3}$.

    A) $b^{-2}$    B) $(b^2)^{-2}$    C) $b^{-7/3}$
    D) $b^{11/3}$    E) $b^2$

5. The expression $\sec^2 x + \csc^2 x$ is equivalent to

    A) $\sin^2 x \cos^2 x$    B) $\dfrac{1}{\sin^2 x \cos^2 x}$
    C) $1 + \tan^2 x$    D) $1 - \tan^2 x$
    E) $\sin^2 x + \cos^2 x$

6. Express $-\sqrt{3} + i$ in polar notation.

    A) $2(\cos 150° + i \sin 150°)$
    B) $2(\cos 270° + i \sin 270°)$
    C) $2(\cos 30° + i \sin 30°)$
    D) $\sqrt{2}(\cos 150° + i \sin 150°)$
    E) $\sqrt{2}(\cos 30° + i \sin 30°)$

7. The expression $\cos 262°$ is equal to

    A) $\sin 8°$    B) $-\cos 8°$    C) $\sin 82°$
    D) $-\cos 82°$    C) $\dfrac{1}{\sec 82°}$

8. Which function is represented in the accompanying graph?

    A) $y = \sin 2x$    B) $y = 2 \sin x$
    C) $y = \cos 2x$    D) $y = 2 \cos x$
    E) $y = 2 \cos 2x$

9. The graph below represents which of the following?

    A) $x < 1$ or $x > 2$    B) $x < 2$ or $x \geqslant 1$
    C) $x \leqslant 2$ or $x \geqslant 1$    D) $x \leqslant 1$ or $x > 2$
    E) $x < 1$ or $x \geqslant 2$

10. The parametric equations $x = 4 + 3\tan\theta$ and $y = -1 + 2 \sec \theta$ represent the graph of a(n)

    A) line    B) parabola    C) circle
    D) ellipse    E) hyperbola

11. If the measure of an interior angle of a regular polygon is 108, how many sides does the polygon have?

    A) 4    B) 5    C) 6    D) 7    E) 11

12. If $A = \text{Arc} \tan \frac{3}{4}$, then $\tan \frac{1}{2}A$ is equal to

    A) $\sqrt{\frac{1}{7}}$    B) $\frac{1}{2}$    C) 3    D) $\frac{1}{3}$    E) $\frac{1}{9}$

13. In $\triangle JSL$, $j = 5$, $\sin J = .35$ and $\sin S = .21$. Find the measure of side $s$.

    A) .3    B) 3    C) $\frac{25}{3}$    D) .5    E) .03

14. $\left(\dfrac{\sqrt{3}+i}{2}\right)^{12} =$

A) $3^6 + 2i$   B) $-1$   C) $1$
D) $2(\cos 30° + i \sin 30°)$
E) $1(\cos 30° + i \sin 30°)$

15. $x = 2 \tan \theta$, $y = \cot \theta$ is the parametric representation of a(n)

A) line   B) circle   C) parabola
D) ellipse   E) hyperbola

16. What is the center of the circle whose equation is given by $x^2 - 2x + y^2 + 6y + 4 = 0$?

A) $(1, 3)$   B) $(-1, -3)$   C) $(-3, 1)$
D) $(6, 2)$   E) $(1, -3)$

17. Find the value of $\tan\left(\text{Arc} \sin \dfrac{\sqrt{3}}{2}\right)$.

A) $\sqrt{3}$   B) $-\dfrac{\sqrt{3}}{2}$   C) $-\sqrt{3}$
D) $\frac{1}{2}$   E) $-\frac{1}{2}$

18. Find the length of side $c$ in $\triangle ABC$ if $a = 3$, $b = 5$ and $\cos C = -\frac{1}{2}$.

A) $\sqrt{19}$   B) $\sqrt{6}$   C) $7$   D) $\sqrt{59}$   E) $8$

19. The expression $\sin 35° \cos 22° + \cos 35° \sin 22°$ is equal to

A) $\sin 13°$   B) $\sin 57°$   C) $\cos 13°$
D) $\cos 57°$   E) $\tan 13°$

20. What is the solution set for the accompanying system of linear equations?

$$2x - y + z = 5$$
$$4x - 3y = 5$$
$$6x + 2y + 2z = 7$$

A) $\{2, 1, 0\}$   B) $\{\frac{1}{2}, -1, 3\}$
C) $\{1, -4, -1\}$   D) $\{2, 1, -1\}$
E) $\{1, 0, 3\}$

Solutions on page 331

# Diagnostic Chart for Cumulative Review Test

## Solutions to Practice Problems in Unit Five

## Practice Problems 17.1

**1. (D)** By cofunctions $\cos 52° = \sin 38°$.

By inverse functions $\cos 52° = \dfrac{1}{\sec 52°}$.

**2. (A)**

$$-140$$
$$+360$$
$$\overline{\phantom{-}220}$$

$\sin(-140°) = \sin 220°$

$\qquad\qquad = -\sin 40°$

or

$\qquad\qquad = -\cos 50°$

$\qquad\qquad\quad \uparrow$

$\qquad\quad$ by cofunctions

**3. (B)** Draw a picture of angle $A$.

$\sec A = \dfrac{\text{hyp}}{\text{adj}} = \dfrac{5}{4}$

$\cos A = \dfrac{1}{\sec} = \dfrac{+4}{5}$

**4. (B)**

$$2\cos^2 \theta - \cos \theta = 0$$
$$\cos \theta(2\cos \theta - 1) = 0$$

$\cos \theta = 0$ $\qquad$ $2\cos \theta - 1 = 0$

$\theta = 90°$ $\qquad\qquad$ $\cos \theta = +\dfrac{1}{2}$

$\quad = 270°$

$\qquad\qquad$ indicated $\qquad$ Ref. $\angle = 60°$

$\qquad\qquad$ First and fourth quadrants

Answer: $60°$

**5. (B)**

$4 + \cos^2 A$

Recall: $\sin^2 A + \cos^2 A = 1$

$\therefore \cos^2 A = 1 - \sin^2 A$

Substitute:

$$4 + \cos^2 A = 4 + (1 - \sin^2 A)$$
$$= 5 - \sin^2 A$$

324

**6. (B)**

$$2 \cos x - 1 = 0$$

$$\cos x = \frac{+1}{2}$$

cos is "+" in        Ref. $\angle = 60°$
First and fourth quadrants

Answer: 60° and 300°

**7. (C)**

$$\frac{7\pi}{6} \times \frac{180}{\pi} = 210$$

**8. (B)**

$$6 \tan x = 6$$

$$\tan x = +1$$

Quad        Ref. $\angle = 45°$
First and third quadrants

$$45° = \frac{\pi}{4}$$

$$225° = \frac{5\pi}{4}$$

**9. (A)** These are cofunctions.

$$(2x + 30) + (3x + 20) = 90$$
$$5x = 40$$
$$x = 8$$

**10. (A)** $\dfrac{1}{1 - \cos x}$ is undefined when

$$1 - \cos x = 0$$
$$\cos x = 1$$
$$x = 0°$$

# Practice Problems 17.2

**1. (E)** At $x = 0$ $y$ is a maximum
∴cos not sin.
Amplitude: 2
Period: $-\pi$ to $\pi = 2\pi$   ∴frequency: 1
∴$y = 2 \cos x$

**2. (A)** $y = a \cos(bx + c)$

Amplitude: $|a|$

$y = -3 \cos(2x - \pi)$

Amplitude: $|-3| = 3$

**3. (B)** Amplitude $= \frac{1}{2}|$max value − min value$|$
$= \frac{1}{2}|2 - (-2)| = 2$

**4. (E)** For $a \sin(bx + c)$

the period $= \dfrac{2\pi}{|b|} = \dfrac{2\pi}{2} = \pi$

**5. (C)** If $y = a \sin(bx + c)$

the phase shift $= \dfrac{c}{b} = \dfrac{-\dfrac{\pi}{3}}{2} = \dfrac{-\pi}{6}.$

Shift is $\dfrac{\pi}{6}$. The minus indicates that the shift is to the right.

**6. (E)** The tangent function has no amplitude.

**7. (D)** Examine the graphs. All can have the values 1 and $-1$.

**8. (D)** If $y = a \sec(bx + c)$

the period $= \dfrac{2\pi}{|b|} = \dfrac{2\pi}{\dfrac{3\pi}{4}} = \dfrac{8\pi}{3\pi} = \dfrac{8}{3}.$

**9. (B)** At $x = 0$ $y$ is a max $\therefore \cos$ not sin.
Amplitude: 1

Half of a cycle is $2\pi$; a complete cycle is $4\pi$

$\therefore$ period $= 4\pi,$   $P = \dfrac{2\pi}{\text{freq}}$   $\therefore$ freq $= \frac{1}{2}$

freq $= \frac{1}{2}$

amplitude   frequency

**10. (D)** Examine the graph.

Decreases from 1 to 0

## Practice Problems 17.3

**1. (D)**

$\tan 2A = \dfrac{2 \tan A}{1 - \tan^2 A} = \dfrac{2(\frac{1}{3})}{1 - (\frac{1}{3})^2}$

$= \dfrac{\frac{2}{3}}{\frac{8}{9}}$

$= \dfrac{3}{4}$

**2. (C)**

$\sin \frac{1}{2}\theta = \pm \sqrt{\dfrac{1 - \cos \theta}{2}}$

$= \sqrt{\dfrac{1 - \frac{7}{25}}{2}}$

$= \sqrt{\dfrac{\frac{18}{25}}{2}} = \sqrt{\dfrac{18}{50}} = \sqrt{\dfrac{9}{25}}$

$= \dfrac{3}{5}$

Since $\theta$ is positive acute (first quadrant), $\sin \frac{1}{2}\theta$ is '+'.

**3. (E)** Since $\cos \theta > 0$ and $\tan \theta < 0$, $\theta$ is in the fourth quadrant.

$3^2 + x^2 = 7^2$
$x^2 = 40$
$x = 2\sqrt{10}$

$\sin 2\theta = 2 \sin \theta \cos \theta$

$= 2\left(-\dfrac{2\sqrt{10}}{7}\right)\left(\dfrac{3}{7}\right) = \dfrac{-12\sqrt{10}}{49}$

$\theta$ is in the fourth quadrant
$\sin \theta = "-"$ $\cos \theta = "+"$

**4. (A)**

$\cos 2x = 1 - 2 \sin^2 x$

$= 1 - 2\left(\dfrac{5}{6}\right)^2$

$= 1 - 2\left(\dfrac{25}{36}\right) = 1 - \dfrac{50}{36}$

$= \dfrac{-14}{36} = \dfrac{-7}{18}$

# Practice Problems 17.4

1. **(A)** $\theta = \text{Arc tan}\left(-\dfrac{\sqrt{3}}{3}\right)$

$\theta$ is a negative angle in the fourth quadrant.

Ref. $\angle = 30°$

$\theta = -30°$

2. **(D)** Let $\theta = \text{Arc tan } \tfrac{3}{4}$. Find $\cos \theta$.
$\theta$ is a first quadrant angle.

Draw a diagram of $\tan \theta = \tfrac{3}{4}$

Find $\cos \theta$: $\cos \theta = \tfrac{4}{5}$.
Since the angle is in the first quadrant
$\underbrace{\cos(\text{Arc tan } \tfrac{3}{4})}_{\theta} = \cos \theta = +\tfrac{4}{5}$.

3. **(B)** $\text{Arc sin}\left(\dfrac{\sqrt{3}}{2}\right)$ is a first quadrant angle.

Ref. $\angle = 60°$

Arc tan 1 is a first quadrant angle.

Ref. $\angle = 45°$

$\text{Arc sin } \dfrac{\sqrt{3}}{2} + \text{Arc tan } 1 = 60° + 45° = 105°$

4. **(D)** Let $\theta = \text{Arc cos } \tfrac{1}{3}$. This is a first quadrant angle.
Draw a diagram to represent $\cos \theta = \tfrac{1}{3}$.

$1^2 + x^2 = 3^2$

$x^2 = 8$

$x = 2\sqrt{2}$

$\sin (2 \overbrace{\text{Arc cos } \tfrac{1}{3}}^{\theta}) = \sin 2\theta$

$= 2 \sin \theta \cos \theta$

$= 2\left(\dfrac{2\sqrt{2}}{3}\right)\left(\dfrac{1}{3}\right)$

$= \dfrac{4\sqrt{2}}{9}$

# Practice Problems 17.5

**1. (A)** Use the law of sines.

$$\frac{n}{\sin N} = \frac{l}{\sin L} \quad \text{or} \quad \frac{\sin N}{n} = \frac{\sin L}{l}$$

$$\frac{.8}{4} = \frac{\sin L}{l}$$

$$\frac{\sin L}{l} = \frac{1}{5}$$

**2. (B)** Use the law of cosines.

$$a^2 = b^2 + c^2 - 2bc \cos A$$

$$5^2 = 8^2 + 9^2 - 2(8)(9) \cos A$$

$$25 = 64 + 81 - 144 \cos A$$

$$25 = 145 - 144 \cos A$$

 The 144 is multiplied by cos $A$. You cannot subtract the 144 from the 145

$$
\begin{array}{r}
25 = \phantom{-}145 - 144 \cos A \\
-145 \quad -145 \phantom{xxxxxx} \\
\hline
\dfrac{-120}{-144} = \dfrac{-144 \cos A}{-144}
\end{array}
$$

$$\cos A = \frac{5}{6}$$

**3. (B)** Use the law of sines.

$$180° - (68° + 35°) = 77°$$

$$\frac{a}{\sin A} = \frac{c}{\sin C} \Rightarrow \frac{a}{\sin 35°} = \frac{20}{\sin 77°}$$

$$\frac{a}{.5736} = \frac{20}{.9563}$$

$$a = \frac{.5736(20)}{.9563} = 11.996$$

$$a = 12$$

**4. (D)**

$$a^2 = b^2 + c^2 - 2bc \cos A$$

$$x^2 = 6^2 + 8^2 - 2(6)(8) \cos 38°$$

$$= 36 + 64 - 96(.7880)$$

$$= 100 - 75.65 = 24.65$$

$$x = \sqrt{24.65} = 4.93$$

Answer: $x = 5$

# Practice Problems 18.1

**1. (D)** $(-\sqrt{3}, 1)$

Point is in the second quadrant.

$(-\sqrt{3}, 1)$

$$r^2 = (-\sqrt{3})^2 + 1^2 = 4$$

$$r = \pm 2$$

$$\theta = \text{Tan}^{-1} \frac{1}{-\sqrt{3}} =$$

$$\theta = \text{Ref.} \angle = 30°$$

From the graph, the point is in the second quadrant

$\therefore \theta = 150°$ and the point is $(2, 150°)$. We could also refer to the point as $(-2, 330°)$.

**2. (D)** $r \sin \theta = -2$
From the conversion table
$r \sin \theta = y$
$\therefore y = -2$

**3. (A)** $x + yi = 1 + i\sqrt{3}$  $x = 1$

$(1, \sqrt{3})$    $y = \sqrt{3}$

$r^2 = 1^2 + (\sqrt{3})^2 = 4$  $r = \pm 2$

**1st Quad**

$\theta = \text{Tan}^{-1} \dfrac{\sqrt{3}}{1} = 60°$

$\theta = \text{reference angle} = 60°$
From the graph, the point is in the first quadrant
$\therefore \theta = 60°$ and the point is $(2, 60°)$. We could also
refer to the point as $(-2, 240°)$

$x + yi = r(\cos \theta + i \sin \theta)$

$1 + i\sqrt{3} = 2(\cos 60° + i \sin 60°)$

or

$= -2(\cos 240° + i \sin 240°)$

**4. (E)**

$\cos 120° = -\cos 60° = -\dfrac{1}{2}$

$\sin 120° = +\sin 60° = \dfrac{\sqrt{3}}{2}$

$x = r \cos \theta = 2\sqrt{6}\left(-\dfrac{1}{2}\right)$

$\qquad = -\sqrt{6}$

$y = r \sin \theta = 2\sqrt{6}\left(\dfrac{\sqrt{3}}{2}\right)$

$\qquad = \sqrt{18} = 3\sqrt{2}$

$r(\cos \theta + i \sin \theta) = x + yi$

$\qquad = -\sqrt{6} + 3\sqrt{2}i$

# Practice Problems 19.1

**1. (C)**

$e^{i(\pi/2)} = \cos \dfrac{\pi}{2} + i \sin \dfrac{\pi}{2}$

$\qquad = 0 + i(1) = i$

**2. (B)**

$\sqrt{3} - i = x + yi$  $x = \sqrt{3}$  $y = -1$

$r = \sqrt{x^2 + y^2} = \sqrt{3 + 1} = 2$

$\tan \theta = \dfrac{y}{x} = \dfrac{-1}{\sqrt{3}}$  $\theta = 330°$

$\sqrt{3} - i = 2(\cos 330° + i \sin 330°)$

$(\sqrt{3} - i)^5 = [2(\cos 330° + i \sin 330°)]^5$

$\qquad = 2^5(\cos 5 \cdot 330° + i \sin 5 \cdot 330°)$

Note: $5(330°) = 1650°$
subtracting four multiples
of 360° yields:   $1650°$
$\underline{\quad -1440°}$
$\qquad\quad 210°$

$= 2^5(\cos 210° + i \sin 210°)$

$= 32\left(\dfrac{-\sqrt{3}}{2} + i\left(-\dfrac{1}{2}\right)\right)$

$= 16(-\sqrt{3} - i)$

$= -16(\sqrt{3} + i)$

$\cos 210° = -\cos 30° = -\dfrac{\sqrt{3}}{2}$

$\sin 210° = -\sin 30° = -\dfrac{1}{2}$

3. **(B)**

$2 + 2i\sqrt{3} = x + yi \quad x = 2, \quad y = 2\sqrt{3}$

$r = \sqrt{x^2 + y^2} = \sqrt{2^2 + (2\sqrt{3})^2}$

$\quad = \sqrt{4 + 12} = 4$

$\tan \theta = \dfrac{y}{x} = \dfrac{2\sqrt{3}}{2} = \sqrt{3}, \quad \theta = 60°$

$2 + 2i\sqrt{3} = 4(\cos 60° + i \sin 60°)$

$(2 + 2i\sqrt{3})^{1/2} = [4(\cos 60° + i \sin 60°)]^{1/2}$

$k = 0 \qquad = 4^{1/2}\left(\cos \dfrac{60°}{2} + i \sin \dfrac{60°}{2}\right)$

$\qquad = 2\left(\dfrac{\sqrt{3}}{2} + \dfrac{1}{2}\right)i = \sqrt{3} + 1i$

$k = 1 \qquad = 4^{1/2}\left(\cos \dfrac{60° + 360°}{2} + i \sin \dfrac{60° + 360°}{2}\right)$

$\qquad = 2\left(-\dfrac{\sqrt{3}}{2} - \dfrac{1}{2}\right) = -\sqrt{3} - 1i$

4. **(C)**

$[2(\cos 30° + i \sin 30°)][7(\cos 10° + i \sin 10°)]$

$\quad = 2 \cdot 7[\cos(30° + 10°) + i \sin(30° + 10°)]$

$\quad = 14[\cos 40° + i \sin 40°]$

## Practice Problems 20.1

1. **(B)**

$y = t + 1$

$t = y - 1$

$x = t^2 - 2$

$x = (y - 1)^2 - 2$

$x + 2 = (y - 1)^2$

$x + 2 = y^2 - 2y + 1$

$x = y^2 - 2y - 1$

2. **(E)**

$y = \tan \theta \qquad x = 2 \cot \theta$

$y = \dfrac{\sin \theta}{\cos \theta} \qquad x = \dfrac{2 \cos \theta}{\sin \theta}$

$\dfrac{\cos \theta}{\sin \theta} = \dfrac{1}{y} \Rightarrow x = 2\left(\dfrac{1}{y}\right)$

$x = \dfrac{2}{y}$

$xy = 2$

This is a hyperbola.

3. **(E)**

$x = e^t \qquad y = e^{-t} = \dfrac{1}{e^t}$

$y = \dfrac{1}{e^t} \Rightarrow e^t = \dfrac{1}{y}$

$x = e^t = \dfrac{1}{y}$

$xy = 1$

This is a hyperbola.

4. **(B)**

$y = 3 \sin \phi \qquad x = 3 \cos \phi$

$\sin \phi = \dfrac{y}{3} \qquad \cos \phi = \dfrac{x}{3}$

$\sin^2 \phi + \cos^2 \phi = 1 \Rightarrow \left(\dfrac{y}{3}\right)^2 + \left(\dfrac{x}{3}\right)^2 = 1$

$x^2 + y^2 = 9$

This is a circle.

# Cumulative Review Test

**1. (D)** GCF: $a$

$$\frac{ax+ay+a^2+a}{a}=x+y+a+1$$

$$ax+ay+a^2+a=a(x+y+a+1)$$

**2. (C)**

$$\frac{r_1(\cos\theta_1+i\sin\theta_1)}{r_2(\cos\theta_2+i\sin\theta_2)}$$

$$=\frac{r_1}{r_2}(\cos(\theta_1-\theta_2)+i\sin(\theta_1-\theta_2))$$

$$=\frac{20}{5}(\cos(15°-3°)+i\sin(15°-3°))$$

$$=4(\cos12°+i\sin12°)$$

**3. (D)**

$$\tan810°=\tan90°$$

$$=\frac{\sin90°}{\cos90°}=\frac{1}{0}=\text{Undefined}$$

$$\begin{array}{r}810°\\-360°\\\hline450°\\-360°\\\hline90°\end{array}$$

**4. (C)**

$$b^{-3}\cdot b^{2/3}=b^{-3+(2/3)}=b^{-(9/3)+(2/3)}$$

$$=b^{-7/3}$$

**5. (B)**

$$\sec^2x+\csc^2x=\frac{1}{\cos^2x}+\frac{1}{\sin^2x}$$

$$=\frac{\sin^2x+\cos^2x}{\sin^2x\cos^2x}$$

$$=\frac{1}{\sin^2x\cos^2x}$$

**6. (A)**

$$x+yi=-\sqrt{3}+1i\quad x=-\sqrt{3},\quad y=1$$

$$r=\sqrt{x^2+y^2}=\sqrt{3+1}=2$$

$$\tan\theta=\frac{y}{x}=\frac{1}{-\sqrt{3}}\quad\theta_{\text{ref}}=30°\quad\theta=150°$$

$$-\sqrt{3}+1i=2(\cos150°+i\sin150°)$$

**7. (D)**

$$\cos262°=-\cos82°$$

$$=-\sin8°\ (\text{by cofunction rules})$$

**8. (D)**

is a cos curve

with amplitude 2 and period $2\pi$

∴ frequency $=1$.

$$y=2\cos x$$

**9. (E)**

**10. (E)**

$$x = 4 + 3\tan\theta \qquad y = -1 + 2\sec\theta$$

$$\tan\theta = \frac{x-4}{3} \qquad \sec\theta = \frac{y+1}{2}$$

$$\tan^2\theta + 1 = \sec^2\theta$$

$$\left(\frac{x-4}{3}\right)^2 + 1 = \left(\frac{y+1}{2}\right)^2$$

$$\left(\frac{x^2 - 8x + 16}{9}\right) + 1 = \frac{y^2 + 2y + 1}{4}$$

Multiply the entire equation by 36 to clear the fractions.

$$4x^2 - 32x + 64 + 36 = 9y^2 + 18y + 9$$

$$4x^2 - 32x - 9y^2 - 18y + 91 = 0$$

This is a hyperbola.

**11. (B)**

Each interior angle $= 108°$
Each exterior angle $= 180° - 108° = 72°$
Each exterior angle $= \dfrac{360}{n} = 72 \qquad n = 5$

5 sides

**12. (D)**

$$\tan\tfrac{1}{2}A = \sqrt{\frac{1 - \cos A}{1 + \cos A}}$$

$$= \sqrt{\frac{1 - \frac{4}{5}}{1 + \frac{4}{5}}}$$

$$= \sqrt{\frac{\frac{1}{5}}{\frac{9}{5}}} = \sqrt{\frac{1}{9}} = \frac{1}{3}$$

**13. (B)**

$$\frac{j}{\sin J} = \frac{s}{\sin S}$$

$$\frac{5}{.35} = \frac{s}{.21} \qquad s = \frac{.21(5)}{.35}$$

$$= 3$$

**14. (C)**

$$\frac{\sqrt{3}+i}{2} = \frac{\sqrt{3}}{2} + \frac{1}{2}i \qquad x = \frac{\sqrt{3}}{2},\ y = \frac{1}{2}$$

$$r = \sqrt{x^2 + y^2} = \sqrt{\tfrac{3}{4} + \tfrac{1}{4}} = 1$$

$$\tan\theta = \frac{y}{x} = \frac{1}{\sqrt{3}}$$

$$\theta_{\text{ref}} = 30°,\ \theta = 30°$$

$$\frac{\sqrt{3}+i}{2} = 1(\cos 30° + i\sin 30°)$$

$$\left(\frac{\sqrt{3}+i}{2}\right)^{12} = 1^{12}(\cos 12\cdot 30° + i\sin 12\cdot 30°)$$

$$= 1(\cos 360° + i\sin 360°)$$

$$= 1(1 + i(0)) = 1$$

**15. (E)**

$$\tan\theta = \frac{x}{2} \qquad \cot\theta = y$$

$$\tan\theta = \frac{1}{\cot\theta} \qquad \therefore \frac{x}{2} = \frac{1}{y}$$

$$xy = 2$$

This is a hyperbola.

**16. (E)**

$$x^2 - 2x \qquad\qquad + y^2 + 6y \qquad\qquad = -4$$

$$\underbrace{x^2 - 2x + (-1)^2}_{(x-1)^2} + \underbrace{y^2 + 6y + (+3)^2}_{+(y+3)^2} = -4 + (-1)^2 + (3)^2$$

$$= 6$$

center: $(1, -3)$ \qquad radius: $\sqrt{6}$

17. **(A)**      First quadrant

$$\tan x = \frac{\sqrt{3}}{1} = \sqrt{3}$$

18. **(C)**
$$c^2 = a^2 + b^2 - 2ab \cos c$$
$$= 9 + 25 - 30(-\tfrac{1}{2})$$
$$= 34 + 15 = 49$$
$$c = 7$$

19. **(B)** The identity

$$\sin(x+y) = \sin x \cos y + \sin y \cos x,$$
where $x = 35°$ and $y = 22°$.

$$\sin 35° \cos 22° + \sin 22° \cos 35°$$
$$= \sin(35° + 22°) = \sin 57°$$

20. **(B)**

Using equations 1 with 3 and equation 2 alone eliminate $z$

| | |
|---|---|
| $2x - y + z = 5$ | $4x - 3y = 5$ |
| $6x + 2y + 2z = 7$ | |
| Multiply the first equation by $-2$ and add it to the second equation to eliminate $z$. | The variable $z$ is already removed |

$$-4x + 2y - 2z = -10$$
$$6x + 2y + 2z = \phantom{-}7$$

$$\boxed{2x + 4y \phantom{-0} = -3} \qquad \boxed{4x - 3y = 5}$$

$$2x + 4y = -3$$
$$4x - 3y = \phantom{-}5$$

Multiply first equation by $-2$ and add to the second to eliminate $x$

$$-4x - 8y = \phantom{-}6$$
$$\underline{4x - 3y = \phantom{-}5}$$
$$-11y = \phantom{-}11$$

$$\boxed{y = -1}$$

Substitute $y = -1$ into one of the 2 variable equations

$$2x + 4y = -3$$
$$2x + 4(-1) = -3$$
$$2x - 4 = -3$$
$$2x = 1$$

$$\boxed{x = \tfrac{1}{2}}$$

Substitute $y = -1$ and $x = \tfrac{1}{2}$ into one of the original 3 variable equations.

$$2x - y + z = 5$$
$$2(\tfrac{1}{2}) - (-1) + z = 5$$
$$1 + 1 + z = 5$$
$$2 + z = 5$$

$$\boxed{z = 3}$$

# UNIT SIX

# 21. Variation

## 21.1 Basics and Beyond

### Types of Variation

| Type | Equation of Variation | Using Ratios | Alternate Phrase | Sentence |
|------|----------------------|--------------|------------------|----------|
| **Direct** | $x = Ky$ | $\dfrac{x_1}{y_1} = \dfrac{x_2}{y_2} \cdots = K$ | proportional | $x$ varies directly as $y$ |
| **Direct (Square)** | $x = Ky^2$ | $\dfrac{x_1}{y_1^2} = \dfrac{x_2}{y_2^2} \cdots = K$ | proportional | $x$ varies directly as the square of $y$ |
| **Inverse** | $x = \dfrac{K}{y}$ | $x_1 y_1 = x_2 y_2 \cdots = K$ | indirectly | $x$ varies inversely as $y$ |
| **Inverse (Square)** | $x = \dfrac{K}{y^2}$ | $x_1 y_1^2 = x_2 y_2^2 = \cdots = K$ | indirectly | $x$ varies inversely as the square of $y$ |
| **Joint** | $x = Kyz$ | $\dfrac{x_1}{y_1 z_1} = \dfrac{x_2}{y_2 z_2} \cdots = K$ | | $x$ varies jointly as $y$ and $z$ |
| **Combined** | $x = \dfrac{Kyz}{w}$ | $\dfrac{x_1 w_1}{y_1 z_1} = \dfrac{x_2 w_2}{y_2 z_2} \cdots = K$ <br> note: **K** is a constant | | $x$ varies jointly with $y$ and $z$ and inversely as $w$ |

**Examples:**

1. Determine if the table expresses a direct variation, indirect variation or neither.

| $x$ | $y$ |
|-----|-----|
| 4 | 8 |
| $\frac{1}{2}$ | 1 |
| 1 | 2 |
| $\frac{1}{3}$ | $\frac{2}{3}$ |
| 50 | 100 |

**Solutions:**

1. Test $x_1 y_1 = x_2 y_1 = K$

   $4(8) \neq \frac{1}{2}(1)$   Not inverse

   Test $\dfrac{x_1}{y_1} = \dfrac{x_2}{y_2} \cdots = K$

   $\dfrac{4}{8} = \dfrac{\frac{1}{2}}{1} = \dfrac{1}{2} = \dfrac{\frac{1}{3}}{\frac{2}{3}} = \dfrac{50}{100} = \dfrac{1}{2}$

   Variation is direct

2. If $x$ varies jointly as $y$ and the $\sqrt{z}$, and if $x=3$ when $y=\frac{1}{4}$ and $z=16$, find $x$ when $y=2$ and $z=25$.

2. $$\frac{x_1}{y_1\sqrt{z_1}}=\frac{x_2}{y_1\sqrt{z_2}}$$

$$\frac{3}{\frac{1}{4}\sqrt{16}}=\frac{x}{2\sqrt{25}}$$

$$\frac{3}{1}=\frac{x}{2\sqrt{25}}$$

$$\frac{3}{1}=\frac{x}{10}$$

$$x=30$$

# Practice Problems 21.1

1. If $x$ varies inversely as the square of $y$, and if $x=3$ when $y=4$, find $x$ when $y=2$.

A) 96   B) 48   C) 24   D) 12   E) $\sqrt{12}$

2. If $y$ varies inversely as $x$, when $x$ is multiplied by 2, $y$ would

A)  be multiplied by 2
B)  be divided by 2
C)  be multiplied by 4
D)  be multiplied by $\frac{1}{2}$
E)  remain unchanged

3. If $y$ varies directly as $x$ and $y=\frac{1}{4}$ when $x=\frac{1}{2}$, find $y$ when $x=4$.

A) $\frac{1}{2}$   B) 2   C) $\frac{1}{4}$   D) 4   E) $\frac{1}{8}$

4. If $x$ varies jointly as $y$ and $z$ and inversely as the square of $w$, and if $x=3$ when $z=6$, $y=8$ and $w=4$, find $x$ when $z=10$, $y=1$ and $w=2$.

A) $\frac{5}{4}$   B) 43.2   C) 28.8   D) $\frac{10}{4}$   E) 5

Solutions on page 359

# 22. Verbal Problems

## 22.1 Fundamentals

This section is relegated to a brief review of the basic verbal problems.

| Type | Problem | Solution |
|---|---|---|
| **Number** | The greater of two numbers is 5 more than twice the smaller. If the sum of the numbers is 17, find the greater number. | Let $x$ = greater number<br>$y$ = smaller number<br>$x + y = 17$<br>$x = 2y + 5$<br><br>$x + y = 17$<br>$(2y + 5) + y = 17$<br>$3y = 12$<br>$y - 4$<br>$x = \boxed{13}$ |
| **Consecutive Integer** | The sum of three consecutive integers is 57. Find the integers. | Let $x$ = first integer<br>$x + 1$ = second integer<br>$x + 2$ = third integer<br>$x + (x+1) + (x+2) = 57$<br><br>$3x + 3 = 57$<br>$3x = 54$<br>$x = \boxed{18}$<br>$x + 1 = \boxed{19}$<br>$x + 2 = \boxed{20}$ |
| **Consecutive Integers (odd/even)** | Find the second of three consecutive odd integers if the sum of the first and third is 38. | Let $x$ = first odd integer<br>$x + 2$ = second odd integer<br>$x + 4$ = third odd integer<br>$x + (x+4) = 38$<br><br>$2x + 4 = 38$<br>$2x = 34$<br>$x = 17$<br>$x + 2 = \boxed{19}$ |

| | | |
|---|---|---|
| **Coin** | Joan has 45 coins which are worth $3.50. If the coins are nickels and dimes only, how many dimes does she have? | Let $x$ = number of nickels<br>$45 - x$ = number of dimes<br><br>$$\text{Type} \quad \text{Qty} \times \begin{array}{c}\text{Unit}\\(\text{Value})\end{array} = \text{Worth}$$<br><br>| Type | Qty | (Value) | Worth |<br>|---|---|---|---|<br>| Nickels | $x$ | 5¢ | $5x$¢ |<br>| Dimes | $45-x$ | 10¢ | $(45-x)10$¢ |<br><br>Total worth $= 5x + (45 - x)10 = 350$ (in cents)<br>$$5x - 10x + 450 = 350$$<br>$$-5x = -100$$<br>$$x = 20 \text{ nickels}$$<br><br>$45 - x = \boxed{25 \text{ dimes}}$ |
| **Age** | Mr. Unger is 24 years older than his son Jerry. In 8 years, Mr. Unger will be twice as old as Jerry will be then. How old is Mr. Unger now? | Let $x$ = Jerry's age now<br>$x + 24$ = Mr. Unger's age now<br><br>| | Age now | Age in 8 yrs |<br>|---|---|---|<br>| Mr. Unger | $x + 24$ | $x + 24 + 8$ |<br>| Jerry | $x$ | $x + 8$ |<br><br>In 8 yrs $(x + 24 + 8) = 2(x + 8)$<br>$$x + 32 = 2x + 16$$<br>$$x = 16, \; x + 24 = \boxed{40}$$ |
| **Investment** | Joan invests $7200, part at 4% and the rest at 5%. If the annual income from both investments was the same, find her total income from each investment. | Let $x$ = amount invested at 4%<br>$7200 - x$ = amount invested at 5%<br><br>| Principal | × Rate | = Income |<br>|---|---|---|<br>| $x$ | .04 | $.04x$ |<br>| $7200 - x$ | .05 | $.05(7200 - x)$ |<br><br>$$.04x = .05(7200 - x)$$<br>$$9x = 5(7200)$$<br>$$x = 4000$$<br>Investment income $= .04(4000) = \boxed{160}$ |

| **Mixture** | Shirley mixes nuts worth 36¢ a pound with nuts worth 52¢ a pound to make a 300 pound mixture worth 40¢ a pound. How many pounds of the 52¢ nuts did she use? | Let $x$ = amt of the 52¢ nuts<br>$300 - x$ = amt of the 36¢ nuts |

(¢ per pound) × (# of pounds) = Cost

| 36¢ | $x$ | $36x$ |
|-----|-----|-------|
| 52¢ | $300 - x$ | $52(300 - x)$ |
| 40¢ | $300$ | $40(300)$ |

$$36x + 52(300 - x) = 40(300)$$
$$36x + 52(300) - 52x = 40(300)$$
$$52(300) - 40(300) = 52x - 36x$$
$$\frac{\overset{3}{\cancel{12}}(300)}{\underset{4}{\cancel{16}}} = \frac{16x}{16}$$
$$\frac{900}{4} = x$$
$$225 = x$$
$$300 - x = \boxed{75}$$

| **Motion (same and opposite direction)** | At 10 a.m. two cars started traveling toward each other from towns 287 miles apart. They passed each other at 1:30 p.m. If the rate of the faster car exceeds the rate of the slower car by 6 m.p.h., find the rate in miles per hour of the faster car. | |

Let $x$ = rate of slower car
$x + 6$ = rate of faster car

Rate × Time = Distance

| $x$ | $3\frac{1}{2}$ hrs | $(3\frac{1}{2})(x)$ |
|-----|-----|-------|
| $x + 6$ | $3\frac{1}{2}$ hrs | $(3\frac{1}{2})(x + 6)$ |

10 a.m. to 1:30 p.m. was $3\frac{1}{2}$ hrs

$$\tfrac{7}{2}(x) + \tfrac{7}{2}(x + 6) = 287$$
$$\tfrac{7}{2}x + \tfrac{7}{2}x + \tfrac{7}{2}(6) = 287$$
$$7x + 21 = 287$$
$$7x = 266$$
$$x = 38$$
$$x + 6 = \boxed{44}$$

| | | |
|---|---|---|
| **Motion (overtake)** | Susie left her home at 11 a.m. traveling at 30 m.p.h. At 1 p.m. Norm left home and started after her on the same road at 45 m.p.h. At what time did Norm catch up to Susie? | Let $t =$ time Susie traveled<br>$t-2 =$ time Norm traveled |

Rate × Time = Distance

| | Rate | Time | Distance |
|---|---|---|---|
| Susie | 30 | $t$ | $30t$ |
| Norm | 45 | $t-2$ | $45(t-2)$ |

To catch up means to have gone the same distance.

$$30t = 45(t-2)$$
$$90 = 15t$$
$$6 = t$$

$\therefore$ 11 a.m. + 6 hrs = 5 p.m.

---

| | |
|---|---|
| **Work** | Joan can paint a house in 6 hours. Her younger sister Karen can do the same job in 9 hours. In how many hours can they do the job if they work together? |

Joan's rate $= \frac{1}{6}$   Karen's rate $= \frac{1}{9}$

Let $x =$ amount of hours they worked together

Rate × Time = Did

| | Rate | Time | Did |
|---|---|---|---|
| Joan | $\frac{1}{6}$ | $x$ | $\frac{x}{6}$ |
| Karen | $\frac{1}{9}$ | $x$ | $\frac{x}{9}$ |

$$\frac{x}{6} + \frac{x}{9} = 1 \quad \text{(one job)}$$

$$\frac{3x + 2x}{18} = 1$$

$$5x = 18$$

$$\boxed{x = \frac{18}{5} \text{ hrs}}$$

---

| | |
|---|---|
| **Percent** | If 25% of a number is 80, find the number. |

Let $x =$ the number
Note: % means division by 100, "of" implies multiplication

$$\frac{25}{100} \cdot x = 80$$

$$x = \frac{8000}{25} = \boxed{320}$$

| | | |
|---|---|---|
| **Percent (cont.)** | Of 140 seniors, 35 were on the wrestling team. What percent of the seniors were on the wrestling team? | $\% = \dfrac{\text{part}}{\text{whole}} \times 100$ <br><br> $= \dfrac{35}{140} \times 100 = \dfrac{1}{4} \times 100 = \boxed{25\%}$ |
| | 6 is 20% of what number? | Let $x =$ the number <br><br> $6 = \frac{20}{100} \cdot x$ <br><br> $\frac{600}{20} = x$ <br><br> $\boxed{30 = x}$ |
| **Perimeter** | If the sides of a pentagon are $x$, $x$, $x+1$, $2x+1$ and $5x-7$, and the perimeter is 45, find the largest side. | $(x)+(x)+(2x+1)+(x+1)+(5x-7)=45$ <br><br> $10x-5=45$ <br><br> $x=5$ <br><br> The largest side is $5x-7=25-7=\boxed{18}$ |
| **Area** | The length of a rectangle is 7 less than 3 times the width. If the area is 6, find the width of the rectangle. | Let $x =$ width of rectangle <br> $3x-7 =$ length of rectangle <br> <br><br> $A = lw = (3x-7)(x) = 6$ <br> $3x^2 - 7x - 6 = 0$ <br> $(3x+2)(x-3) = 0$ <br> $3x+2=0 \mid x-3=0$ <br> $x = -\frac{2}{3} \mid \boxed{x=3}$ <br> Reject |
| **Digit** | The sum of the digits of a two-digit number is 9. If the number with the digits reversed is subtracted from 3 times the original number the difference is 9. Find the number. | Let $t =$ ten digits <br> $u =$ units digit <br><br> $t+u =$ sum of the digits <br> $10t+u =$ original number <br> $10u+t =$ number with the digits reversed <br> ——————— <br> $t+u=9$ <br> $3(10t+u)-(10u+t)=9$ <br> ——————— <br> $30t+3u-10u-t=9$ <br> $29t-7u=9 \qquad t+u=9$ <br> $29(9-u)-7u=9 \qquad \therefore t=9-u$ <br> $261-29u-7u=9$ <br> $-36u=-252$ <br> $u=7$ <br> $t=2$ <br> Number $=\boxed{27}$ |

| | | |
|---|---|---|
| **Ratio** | An angle which measures 150° is divided into three parts whose ratios are 2:5:8. Find the measure of the three parts. | Let $2x$ = the measure of the part of ratio 2<br>$5x$ = the measure of the part of ratio 5<br>$8x$ = the measure of the part of ratio 8<br><br>$$2x + 5x + 8x = 150$$<br>$$15x = 150$$<br>$$x = 10$$<br><br>$2x = 20$<br>$5x = 50$<br>$8x = 80$ |
| **Current** | A boat can travel 12 m.p.h. in still water. If it can travel 22 miles down a stream in the same time that it can travel 11 miles up the stream, what is the rate of the stream? | $s$ = rate of the stream |

Distance ÷ Rate = Time

| | Distance | Rate | Time |
|---|---|---|---|
| Up | 11 | $12 - s$ | $\dfrac{11}{12 - s}$ |
| Down | 22 | $12 + s$ | $\dfrac{22}{12 + s}$ |

$$D = RT \qquad \frac{D}{R} = T$$

$$\frac{11}{12 - s} = \frac{22}{12 + s}$$

$$11(12 + s) = 22(12 - s)$$

$$11(12) + 11s = 22(12) - 22s$$

$$33s = 22(12) - 11(12) = 11(12)$$

$$33s = 11(12)$$

$$s = \frac{11(12)}{33} \qquad S = \boxed{4}$$

| | |
|---|---|
| **Percent Mixture** | If a 40% acid solution is mixed with a 60% acid solution how much of each is necessary to yield 60 gallons of a 50% acid solution? |

$$\binom{\text{Amount}}{\text{of solution}} \times (\%\ \text{acid}) = \binom{\text{Amount}}{\text{of acid}}$$

| | | |
|---|---|---|
| $x$ | .40 | $.40x$ |
| $60 - x$ | .60 | $.60(60 - x)$ |
| 60 | .50 | $.50(60)$ |

Let $x$ = amount of 40% solution
$60 - x$ = amount of 60% solution

$$.40x + .60(60 - x) = .50(60)$$

Multiply equation by 100 to clear the decimals.

$$40x + 60(60 - x) = 50(60)$$
$$4x + 360 - 6x = 300$$
$$-2x = -60$$
$$x = 30$$
$$60 - x = 30$$

$\boxed{\text{30 gallons of each}}$

# Practice Problems 22.1

1. The tens digit of a two-digit number is three times the units digit. The number obtained by reversing the digits is 18 less than the original number. Find the original number.

   A) 84   B) 31   C) 26   D) 13   E) 93

2. How many ounces of pure acid must be added to 30 ounces of a 20% solution of acid to make it a 50% solution?

   A) 18   B) 30   C) 28   D) 13   E) 180

3. A new machine can package in 3 hours the whole day's production. An older machine takes 5 hours to do the same job. If both machines are run concurrently, how many hours will it take for them to package a day's output?

   A) $\frac{5}{3}$   B) $\frac{3}{5}$   C) $\frac{8}{15}$   D) $\frac{15}{8}$   E) 8

4. A car leaves town A at 1:00 p.m. traveling at 45 m.p.h. A second car leaves town A at 2:00 p.m. traveling the same route at 55 m.p.h. How many hours will it take the second car to overtake the first car?

   A) 1   B) $5\frac{1}{2}$   C) 10   D) $4\frac{1}{2}$   E) $3\frac{3}{4}$

5. A man invests $6000 for one year, part in a certificate which pays 16%, and the remainder in bonds which pay 9%. If the annual income from both investments is the same, find the amount invested at 16%.

   A) 2160   B) 3840   C) 2750
   D) 3250   E) 3000

6. Find the largest of three positive consecutive odd integers such that the square of the smallest is 9 more than the sum of the other two.

   A) 3   B) 5   C) 7   D) 9   E) 11

Solutions on page 360

# 23. Arithmetic and Geometric Progressions and Series

## 23.1 Essentials

**Arithmetic Progression:** A progression of terms such that the difference between any term and the term immediately preceeding it is a constant.

**Arithmetic Series:** The sum of the terms of an arithmetic progression.

**Geometric Progression:** A progression of terms such that the ratio between any term and the term immediately preceeding it is a constant.

**Geometric Series:** The sum of the terms of a geometric progression.

$$\underbrace{a_1=a}_{\text{1st term}}, \underbrace{a_2}_{\text{2nd term}}, \underbrace{a_3}_{\text{3rd term}}, \underbrace{a_4}_{\text{4th term}}, \ldots \underbrace{a_k}_{k\text{th term}}, \underbrace{a_{k+1}}_{k+1\text{ term}}, \ldots \underbrace{a_{n-1}}_{n-1\text{ term}}, \underbrace{a_n=l}_{n\text{th term}}$$

The $a$'s represent the values of the respective terms.

$a$ (without a subscript) usually represents the value of $a_1$ the first term.

$l$ usually represents the value of $a_n$ or the last term.

$n$ represents the number of terms involved.

| | Arithmetic | Geometric |
|---|---|---|
| **Common Difference** | $d=a_{k+1}-a_k$ | — |
| **Common Ratio** | — | $r=\dfrac{a_{k+1}}{a_k}$ |
| **Mean** | Given two real numbers $a$ and $b$<br><br>**Arithmetic Mean** $=\dfrac{a+b}{2}$ | Given two real numbers $a$ and $b$<br><br>**Geometric Mean** $=\pm\sqrt{ab}$<br><br>$\sqrt{ab}$ if $a$ and $b$ are positive<br>$-\sqrt{ab}$ if $a$ and $b$ are negative |
| **Last Term** $(a_n)$ | $l=a_n=a+(n-1)d$ | $l=a_n=a\cdot r^{n-1}$ $(n\geqslant 1)$ |
| **Sum of the First $n$ Terms** | $S=\dfrac{n}{2}(a+l)$<br><br>$=\dfrac{n}{a}(2a+(n-1)d)$ | $S=\dfrac{a-rl}{1-r}$ $(r\neq 1)$<br><br>$=\dfrac{a(1-r^n)}{1-r}$ |

346

| | Arithmetic | Geometric |
|---|---|---|
| **Sum of an Infinite Series** | — | $S_\infty = \dfrac{a}{1-r} \qquad (-1 < r < 1)$ |
| **Sample: Using Letters** | $a, a+d, a+2d, a+3d, \ldots a+(n-1)d$ <br><br> $\displaystyle\sum_{k=1}^{n} a+(k-1)d$ | $a, ar, ar^2, ar^3 \cdots ar^{n-1}$ <br><br> $\displaystyle\sum_{k=1}^{n} ar^{k-1}$ |

| | |
|---|---|
| $\sum$ is the summation symbol. <br><br> $\Sigma$ is the capital greek letter sigma. | $\prod$ is the product symbol. <br><br> $\Pi$ is the capital greek letter pi. |

$\displaystyle\sum_{k=a}^{b} (3k+7)$ means:    Substitute the integers consecutively from $a$ to $b$ inclusive for the value of $k$. Once all the terms have been established—add them.

$\displaystyle\sum_{k=1}^{5} (2k-1)$

$$= [2(1)-1] + [2(2)-1] + [2(3)-1] + [2(4)-1] + [2(5)-1]$$
$$= 1+3+5+7+9 = 25$$

(Note: This is an arithmetic progression.)

$\displaystyle\prod_{k=a}^{b} (3k+7)$ means:    Substitute the integers consecutively from $a$ to $b$ inclusive for the value of $k$. Once all the terms have been established—multiply them.

$\displaystyle\prod_{k=1}^{4} (2k-1)$

$$= [2(1)-1] \cdot [2(2)-1] \cdot [2(3)-1] \cdot [2(4)-1]$$

$$= 1 \cdot 3 \cdot 5 \cdot 7 = 105$$

---

**To convert a never ending repeating decimal to $\dfrac{a}{b}$ form.**

---

This process can be achieved using an infinite geometric series, but there is an easier procedure.

| Step | Description | Example <br> Convert $.33\overline{333}$ to $\dfrac{a}{b}$ form |
|---|---|---|
| 1 | Let $x = $ the repeating decimal | $x = .333\overline{3}$ |
| 2 | Represent $10x$ | $10x = 3.333\overline{3}$ |
| 3 | The goal is to find any two $10^n$ multiples of $x$, i.e., $10x$, $100x$, $1000x$, etc., which repeats exactly the same endless decimal ($n$ is an integer) | $10x$ has decimal $.333\overline{3}$ <br> $x$ has decimal $.333\overline{3}$ <br> These are the same! |
| 4 | Subtract these $10^n$ multiples of $x$ to eliminate the decimals | $10x = 3.333\overline{3}$ <br> $\underline{\quad x = \phantom{3}.333\overline{3}}$ <br> $9x = 3$ |
| 5 | Solve for $x$ and reduce | $x = \frac{3}{9} = \frac{1}{3}$ |

**Examples:**

1. Find the 21st term of the progression

$$1, 4, 7, 10 \ldots$$

2. Find the sum of the first 20 terms of the series

$$1 + 4 + 7 + 10 + \ldots$$

3. Find the two geometric means between 3 and 24.

4. Express $.16666\overline{6}$ in $\frac{a}{b}$ form.

**Solutions:**

1. $d = 4 - 1 = 3, \quad a = 1, \quad n = 21$

$$a_n = a_{21} = l = a + (n-1)d$$
$$l = 1 + (21 - 1)3$$
$$= 1 + 20(3)$$
$$= 61$$

2. $S_{20} = \dfrac{n}{2}(a + l), \quad d = 4 - 1 = 3$

$$l = a + (n - 1)d$$
$$l = 1 + (20 - 1)3$$
$$= 1 + 19(3)$$
$$= 58$$
$$S_{20} = \frac{20}{2}(1 + 58) = 590$$

3. $\underline{3} \quad \underline{\phantom{-}} \quad \underline{\phantom{-}} \quad \underline{24}, \quad n = 4$

$$a_4 = l = ar^{n-1}$$
$$24 = 3r^3$$
$$r^3 = 8$$
$$r = 2$$
$$\therefore \quad \underline{3} \quad \underline{3 \cdot 2} \quad \underline{3 \cdot 2^2} \quad \underline{24}$$

$$6 \text{ and } 12$$

4. Method 1

$$x = \quad .16666\overline{6}$$
$$\left. \begin{matrix} 10x = \quad 1.66666\overline{6} \\ 100x = 16.6666\overline{6} \end{matrix} \right\} \quad \begin{matrix} 100x = 16.6666\overline{6} \\ \underline{10x = \quad 1.6666\overline{6}} \\ 90x = 15 \\ x = \dfrac{15}{90} = \dfrac{1}{6} \end{matrix}$$

Method 2

$$x = \underbrace{.1 + .06 + .006 + .0006 \ldots}_{\text{This is a geometric series.}}$$

$$r = \frac{.006}{.06} = .1$$

$$S_\infty = \frac{a}{1 - r} = \frac{.06}{1 - .1} = \frac{.06}{.9} = \frac{6}{90} = \frac{1}{15}$$

$$x = .1 + \underbrace{\frac{1}{15}}_{S_\infty} = \frac{1}{10} + \frac{1}{15} = \frac{25}{150} = \frac{1}{6}$$

# Practice Problems 23.1

1. Find the 18th term of the arithmetic progression

$$2, 7, 12, 17\ldots.$$

A) 90  B) 98  C) 78  D) 87  E) 92

2. Find the 8th term of the geometric progression

$$5, 1, \tfrac{1}{5}, \tfrac{1}{25}\ldots.$$

A) $\frac{1}{15625}$  B) $\frac{1}{625}$  C) $\frac{1}{125}$  D) 6.4
E) $\frac{1}{3125}$

3. Find the sum of the first 20 terms of the series

$$2+6+10+14+\ldots.$$

A) 810  B) 850  C) 860
D) 840  E) 800

4. Find the two arithmetic means between 3 and 18.

A) 10 and 11  B) 8 and 13  C) $10\frac{1}{2}$ only
D) 6 and 12  E) 8 and $-8$

5. Find the sum of the first 10 terms of the geometric series

$$2+4+8+16+\ldots.$$

A) $-2048$  B) $-2046$  C) 2048
D) 2046  E) 1023

6. Write $.1111\bar{1}$ in $\dfrac{a}{b}$ form.

A) $\frac{11}{100}$  B) $\frac{111}{1000}$  C) $\frac{1}{9}$  D) $\frac{1}{90}$  E) $\frac{10}{9}$

7. Find the sum of the infinite geometric series

$$\tfrac{1}{2}+\tfrac{1}{4}+\tfrac{1}{8}+\tfrac{1}{16}+\ldots.$$

A) 1  B) 2  C) 3  D) $\frac{19}{20}$  E) $\frac{153}{154}$

8. $\displaystyle\sum_{k=1}^{99}(k+3)=$

A) 396  B) 5247  C) 1030
D) 5427  E) 10197

9. Find the result of the infinite product

$$17^{1/2}\cdot 17^{1/4}\cdot 17^{1/8}\cdot 17^{1/16}\ldots 17^{1/2n}\ldots.$$

A) 17  B) $17^{\frac{111}{112}}$  C) 17171717
D) $17^{17}$  E) 1

Solutions on page 361

# 24. Permutations, Combinations, Probability and the Binomial Theorem

## 24.1 Permutations

**Permutation:** An *arrangement* of items in some *specified order.*

**Fundamental Counting Principle:** If one activity can be done in any of $g$ ways and once completed a second independent activity can be done in $h$ ways, then both activities can be done in the stated order in $g \cdot h$ ways.

| | |
|---|---|
| Permutation of $n$ things taken $n$ at a time | $_nP_n = n! = n(n-1)(n-2)\ldots(3)(2)(1)$ |
| Permutation of $n$ things taken $r$ at a time | $_nP_r = P(n,\ r) = {}^nP_r = P_r^n = P_{n,r} =$ $\underbrace{n(n-1)(n-2)\ldots}_{r \text{ factors}} = \dfrac{n!}{(n-r)!}$ |
| The number of distinguishable permutations of $n$ objects of which $a$ are alike | $\dfrac{n!}{a!}$ |
| The number of circular permutations of $n$ things | $(n-1)!$ |
| **Math Notes** | $0! = 1$ $\qquad n! = n \cdot (n-1)!$ |

**Examples:**

1. A diner has 3 choices of appetizer, 2 choices of soup, 4 choices of entrée and 10 choices of dessert. How many different four-course meals are available?

2. How many different 7-letter permutations can be formed from the word "college"?

**Solutions:**

1. By the fundamental counting principle:

$$\underbrace{3}_{\text{appetizer}} \cdot \underbrace{2}_{\text{soup}} \cdot \underbrace{4}_{\text{entrée}} \cdot \underbrace{10}_{\text{dessert}} = 240$$

2. College has 7 letters
   2 l's
   2 e's

   Number of distinguishable permutations

   $$= \frac{7!}{2!2!} = 1260$$

   for the    for the
   2-l's      2-e's

3. In how many arrangements can 8 people be seated around a circular table?

4. How many different 6-letter permutations are possible from the letters in the word "waiter" with a "w" or "e" as the first letter?

3. Circular permutations $= (n-1)! = 7! = 5040$

4.

$$\frac{2}{\uparrow} \cdot \frac{5}{\uparrow} \cdot \frac{4}{\uparrow} \cdot \frac{3}{\phantom{.}} \cdot \frac{2}{\phantom{.}} \cdot \frac{1}{\phantom{.}} = 240$$

w or e first
2 possibilities

having picked two letters, 4 remain

having picked a first letter, 5 letters are left

# Practice Problems 24.1

1. How many different 4-digit permutations are there from the numeral 1981?

   A) 10  B) 19  C) 12  D) 24  E) 48

2. There are 5 boys and 3 girls in a math class. In how many ways can the 8 students be arranged in a line?

   A) 40320  B) 1024  C) 56
   D) 20  E) 8

3. There are 5 boys and 3 girls in a math class. In how many different ways may the 8 students be arranged in a line if all 3 girls are to be in line ahead of the 5 boys.

   A) 40320  B) 720  C) 620
   D) 500  E) 210

4. There are 4 different roads from town A to town B, 6 different roads from town B to town C and 2 different roads from town C to town D. How many different direct routes are available to travel from town A to town D, through towns B and C?

   A) 6  B) 12  C) 24  D) 36  E) 48

Solutions on page 362

## 24.2 Combinations

**Combination: A selection** (picking) of items regardless of order.

| Combinations of $n$ things taken $r$ at a time | $\displaystyle {}_nC_r = \binom{n}{r} = {}^nC_r = C(n, r)$ $= C_{n,r}$ $= \dfrac{{}_nP_r}{r!} = \dfrac{n!}{r!(n-r)!} \qquad r \geqslant n > 0$ |
|---|---|
| **Math Notes** | ${}_nC_r = {}_nC_{n-r}$ $\qquad\qquad$ ${}_nC_n = {}_nC_0 = 1$ <br> ● **At least:** means equal to or greater than <br> ● **At most:** means less than or equal to |

**Example:**

A vase contains 4 yellow roses, 3 pink roses and 5 red roses.

a) How many combinations of 3 roses are possible?
b) How many combinations of 3 roses will be of one color only?
c) How many combinations of 3 roses will contain exactly one rose of each color?
d) How many combinations of 3 roses will contain at least 2 yellow roses?

**Solutions:**

1. a) This is selecting any 3 of the 12 $(4+3+5)$ roses

$$_{12}C_3 = \frac{12 \cdot 11 \cdot 10}{3 \cdot 2} = \boxed{220}$$

b)    $_4C_3$    or    $_3C_3$    or    $_5C_3$
   selecting 3     selecting 3     selecting 3
   yellow         pink          red
   out of 4       out of 3       out of 5

   "or" indicates $+$ (adding)
   "and" indicates $\cdot$ (multiplication)
   $_4C_3 + {}_3C_3 + {}_5C_3 = 4 + 1 + 10 = \boxed{15}$

c)    $_4C_1$    and    $_3C_1$    and    $_5C_1$
   selecting 1      1 pink       1 red
   yellow         out of 3     out of 5
   out of 4

     $= {}_4C_1 \cdot {}_3C_1 \cdot {}_5C_1$
     $= 4 \cdot 3 \cdot 5 = \boxed{60}$

d) At least 2 yellow roses:

   2 yellow and 1 non-yellow

   $= \quad {}_4C_2 \quad \cdot \quad {}_8C_1 \quad = 6 \cdot 8 = 48$
     2 yellow   1 non-yellow
     out of 4    out of 8
              non yellow
  or

3 yellow:     $_4C_3$     $= 4$
         3 yellow
         out of 4

    $({}_4C_2 \cdot {}_8C_1) + {}_4C_3 = 48 + 4 = 52$

# Practice Problems 24.2

1. There are 5 boys and 3 girls in a class. How many 6-member committees can be formed consisting of exactly 4 boys and 2 girls?

   A) 720   B) 48   C) 24   D) 15   E) 8

2. A vase contains 6 yellow, 4 red and 2 pink roses. How many selections of 4 roses will have 2 yellow, 1 red and 1 pink?

   A) 2   B) 4   C) 48   D) 100   E) 120

3. Evaluate $_{30}C_{28}$.

   A) 5040   B) 840   C) 435
   D) 345   E) 58

4. From a regular deck of 52 cards, in how many ways can 5 cards be selected containing all four aces?

   A) 0   B) 1   C) 4   D) 48   E) 120

Solutions on page 362

## 24.3 Probability

| | |
|---|---|
| **Probability** of an event: | $\dfrac{\textbf{Number of favorable or successful outcomes}}{\textbf{Total number of all alternative ways}}$ |

| | |
|---|---|
| Probability of **A or B** (**inclusive** events) | $P(A) + P(B) - P(A \text{ and } B)$ |
| Probability of **A or B** (**exclusive** events) | $P(A) + P(B)$ |
| Probability of **A and B** (**independent** events) | $P(A) \cdot P(B)$ |
| Probability of **A given B** (**dependent** events) | $P(A \text{ given } B) = \dfrac{P(A \text{ and } B)}{P(B)}$ |

**Exclusive:** not more than one of the events can occur in a single trial

**Inclusive:** more than one event can occur in a single trial

**Independent:** if one event does not affect the other event

| | |
|---|---|
| **Probability of r successes out of n attempts**<br>$p$ = probability of success in a single trial<br>$q$ = probability of failure $(1-p)$ | $_nC_r p^r q^{n-r}$ |
| Probability (**Certainty**) = 1<br>Probability (**Never or Impossibility**) = 0 | $P(A \cup B)$ = probability (**A or B**)<br>$P(A \cap B)$ = probability (**A and B**)<br>$P(A/B)$ = probability (**A given B**) |

**Examples:**

1. With one fair die, find the probability of throwing 3 fours in 5 attempts.

2. If three cards are selected from an ordinary deck, find the probability that all three are black.

3. Given a parallelogram, a rhombus, a rectangle and a square, if one of these quadrilaterals is selected at random, what is the probability that its diagonals bisect each other?

**Solutions:**

1. Probability of a four = $\frac{1}{6}$.

$$p = p(\text{success}) = \tfrac{1}{6} \quad q = p(\text{failure}) = \tfrac{5}{6}$$
$$n = 5 \text{ attempts}, \quad r = 3 \text{ successes}$$
$$_5C_3(\tfrac{1}{6})^3(\tfrac{5}{6})^2 = (10)(\tfrac{1}{216})(\tfrac{25}{36}) = (\tfrac{125}{3888})$$

2. Number of ways to succeed = $_{26}C_3$, selecting 3 black out of 26.

Total number of ways = $_{52}C_3$, selecting 3 cards out of 52.

$$P = \frac{_{26}C_3}{_{52}C_3} = \frac{26 \cdot 25 \cdot 24}{3 \cdot 2 \cdot 1} \cdot \frac{3 \cdot 2 \cdot 1}{52 \cdot 51 \cdot 50} = \frac{\cancel{26} \cdot 25 \cdot 24}{\cancel{3 \cdot 2 \cdot 1}} \cdot \frac{\cancel{3 \cdot 2 \cdot 1}}{\underset{(2)}{52} \cdot 51 \cdot \underset{(2)}{50}}$$

$$= \frac{24}{4(51)} = \frac{2}{17}$$

3. All parallelograms have diagonals that bisect each other.

∴ All four are successes.

$P$(diagonals bisecting each other) = certainty = 1

4. What is the probability of 6 heads out of 10 tosses of a fair coin?

4. $P(\text{head}) = \frac{1}{2}$    $P(\text{tail}) = \frac{1}{2}$

$p = p(\text{success}) = p(\text{head}) = \frac{1}{2}$

$q = p(\text{failure}) = \frac{1}{2}$

$r = 6$ success

$n = 10$ attempts

$$_{10}C_6 (\tfrac{1}{2})^6 (\tfrac{1}{2})^4 = (210)(\tfrac{1}{64})(\tfrac{1}{16}) = \tfrac{105}{512}$$

# Practice Problems 24.3

1. From a standard deck of 52 cards, two cards are drawn at random without replacement. What is the probability that both cards are aces?

   A) $\frac{6}{2652}$  B) $\frac{4}{2652}$  C) $\frac{12}{2652}$  D) $\frac{4}{52}$
   E) $\frac{4}{48}$

2. If a card is randomly selected from a standard deck of 52 cards, what is the probability of picking a black king or a club?

   A) $\frac{1}{52}$  B) $\frac{13}{2652}$  C) $\frac{14}{52}$  D) $\frac{15}{52}$
   E) $\frac{12}{2652}$

3. On a loaded coin the $P(\text{head}) = \frac{1}{3}$. Find the probability of 2 heads on 7 throws of this coin.

   A) $_7C_2(\frac{1}{3})^2(\frac{2}{3})^5$  B) $_7C_2(\frac{1}{3})^5(\frac{2}{3})^2$  C) $_7C_2(\frac{1}{3})^7(\frac{2}{3})^2$
   D) $_7C_2(\frac{1}{3})^7$  D) $_7C_2(\frac{2}{3})^7$

4. From a group of 3 boys and 4 girls, what is the probability that a randomly selected subgroup of 5 will consist of 3 boys and 2 girls?

   A) $\frac{5}{7}$  B) $\frac{2}{7}$  C) $\frac{1}{3}$  D) $\frac{5}{8}$  E) $\frac{2}{9}$

Solutions on page 362

## 24.4 Binomial Theorem

Binomial theorem provides the **expansion of** $(a+b)^n$.

$$(a+b)^n = a^n + na^{n-1}b + \frac{n(n-1)}{2\cdot 1}a^{n-2}b^2 + \frac{n(n-1)(n-2)}{3\cdot 2\cdot 1}a^{n-3}b^3 + \cdots + b^n$$

(good for all real $n$)

or

$$= \sum_{k=1}^{n+1} {}_nC_{k-1}(a)^{n-(k-1)}(b)^{k-1} \quad [n \text{ is a positive integer}]$$

$k$th term of $(a+b)^n = {}_nC_{k-1}(a)^{n-(k-1)}(b)^{k-1}$

| **Math Notes** | $(a+b)^n$: • has $n+1$ terms |
| --- | --- |
| | • in each term the sum of the exponents is $n$ |
| | • the exponent of $b$ is one less than the number of the term |
| | • ${}_nC_{k-1} = {}_nC_{n-(k-1)}$ |

**Examples:**

1. Find the 3rd term of $(5x + y)^4$.

2. Find the 4th term of $(\sqrt{x} - 2\sqrt[3]{y})^4$.

**Solutions:**

1. $k$th term $= {}_nC_{k-1}(a)^{n-(k-1)}(b)^{k-1}$

   $k = 3$,     $k - 1 = 2$     $\underbrace{(5x}_{a} + \underbrace{y)}_{b}^{4 \leftarrow n}$

   $= {}_4C_2(5x)^{4-2}(y)^2 = \dfrac{4 \cdot 3}{2 \cdot 1}(5x)^2(y)^2$

   $= 6(25x^2)(y^2)$

   $= 150x^2y^2$

2. $k$th term $= {}_nC_{k-1}(a)^{n-(k-1)}(b)^{k-1}$

   $k = 4$     $k - 1 = 3$     $\underbrace{(\sqrt{x}}_{a} \underbrace{- 2\sqrt[3]{y})}_{b}^{4 \leftarrow n}$

   $= {}_4C_3(\sqrt{x})^{4-3}(-2\sqrt[3]{y})^3$

   $= \dfrac{4 \cdot 3 \cdot 2}{3 \cdot 2 \cdot 1}(\sqrt{x})^1(-2\sqrt[3]{y})^3 = 4(\sqrt{x})(-8y)$

   $= -32y\sqrt{x}$

# Practice Problems 24.4

1. What is the last term in the expansion of $(2x - 3y)^4$?

   A) $-81y^4$   B) $81y^4$   C) $3y^4$
   D) $-54xy^3$   E) $324y^4$

2. Find the middle term of $(2x^2 - y)^4$.

   A) ${}_4C_2(2x^2)^2(-y)^2$   B) ${}_4C_2(2x)^2(y)^2$
   C) ${}_4C_3(2x^2)^3(-y)$   D) ${}_4C_3(2x)^2(-y)^2$
   E) ${}_4C_2(2x^2)^4$

3. Find the 4th term of the expansion $(x + y)^6$.

   A) $30x^3y^4$   B) $15x^4y^2$   C) $20x^3y^3$
   D) $15x^2y^3$   E) $30x^3y^3$

4. What is the middle term of $(x + \frac{1}{x})^6$?

   A) $\dfrac{15}{x^2}$   B) $20x^3$   C) $20x$   D) $20$

   E) $\dfrac{20}{x}$

Solutions on page 363

# Cumulative Review Test

1. The solution set of the equation
$$2x^3 - 3x^2 - 11x + 6 = 0$$
is

   A) $\{-\frac{1}{2}, -2, 5\}$   B) $\{\frac{1}{2}, -2, 3\}$   C) $\{-1, 2, 3\}$
   D) $\{1, -2, \frac{1}{2}\}$   E) $\{-3, -2, -1\}$

2. If $x$ varies directly as the square of $y$ and $x = 100$ when $y = 5$, find the value of $x$ when $y = 4$.

   A) 64   B) 75   C) 80   D) 210   E) 125

3. Find the sum of the infinite series
$$4 + 2 + 1 + \tfrac{1}{2} + \tfrac{1}{4} + \ldots .$$
   A) 7.79   B) 7.93   C) 7.9   D) 8   E) $8\frac{1}{2}$

4. The sum of the digits of a two-digit number is 10. If the digits are reversed and the resulting number is divided by the original number, the quotient is 1 and the remainder is 36. Find the original number.

   A) 28   B) 37   C) 73   D) 19   E) 91

5. The last term in the expansion $(2x + 3y)^4$ is

   A) $-81y^4$   B) $81y^4$   C) $3y^4$
   D) $-54xy^3$   E) $8x^4$

6. Evaluate $\dfrac{|a-3|}{4} + a^0 + a^{1/2}$ when $a = 4$.

   A) $3\frac{1}{4}$   B) $2\frac{1}{4}$   C) $2\frac{3}{4}$   D) $3\frac{3}{4}$   E) $1\frac{3}{4}$

7. What is the numerical value of $\sin \dfrac{\pi}{6} + \cos \dfrac{\pi}{2}$?

   A) $\dfrac{1}{2}$   B) 1   C) $\dfrac{3}{2}$   D) $\dfrac{\sqrt{3}}{2}$   E) $\dfrac{2+\sqrt{3}}{2}$

8. How many liters of water must be evaporated from 84 liters of a 20% salt solution to raise it to a 35% salt solution?

   A) 14   B) 16   C) 24   D) 36   E) 34

9. Figure I is a rectangle, II is a square, III is a right triangle, IV is an acute scalene triangle and V is a parallelogram. The lengths of particular segments of the figures are indicated. Which figure has the greatest area?

   A) I   B) II   C) III   D) IV   E) V

10. Which value of $x$ does not satisfy the equation $\tan^2 x = 1$?

   A) 45°   B) 135°   C) 225°
   D) 300°   E) 315°

11. If $x$ varies directly as $y$ and $x = 16$ when $y = 12$, find $x$, when $y = 21$.

   A) 28   B) 24   C) $\frac{64}{7}$   D) $\frac{63}{4}$   E) 62

12. Find the third term of the expansion $(2x - y)^4$.

   A) $12x^2y^2$   B) $-12x^2y^2$
   C) $24x^2y^2$   D) $4x^3y$
   E) $4xy^3$

13. Find the common ratio of the infinite geometric progression whose first term is 3 and whose sum is 6.

    A) $\frac{1}{2}$   B) 2   C) $-\frac{1}{2}$   D) $-2$   E) $-\frac{3}{2}$

14. In how many different ways can the letters of the word "cosine" be arranged?

    A) 24   B) 6!   C) $6^6$   D) $6^2$   E) $2^6$

15. Find the 20th term of the arithmetic progression 20, 16, 12, 8 . . . .

    A) $-57$   B) $-56$   C) $-60$
    D) $-64$   E) $-68$

16. Which graph does not represent a function?

A)   B)   C)

D)   E)

17. A committee of 4 is to be chosen from 3 boys and 4 girls. How many different ways are there to choose committees which will contain exactly 2 boys and 2 girls?

    A) 18   B) 35   C) 12   D) 7   E) 6

18. The product of two numbers is 54 and their arithmetic mean is $\frac{15}{2}$. Find the larger number.

    A) 5   B) 6   C) 7   D) 8   E) 9

19. The area of a rectangle is $30m^2$. If the length of the rectangle is increased by 4 inches and the width is decreased by 1 inch, the area of the rectangle becomes 33 square inches. Find the width of the original rectangle.

    A) 8   B) $\frac{15}{4}$   C) 3   D) 10   E) $\frac{11}{4}$

20. The first term of an arithmetic progression is 8 and the fifth term is 2. Find the common difference.

    A) $-\frac{3}{2}$   B) $-2$   C) $-1$   D) $-\frac{1}{2}$   E) $\frac{1}{2}$

Solutions on page 364

# Diagnostic Chart for Cumulative Review Test

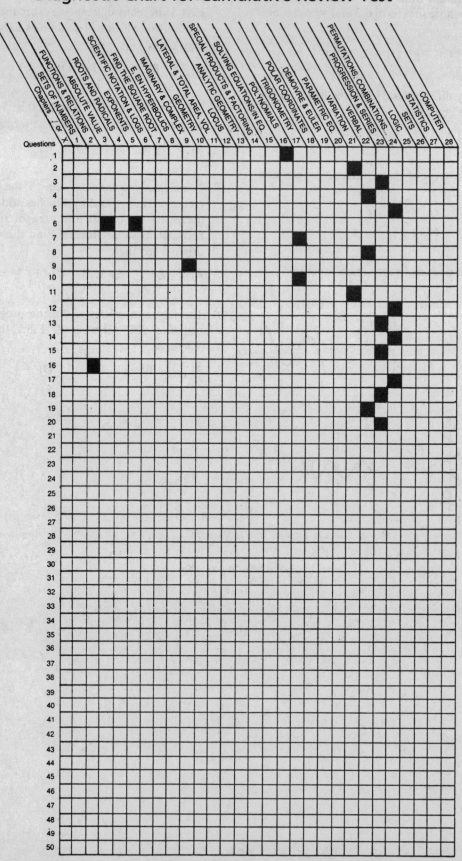

# Solutions to Practice Problems in Unit Six

## Practice Problems 21.1

1. **(D)** $x_1 y_1^2 = x_2 y_2^2 = k$

$$3(4)^2 = x_2(2)^2 = k$$
$$48 = 4x$$
$$12 = x$$

2. **(B)** $y = \dfrac{k}{x}$

When $x = 1$  $y = k$

When $x = 2$ (doubled)

$$y = \frac{k}{2}$$

If $x$ is doubled, $y$ is halved
$y$ is divided by 2.

3. **(B)** $\dfrac{x_1}{y_1} = \dfrac{x_2}{y_2} = K$

$$\frac{\frac{1}{2}}{\frac{1}{4}} = \frac{4}{y}$$

$$\frac{1}{2} \cdot \frac{4}{1} = \frac{4}{y}$$

$$2 = \frac{4}{y}$$

$$2y = 4$$

$$y = 2$$

4. **(D)** $\dfrac{x_1 w_1^2}{y_1 z_1} = \dfrac{x_2 w_2^2}{y_2 z_2}$

$$\frac{3(4)^2}{8(6)} = \frac{x(2)^2}{1(10)}$$

$$\frac{3(16)}{8(6)} = \frac{48}{48} = \frac{4x}{10}$$

$$1 = \frac{4x}{10}$$

$$x = \frac{10}{4}$$

359

# Practice Problems 22.1

**1. (B)** Let $t$ = tens digit

$u$ = units digit

$t = 3u$

$(10u + t) + 18 = 10t + u$

$10u + t + 18 = 10t + u$

$9t - 9u = 18$

$t - u = 2$

$(3u) - u = 2$

$2u = 2$

$u = 1$

$t = 3$

$$\boxed{31}$$

**2. (A)**

Let $x$ = amount of pure acid

| $\left(\begin{array}{c}\text{Amount}\\\text{of solution}\end{array}\right)$ × | (% acid) = | Amount acid |
|---|---|---|
| $x$ | 1.00 | $1.00x$ |
| 30 | .20 | 6 |
| $30 + x$ | .50 | $.50(30 + x)$ |

$1.00x + 6 = .50(30 + x)$

Multiply by 100 to clear the decimals.

$100x + 600 = 50(30 + x)$

$100x + 600 = 1500 + 50x$

$50x = 900$

$x = 18$

**3. (D)**

Let $t$ = time working together

| | Rate × | Time = | Work accomplished |
|---|---|---|---|
| New | $\dfrac{1}{3}$ | $t$ | $\dfrac{t}{3}$ |
| Old | $\dfrac{1}{5}$ | $t$ | $\dfrac{t}{5}$ |

New rate = $\frac{1}{3}$     Old rate = $\frac{1}{5}$

$\dfrac{t}{3} + \dfrac{t}{5} = 1$   (did 1 job)

$5t + 3t = 15$

$8t = 15$

$t = \dfrac{15}{8}$

**4. (D)**

Let $t$ = time traveled by first car

$t - 1$ = time traveled by second car

| | Rate × | Time = | Distance |
|---|---|---|---|
| First car | 45 | $t$ | $45t$ |
| Second car | 55 | $t - 1$ | $55(t - 1)$ |

$45t = 55(t - 1)$

$10t = 55$

$t = 5.5$ hrs

It will take the second car $t - 1 = 4.5$ hrs to overtake the first car.

**5. (A)**

Let $x$ = amount invested at 16%

| Principal × | Rate = | Interest |
|---|---|---|
| $x$ | .16 | $.16x$ |
| $6000 - x$ | .09 | $.09(6000 - x)$ |

$.16x = .09(6000 - x)$

Multiply by 100 to clear the decimals.

$16x = 9(6000 - x)$

$25x = 54000$

$x = 2160$

**6. (D)**

Let $x$ = first positive odd integer

$x + 2$ = second odd integer

$x + 4$ = third odd integer

$(x)^2 = (x + 2) + (x + 4) + 9$

$x^2 = 2x + 15$

$x^2 - 2x - 15 = 0$

$(x - 5)(x + 3) = 0$

$x = 5$    |    $x = -3$

$x + 2 = 7$    |    Reject (not positive)

$x + 4 = 9$

Answer: The largest is 9.

# Practice Problems 23.1

**1. (D)**

$$a_{18} = l = a + (n-1)d$$
$$a = 2, \quad d = 7-2 = 5, \quad n = 18$$
$$l = 2 + (18-1)5 = 2 + 85 = 87$$

**2. (A)**

$$a_8 = l = a \cdot r^{n-1}$$
$$a = 5 \quad r = \tfrac{1}{5} \quad n = 8$$
$$l = 5\left(\frac{1}{5}\right)^{8-1} = 5\left(\frac{1}{5}\right)^7 = \frac{1}{5^6} = \frac{1}{15625}$$

**3. (E)**

$$S_{20} = \frac{n}{2}(a+l)$$

$$a = 2, \quad d = 6-2 = 4, \quad l = a + (n-1)d$$
$$n = 20 \qquad\qquad = 2 + (20-1)4 = 78$$

$$S = \frac{20}{2}(2+78) = 800$$

**4. (B)** $\dfrac{3}{-} \quad - \quad - \quad \dfrac{18}{-}$

$$a = 3, \quad l = 18, \quad n = 4$$
$$l = a + (n-1)d$$
$$18 = 3 + (4-1)d$$
$$15 = 3d$$
$$5 = d$$

$3, \; 3+5 = 8, \; 3+2(5) = 13, \; 18$

The two arithmetic means are 8 and 13.

$\underline{3} \quad \underline{8} \quad \underline{13} \quad \underline{18}$

**5. (D)**

$$S_{10} = \frac{a(1-r^n)}{1-r}$$

$$a = 2, \quad r = \tfrac{4}{2} = 2, \quad n = 10$$

$$S_{10} = \frac{2(1-2^{10})}{1-2} = \frac{2(1-1024)}{-1}$$

$$= \frac{2(-1023)}{-1} = 2046$$

**6. (C)**

$$\left.\begin{array}{l} x = .11111\overline{1} \\ 10x = 1.11111\overline{1} \end{array}\right\}$$

$$10x = 1.111\overline{1}$$
$$\underline{\quad x = \;\; .111\overline{1}\quad}$$
$$9x = 1$$
$$x = \tfrac{1}{9}$$

**7. (A)**

$$S_\infty = \frac{a}{1-r} \qquad a = \tfrac{1}{2}$$

$$r = \frac{\tfrac{1}{4}}{\tfrac{1}{2}} = \tfrac{1}{2}$$

$$= \frac{\tfrac{1}{2}}{1-\tfrac{1}{2}} = \frac{\tfrac{1}{2}}{\tfrac{1}{2}} = 1$$

**8. (B)**

$$\sum_{k=1}^{99}(k+3) = [1+3] + [2+3] + [3+3] \ldots [99+3]$$

$$= 4 + 5 + 6 \ldots 102$$

there are 99 terms.

$$S_{99} = \frac{n}{2}(a+l)$$

$$= \frac{99}{2}(4+102) = 5247$$

**9. (A)** $x^a \cdot x^b = x^{a+b}$

$$17^{1/2} \cdot 17^{1/4} \cdot 17^{1/8} \cdots$$
$$= 17^{\frac{1}{2}+\frac{1}{4}+\frac{1}{8}+\frac{1}{16}\cdots}$$

$$\frac{1}{2} + \frac{1}{4} + \frac{1}{8} + \frac{1}{16} \cdots$$

$$S_\infty = \frac{a}{1-r} = \frac{\tfrac{1}{2}}{1-\tfrac{1}{2}} = 1$$

$$= 17^1 = 17$$

# Practice Problems 24.1

1. **(C)** Distinguishable permutations of $n$ items of which $a$ are alike $= \dfrac{n!}{a!}$.

1981   $n = 4$ numbers
There are 2-1's.

$$= \frac{4!}{2!} = \frac{24}{2} = 12$$

2. **(A)**

$$\underline{8} \cdot \underline{7} \cdot \underline{6} \cdot \underline{5} \cdot \underline{4} \cdot \underline{3} \cdot \underline{2} \cdot \underline{1} = 40320$$

$8! = 40320$

3. $\underbrace{\underline{3} \cdot \underline{2} \cdot \underline{1} \cdot \underline{5} \cdot \underline{4} \cdot \underline{3} \cdot \underline{2} \cdot \underline{1}} = 720$

any of the 5 remaining boys
the remaining girl
any of   one girl having been seated,
3 girls   any of 2 remaining girls.

4. **(E)**
$4 \cdot 6 \cdot 2 = 48$ different routes

# Practice Problems 24.2

1. **(D)** 5 boys and 3 girls

$$\underset{\substack{\text{select 4} \\ \text{boys out of 5}}}{{}_5C_4} \cdot \underset{\substack{\text{select 2} \\ \text{girls out of 3}}}{{}_3C_2} = 5 \cdot 3 = 15$$

2. **(E)** 6 yellow, 4 red, 2 pink

$$\underset{\substack{\text{select 2} \\ \text{yellow out} \\ \text{of 6}}}{{}_6C_2} \cdot \underset{\substack{\text{select 1} \\ \text{red out} \\ \text{of 4}}}{{}_4C_1} \cdot \underset{\substack{\text{select 1} \\ \text{pink out} \\ \text{of 2}}}{{}_2C_1} = 15 \cdot 4 \cdot 2 = 120$$

$${}_6C_2 = \frac{6 \cdot 5}{2 \cdot 1} = 15 \quad {}_4C_1 = 4 \quad {}_2C_1 = 2$$

3. **(C)**

$${}_{30}C_{28} = {}_{30}C_2 = \frac{30 \cdot 29}{2 \cdot 1} = 435$$

4. **(D)** 4 aces,   48 non-aces

$$\underset{\substack{\text{select 4} \\ \text{aces out} \\ \text{of 4}}}{{}_4C_4} \cdot \underset{\substack{\text{select 1} \\ \text{non-ace} \\ \text{out of} \\ \text{48 non-aces}}}{{}_{48}C_1} = 1 \cdot 48 = 48$$

# Practice Problems 24.3

1. **(C)**

$P(\text{1st ace}) = \frac{4}{52}$

$P(\text{2nd ace}) = \dfrac{3 \leftarrow \text{only 3 aces left}}{51 \leftarrow \text{only 51 cards left}}$

$P(\text{ace and ace}) = P(\text{1st ace}) \cdot P(\text{2nd ace})$

$$= \frac{4}{52} \cdot \frac{3}{51} = \frac{12}{2652}$$

2. **(C)**

$P(\text{black king}) = \frac{2}{52}$

$P(\text{club}) = \frac{13}{52}$

$P(\text{king of clubs}) = \frac{1}{52}$

$P(A \text{ or } B) = P(A) + P(B) - P(A \text{ and } B)$

$$= \frac{2}{52} + \frac{13}{52} - \frac{1}{52} = \frac{14}{52}$$

3. **(A)**

$p = p(\text{success}) = \frac{1}{3}, \quad n = 7 \text{ attempts}$

$q = p(\text{failure}) = \frac{2}{3}, \quad r = 2 \text{ successes}$

${}_7C_2(\frac{1}{3})^2(\frac{2}{3})^5$

4. **(B)**

$\text{Success} = \quad {}_3C_3 \quad\quad {}_4C_2 \quad = 1 \cdot 6 = 6$

$\quad\quad\quad\quad\quad \uparrow \quad\quad\quad\quad \uparrow$

$\quad\quad\quad\quad \text{select 3} \quad\quad \text{select 2}$

$\quad\quad \text{boys out of 3} \quad \text{girls out of 4}$

All possible ways to pick 5

$= \quad {}_7C_5 \quad = {}_7C_2 = \dfrac{7 \cdot 6}{2 \cdot 1} = 21$

$\quad\quad \uparrow$

$\quad \text{pick 5 out}$

$\quad\quad \text{of 7}$

$P\begin{pmatrix} 3 \text{ boys and 2 girls} \\ \text{out of a subgroup of 5} \end{pmatrix} = \frac{6}{21} = \frac{2}{7}$

# Practice Problems 24.4

1. **(B)** The last term is the 5th term

5th term $= {}_nC_{k-1}(a)^{n-(k-1)}(b)^{k-1}$

$k = 5,$

$\therefore k - 1 = 4 \quad \underbrace{(2x}_{a} \underbrace{- 3y)}_{b}^{4 \leftarrow n}$

$= {}_4C_4(2x)^{4-(4)}(-3y)^4$

$= 1 \cdot 1 \cdot 81y^4$

$= 81y^4$

2. **(A)** There are 5 terms; the 3rd term is in the middle.

3rd term $= {}_nC_{k-1}(a)^{n-(k-1)}(b)^{k-1}$

$k = 3,$

$\therefore k - 1 = 2 \quad \underbrace{(2x^2}_{a} \underbrace{- y)}_{b}^{4 \leftarrow n}$

$= {}_4C_2(2x^2)^2(-y)^2$

3. **(C)**

4th term $= {}_nC_{k-1}(a)^{n-(k-1)}(b)^{k-1}$

$k = 4,$

$k - 1 = 3 \quad = {}_6C_3(x)^{6-3}(y)^3$

$= \dfrac{6 \cdot 5 \cdot 4}{3 \cdot 2 \cdot 1} x^3 y^3$

$= 20x^3 y^3$

4. **(D)** There are 7 terms; the 4th term is in the middle

4th term $= {}_nC_{k-1}(a)^{n-(k-1)}(b)^{k-1}$

$k = 4,$

$\therefore k - 1 = 3 \quad \left(\underbrace{\frac{x}{a}}_{} + \underbrace{\frac{1}{x}}_{b}\right)^{6 \leftarrow n}$

$= {}_6C_3(x)^{6-3}\left(\dfrac{1}{x}\right)^3$

$= \dfrac{6 \cdot 5 \cdot 4}{3 \cdot 2 \cdot 1} x^3 \left(\dfrac{1}{x}\right)^3 = 20$

# Cumulative Review 1-24

**1. (B)** $P(x) = 2x^3 - 3x^2 - 11x + 6$

$$\frac{p}{q} = \frac{\pm 6, \ \pm 3, \ \pm 2, \ \pm 1}{\pm 2}$$

$P(x) = 2x^3 - 3x^2 - 11x + 6$ has 2 sign changes

∴2 or 0 positive roots.

$P(-x) = -2x^3 - 3x^2 + 11x + 6$ has 1 sign change

∴1 negative root.

$P(-1) = -2 - 3 + 11 + 6 \neq 0$

$P(-2) = -16 - 12 + 22 + 6 = 0 \ \checkmark$

$x = -2$ is a root, $x + 2$ is a factor

$$\begin{array}{r|rrrr} -2 & 2 & -3 & -11 & 6 \\ & & -4 & 14 & -6 \\ \hline & 2 & -7 & 3 & \boxed{0} \end{array}$$

$P(x) = 2x^3 - 3x^2 - 11x + 6$

$= (x + 2)(x^2 - 7x + 3)$

$\qquad\quad (2x - 1)(x - 3)$

$= (x + 2)(2x - 1)(x - 3) = 0$

$x = -2 | x = \frac{1}{2} | x = 3$

**2. (A)**

$$\frac{x_1}{y_1^2} = \frac{x_2}{y_2^2} \qquad \frac{100}{5^2} = \frac{x}{4^2}$$

$$x = \frac{(100)(16)}{25} = 64$$

**3. (D)** $S_\infty = \dfrac{a}{1-r} \qquad r = \dfrac{2}{4} = \dfrac{1}{2}$

$$= \frac{4}{1 - \left(\frac{1}{2}\right)} = 8$$

**4. (B)**

$t + u = 10$

$\dfrac{10u + t}{10t + u} = 1$ remainder 36

∴ by cross multiplication $10u + t = (10t + u) + 36$
rearrange the terms and divide by 9

$9u - 9t = 36$

$u - t = 4 \Rightarrow -t + u = 4$

combining the two equations

$$\begin{array}{r} t + u = 10 \\ -t + u = \ \ 4 \\ \hline 2u = 14 \\ u = \ \ 7 \end{array}$$

Since $t + u = 10$ and $u = 7$

$$t + 7 = 10$$
$$t = \ \ 3$$

The number is 37.

**5. (B)** The last term is the 5th term

5th term $= {}_nC_{k-1}(a)^{n-(k-1)}(b)^{k-1}$

$k = 5, \quad \therefore k - 1 = 4$

$\underbrace{(2x}_{a} + \underbrace{3y)}_{b}^4$

5th term $= {}_4C_4(2x)^{4-4}(3y)^4 = 81y^4$

**6. (A)**

$$\frac{|4 - 3|}{4} + 4^0 + 4^{1/2} = \frac{1}{4} + 1 + 2 = 3\tfrac{1}{4}$$

**7. (A)**

$$\sin \frac{\pi}{6} + \cos \frac{\pi}{2} = \frac{1}{2} + 0 = \frac{1}{2}$$

Note: $\sin \dfrac{\pi}{6} = \sin 30° = \dfrac{1}{2}$

$\cos \dfrac{\pi}{2} = \cos 90° = 0$

**8. (D)** Let $x$ = the amount of water evaporated.

| $\begin{pmatrix}\text{Amount} \\ \text{of solution}\end{pmatrix} \times$ | (% salt) = | $\begin{array}{c}\text{Amount} \\ \text{of salt}\end{array}$ |
|---|---|---|
| 84 | .20 | 16.8 |
| $x$ | 0 | 0 |
| $84-x$ | .35 | $.35(84-x)$ |

$$16.8 = .35(84-x)$$
$$16.8 = 29.4 - .35x$$

Multiply by 100 to clear the decimals.

$$1680 = 2940 - 35x$$
$$x = 36$$

**9. (B)**

$A = bh$
$A = 40$

$A = \dfrac{d^2}{2}$

$= \boxed{50}$

$A = \frac{1}{2}bh$
$= 30$

$A = \frac{1}{2}bh$
$= 30$

$A = bh$
$= 40$

**10. (D)**
$$\tan^2 x - 1 = 0$$
$$(\tan x + 1)(\tan x - 1) = 0$$

| $\tan x = 1$ | $\tan x = -1$ |
|---|---|
| $x = 45°$ | $x = 135°$ |
| $x = 225°$ | $x = 315°$ |

**11. (A)**

$$\frac{x_1}{y_1} = \frac{x_2}{y_2}$$

$$\frac{16}{12} = \frac{x}{21}$$

$$x = 28$$

**12. (C)**

$$\text{kth term} = {}_nC_{k-1}(a)^{n-(k-1)}(b)^{k-1}$$

$k = 3$

$\therefore k - 1 = 2$

$\underbrace{(2x}_{a} - \underbrace{y)^{4n}}_{b}$

$$= {}_4C_2(2x)^{4-2}(-y)^2$$

$$= \frac{4 \cdot 3}{2 \cdot 1}(4x^2)(y^2)$$

$$= 24x^2 y^2$$

**13. (A)**

$$S_\infty = \frac{a}{1-r} \Rightarrow 6 = \frac{3}{1-r}$$

$$6 - 6r = 3$$

$$-6r = -3$$

$$r = \tfrac{1}{2}$$

**14. (B)**

$$\underline{6 \cdot 5 \cdot 4 \cdot 3 \cdot 2 \cdot 1} = 6!$$

**15. (B)** This is an arithmetic progression.

$d = -4$
$l = a + (n-1)d$
$l = 20 + (20-1)(-4)$
$\quad = 20 + (19)(-4)$
$\quad = -56$

**16. (D)** Use the vertical line test.

**17. (A)**
${}_3C_2$: pick 2 boys out of 3
${}_4C_2$: pick 2 girls out of 4
$${}_3C_2 \cdot {}_4C_2 = \frac{3 \cdot 2}{2 \cdot 1} \cdot \frac{4 \cdot 3}{2 \cdot 1}$$
$$= 3 \cdot 6 = 18$$

18. **(E)**

Let $x=$ first number

$y=$ second number

$xy=54$

$$\frac{x+y}{2}=\frac{15}{2}$$

$$x+y=15$$

$$xy=54 \quad \therefore x=\frac{54}{y}$$

Substitute $\frac{54}{y}$ for $x$ in $x+y=15$

$$x+y=15$$

$$\frac{54}{y}+y=15$$

$$54+y^2=15y \implies y^2-15y+54=0$$

$$(y-9)(y-6)=0$$

| $y=9$ | $y=6$ |
|---|---|
| $x=6$ | $x=9$ |

19. **(B)**

$$wl=30 \implies w=\frac{30}{l}$$

$$(w-1)(l+4)=33$$

$$wl+4w-l-4=33$$

Substitute $\frac{30}{l}$ for $w$

$$\left(\frac{30}{l}\right)(l)+4\left(\frac{30}{l}\right)-l=37$$

$$30+\frac{120}{l}-l=37$$

Multiply each term by $l$ and set the equation equal to zero.

$$30l+120-l^2=37l$$

$$l^2+7l-120=0$$

$$(l+15)(l-8)=0$$

| $l=-15$ reject | $l=8 \therefore w=\frac{15}{4}$ |
|---|---|

20. **(A)**

$$l=a+(n-1)d$$

$$2=8+(5-1)d$$

$$2=8+4d$$

$$-6=4d$$

$$-\tfrac{3}{2}=d$$

*UNIT SEVEN*

# 25. Logic

## 25.1 Basics and Beyonds

**Symbols—Summary**

| | Name | $p$: It is blue; $q$: It is heavy |
|---|---|---|
| $\sim$ | **Negation**<br>(not) | $\sim p$: It is **not** blue |
| $\wedge$<br>$(\cdot)$ | **Conjunction**<br>(and) | $p \wedge q$: It is blue **and** it is heavy |
| $\vee$ | **Disjunction**<br>(or) | $p \vee q$: It is blue **or** it is heavy |
| $\rightarrow$<br>$(\supset)$ | **Conditional**<br>If —, then —<br>(implies) | $p \rightarrow q$: **If** it is blue, **then** it is heavy   or<br>It is blue **implies** it is heavy |
| $\leftrightarrow$ | **Biconditional**<br>If and only if<br>(iff) | $p \leftrightarrow q$: It is blue **if and only if** it is heavy<br>or **If** it is blue, **then** it is heavy<br>**AND**<br>**If** it is heavy, **then** it is blue |
| | **Tautology** | A logical sentence whose truth value is always **True** |
| | **Contradiction** | A logical sentence whose truth value is always **False** |
| | **Converse**<br>(reverse, backwards) | Statement $p \rightarrow q$: If it is blue, then it is heavy<br>Converse $q \rightarrow p$: If it is heavy, then it is blue |
| | **Inverse*** | Statement $p \rightarrow q$:   If it is blue, then it is heavy<br>Inverse $\sim p \rightarrow \sim q$: If it is **not** blue, then it is **not** heavy |
| | **Contrapositive**<br>(Inverse of the<br>converse) or<br>(converse of<br>the inverse) | Statement $p \rightarrow q$:          If it is blue, then it is heavy<br>Contrapositive $\sim q \rightarrow \sim p$: If it is not heavy, then it is<br>not blue |

*The inverse is obtained by negating the hypothesis and negating the conclusion.

# Truth Table - Summary

| p | q | ~p | ~q | p∧q | p∨q | statement p→q | p↔q | converse q→p | inverse ~p→~q | contrapositive ~q→~p | tautology | contradiction |
|---|---|----|----|-----|-----|---------------|-----|--------------|---------------|----------------------|-----------|---------------|
|   |   | negation (not) |  | conjunction (and) | disjunction (or) | conditional (if_, then_) | biconditional (iff) | converse | inverse | contrapositive | | |
| T | T | F | F | T | T | T | T | T | T | T | T | F |
| T | F | F | T | F | T | F | F | T | T | F | T | F |
| F | T | T | F | F | T | T | F | F | F | T | T | F |
| F | F | T | T | F | F | T | T | T | T | T | T | F |

 Given a statement

- The contrapositive is ALWAYS logically equivalent.
- The truth or falsity of the inverse and converse vary with the particular statement.

 Note: The converse and inverse are logically equivalent.
(One is the contrapositive of the other.)

# Laws of Inference

| | | |
|---|---|---|
| $p \rightarrow q$<br>$p$<br>―――<br>$q$ | $p$ implies $q$<br>Given: $p$<br>―――<br>Conclusion: $q$ | **Law of Detachment**<br>or<br>**Modus Ponens** (M.P.) |
| $p \rightarrow q$<br>―――<br>$\sim q \rightarrow \sim p$ | $p$ implies $q$<br>―――<br>Conclusion: not $q$ implies not $p$ | **Law of Contrapositive** |
| $\sim(\sim p)$<br>―――<br>$p$ | not(not $p$)<br>―――<br>Conclusion: $p$ | **Law of Double Negation** |
| $p \rightarrow q$<br>$\sim q$<br>―――<br>$\sim p$ | $p$ implies $q$<br>Given: not $q$<br>―――<br>Conclusion: not $p$ | **Law of Modus Tollens** (M.T.) |
| $p \rightarrow q$<br>$q \rightarrow r$<br>―――<br>$p \rightarrow r$ | $p$ implies $q$<br>$q$ implies $r$<br>―――<br>Conclusion: $p$ implies $r$ | **Law of Hypothetical Syllogism**<br>(chain rule) |

| | | | | |
|---|---|---|---|---|
| $p \vee q$<br>$\sim q$<br>___<br>$p$ | $p \vee q$<br>$\sim p$<br>___<br>$q$ | $p$ or $q$<br>Given: not $q$<br>___<br>Concl: $p$ | $p$ or $q$<br>Given: not $p$<br>___<br>Concl: $q$ | **Law of Disjunctive Inference**<br>(Disjunctive Syllogism) |
| $p \wedge q$<br>___<br>$p$ | $p \wedge q$<br>___<br>$q$ | $p$ and $q$<br>___<br>Concl: $p$ | $p$ and $q$<br>___<br>Concl: $q$ | **Law of Simplification** |
| $\sim(p \wedge q)$<br>___<br>$\sim p \vee \sim q$ | $\sim(p \vee q)$<br>___<br>$\sim p \wedge \sim q$ | not $(p$ and $q)$<br>___<br>Concl: not<br>$p$ or not $q$ | not $(p$ or $q)$<br>___<br>Concl: not<br>$p$ and not $q$ | **De Morgan's Laws** |
| $p$<br>$q$<br>___<br>$p \cdot q$ | | Given: $p$<br>Given: $q$<br>___<br>Concl: $p$ and $q$ | | **Law of Conjunction** |
| $p$<br>___<br>$p \vee q$ | | Given: $p$<br>___<br>Concl: $p$ or $q$ | | **Law of Addition** |
| $p \rightarrow q$<br>___<br>$p \rightarrow (p \wedge q)$ | | $p$ implies $q$<br>___<br>Concl: $p$ implies $(p$ and $q)$ | | **Law of Absorption** |

# Quantifiers

| | | |
|---|---|---|
| $\forall_x$: | Universal Quantifier | All, No<br>For all $x$<br>Each<br>For each $x$<br>Every, None<br>For every $x$ |
| $\exists_x$: | Existential Quantifier | Some<br>For some<br>At least one<br>There is at least one |

| | |
|---|---|
| The negation of $\forall_x$: $p$ is $\exists_x$: $\sim p$ | The negation of "for all $p$" is "for some not $p$" |
| The negation of $\exists_x$: $p$ is $\forall_x$: $\sim p$ | The negation of "for some $p$" is "for all not $p$" |

**Examples:**

1. Using the laws of inference:

   Given: $S \rightarrow W$
   $\quad\quad\quad P \rightarrow \sim W$
   $\quad\quad\quad S \vee \sim H$
   $\quad\quad\quad H \vee \sim A$
   $\quad\quad\quad A$

   prove $\sim P$.

**Solutions:**

1. 

| Statements | Reasons |
|---|---|
| 1) $S \rightarrow W$ | Given |
| 2) $P \rightarrow \sim W$ | Given |
| 3) $S \vee \sim H$ | Given |
| 4) $H \vee \sim A$ | Given |
| 5) $A$ | Given |
| 6) $H$ | 4,5 Disjunctive inference |
| 7) $S$ | 3,6 Disjunctive inference |
| 8) $W$ | 1,7 Modus ponens |
| 9) $\sim P$ | 2,8 Modus tollens |

2. Is $\sim(p \vee q) \leftrightarrow (\sim p \wedge \sim q)$ a tautology?

2. 

| $p$ | $q$ | $p \vee q$ | $\sim(p \vee q)$ | $\sim p$ | $\sim q$ | $\sim p \wedge \sim q$ | $\sim(p \vee q) \leftrightarrow (\sim p \wedge \sim q)$ |
|---|---|---|---|---|---|---|---|
| T | T | T | F | F | F | F | T |
| T | F | T | F | F | T | F | T |
| F | T | T | F | T | F | F | T |
| F | F | F | T | T | T | T | T |

Yes it is a tautology. Its truth value is always true.

3. What is the negation of "All trees have green leaves"?

3. The negation of $\forall_x : p$ is $\exists_x : \sim p$.

   $\forall_x : p =$ All trees have green leaves

   $\exists_x : \sim p =$ Some trees do not have green leaves

4. By DeMorgan's Law what is logically equivalent to $\sim(p \vee \sim q)$?

4. $\sim(p \vee q) \rightarrow \sim p \wedge \sim q \quad \therefore \sim(p \vee \sim q) \rightarrow \boxed{\sim p \wedge q}$

   Proof:

| $p$ | $q$ | $\sim q$ | $p \vee \sim q$ | $\sim(p \vee \sim q)$ | $\sim p$ | $\sim p \wedge q$ |
|---|---|---|---|---|---|---|
| T | T | F | T | F | F | F |
| T | F | T | T | F | F | F |
| F | T | F | F | T | T | T |
| F | F | T | T | F | T | F |

logically equivalent

# Practice Problems 25.1

1. If $x$ is an integer, which statement is true?

   A) $\forall_x : x^2 + 2x = 0$
   B) $\exists_x : x^2 + 2x = 0$
   C) $\exists_x : x^2 < 0$  D) $\forall_x x^2 \leqslant 0$
   E) $\forall_x : x^2 > 3$

2. Which statement is false when $p$ is false and $q$ is false?

   A) $p \wedge q$  B) $p \rightarrow q$  C) $\sim p \rightarrow \sim q$
   D) $p \leftrightarrow q$  E) $q \rightarrow p$

3. Which is the converse of the statement $p \rightarrow \sim q$?

   A) $p \rightarrow \sim q$  B) $p \rightarrow q$
   C) $\sim p \rightarrow \sim q$  D) $q \rightarrow \sim p$
   E) $\sim q \rightarrow p$

4. The inverse of $\sim p \rightarrow q$ is

   A) $p \rightarrow \sim q$  B) $\sim p \rightarrow \sim q$
   C) $\sim q \rightarrow p$  D) $q \rightarrow \sim p$
   E) $p \rightarrow q$

5. Let $p = x$ is a prime number
   Let $q = x < 10$
   If $x = 11$ which statement is true?

   A) $\sim p \vee q$  B) $p \wedge q$  C) $\sim p \rightarrow q$
   D) $p \rightarrow q$  E) $p \leftrightarrow q$

6. Which is logically equivalent to the statement "If I live, then I love"?

   A) If I don't live, then I don't love
   B) If I love, then I live
   C) If I don't love, then I don't live
   D) If I don't love, then I live
   E) If I don't live, then I love

7. What is the negation of the statement 'Some students do not do their homework"?

   A) Some students do their homework
   B) All students do not do their homework
   C) All students do their homework
   D) No student does his homework
   E) At least one student does his homework

8. The negation of $r \wedge \sim t$ is

   A) $\sim r \vee t$  B) $\sim r \wedge t$
   C) $\sim r \vee \sim t$  D) $\sim r \wedge \sim t$
   E) $r \rightarrow t$

Solutions on page 390

# 26. Sets

## 26.1 Fundamentals

We often consider a collection, or SET, of items, numbers or things.

{ }          Bracket notation. Used to indicate a set. Each number or item in a set is called a **MEMBER** or an **ELEMENT** of the set.

$\in$          Used to indicate set membership. If $x = \{0, 1, 3, 7\}$, then $3 \in x$, that is, 3 is a member of set $x$. $5 \notin x$ means 5 is not a member of set $x$.

{ } **or** $\varnothing$     The empty or null set.

$\subset$          Indicates a subset. If all the elements of a set $B$ are included in set $D$, then $B$ is called a subset of $D$, written $B \subset D$. For example, if $R = \{-4, 0, 2\}$ and $S = \{0, 2\}$, then $S \subset R$. The symbol $\not\subset$ means is not a subset.

$\cup$          Indicates a union. The union of two sets is a new set, each of whose elements are in either or both of the original sets. For example, if $M = \{1, 2, 3, 4\}$ and $N = \{2, 4, 6\}$ then $M \cup N = \{1, 2, 3, 4, 6\}$.

$\cap$          Indicates intersection. The intersection of two sets is a new set, whose elements are only those elements shared by the original sets. For example, if $M = \{1, 2, 3, 4\}$ and $N = \{2, 4, 6\}$ then $M \cap N = \{2, 4\}$.

---

**Introduction to Venn Diagrams**

The circle represents set $X$; the square represents the **universe of available data**, and the space outside the circle but within the square are elements not contained in set $X$ but in the universe of the data.

Sets $X$ and $Y$ have no elements in common. They are disjoint.

The area indicated by $b$ is the intersection of sets $X$ and $Y$ $(X \cap Y)$. $b$ represents the common elements of sets $X$ and $Y$.

The entire shaded area represents $X \cup Y$, the set of elements of $X$ together with the elements of $Y$.

Since set $Y$ is wholly contained within set $X$, set $Y$ is a subset of $X$. $Y \subset X$.

---

374

The elements of a set may be indicated using SET-BUILDER notation.
$\{x/x+5=7\}$ means the set of all numbers $x$ such that $x+5=7$. This set has a single element 2.

$\{x/x>3\}$ means the set of all numbers $x$ which are greater than 3.

If $A=\{x/x^2=16\}$, then the elements of $A$ are all solutions to the equation

$$x^2=16, \ A=\{4, \ -4\}$$

If $B=\{x/|x|=-3\}$, the elements of $B$ are all the solutions to $|x|=-3$. Since there are no solutions, $B=\{ \ \}$ or $B=\varnothing$.

**Examples:**

1. If $Q=\{\text{set of rational numbers}\}$ and $H=\{\text{set of irrational numbers}\}$, then $Q\cup H$ represents what?

2. If $x=\{\text{odd integers}\}$ and $y=\{\text{prime numbers}\}$, find the element which is NOT in the set $X\cap Y$.

**Solutions:**

1. $Q\cup H$ is the union of the set of rational numbers with the set of irrational numbers; this yields the set of real numbers indicated by $R$.

$$Q\cup H=R$$

2. The set $y$ is the numbers 2, 3, 5, 7, 11, 13, 17, 19, 23..., all of which are odd with the exception of 2. The only element of $y$ which is not contained in $x$ is 2.

# Practice Problems 26.1

1. If $R=\{5, 10, 15\}$ and $S=\{5\}$, which statement is true?

   A) $10\in S$  B) $S\notin R$  C) $R\subset S$
   D) $S\subset R$  E) $R\cup S=S$

2. Which is not a subset of $\{-2, -1, 1, 2\}$?

   A) $\{ \ \}$  B) $\{0\}$  C) $\{1\}$
   D) $\{-2, -1, 1, 2\}$  E) $\{-2, -1\}$

3. If $P=\{2, 4\}, Q=\{1, 2, 3, 4, 5\}, R=\{1, 3, 5\}$, which is NOT true?

   A) $P\cup R=Q$  B) $Q\cap R=R$  C) $R\subset Q$
   D) $P\subset R$  E) $P\cap R=\varnothing$

4. If $A=\{x/|x-2|=5\}$, then $A=$

   A) $\{ \ \}$  B) $\{5\}$  C) $\{5, -5\}$
   D) $\{-3, 7\}$  E) $\{3, -7\}$

5. If $M=\{x/x>0\}$ and $N=\{x/x^2+5x-24=0\}$, the $M\cap N=$

   A) $\{3, -8\}$  B) $\{-3, 8\}$  C) $\{8\}$
   D) $\{3\}$  E) $\{ \ \}$

6. If $A=\{\pi, \sqrt{2}, e, \sqrt{17}\}$ and $B=\{\text{all rational numbers}\}$ then $A\cap B$ is

   A) $\{\pi, e\}$  B) $\{\sqrt{2}, \sqrt{17}\}$
   C) $\{\text{whole numbers}\}$
   D) $\{\text{real numbers}\}$  E) $\{ \ \}$

Solutions on page 390

## 26.2 Applications of Venn Diagrams

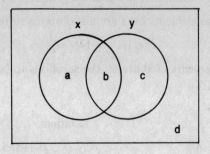

$\odot X$   represents the set of all $x$

$\odot Y$   represents the set of all $y$

    **a**   represents the set of all $x$ that is not also $y$

    **c**   represents the set of all $y$ that is not also $x$

    **b**   represents the set of items that are both $x$ and $y$

    **d**   represents the set of all items that are neither $x$ nor $y$

The set of all items is $a+b+c+d$.

**Examples:**

1. At a dog show there are 23 dogs. 6 have black spots, 14 have brown spots and 7 have neither brown nor black spots. How many dogs have both brown and black spots?

**Solutions:**

1.

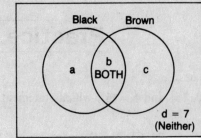

$$a+b=6$$
$$c+b=14$$

$$\text{Total}=\underbrace{a+b}_{6}+c+\underbrace{d}_{7}=23$$

$$\therefore c=10$$

Since $c+b=14$ and $c=10$

$$\therefore b=4$$

4 dogs have both black and brown spots.

2. Of 100 cats, 82 had long hair and 40 had black stripes. If 23 had both long hair and black stripes, how many had neither?

2.

$$a+b=82$$
$$b+c=40$$
$$b=23$$

$$b+c=40$$
$$\underline{b=23}$$
$$\therefore c=17$$

$$\text{Total}=\underbrace{a+b}_{82}+\underbrace{c}_{17}+d=100$$

$$82+17+d=100$$
$$99+d=100$$
$$d=1$$

# Practice Problems 26.2

1. In a pantry there are 28 cans of vegetables. Eight of these have labels with white lettering, 18 have labels with green lettering and 8 have labels with neither white nor green lettering. How many cans have both white and green lettering?

   A) 2   B) 10   C) 8   D) 6   E) 26

2. In a class of 175 students everyone takes math and/or English. If 125 take math and 117 take English. How many take both math and English?

   A) 50   B) 58   C) 67   D) 123   E) 76

3. In a kennel of 100 dogs 40 dogs have only brown spots. Fifty dogs have neither brown nor black spots. How many have black spots (with or without brown spots)?

   A) 10   B) 25   C) 50   D) 60   E) 90

4. In a class of 40 students, twice as many students take English as math. If 12 students take both and 10 take neither, how many students take math?

   A) 6   B) 8   C) 10   D) 12   E) 14

Solutions on page 391

# 27. Statistics

## 27.1 Measures of Central Tendency

| | Given sample values $x_1, x_2, x_3, \ldots, x_n$ | Given sample values $x_1, x_2, x_3, \ldots, x_n$ with respective frequencies $f_1, f_2, f_3, \ldots, f_n$ |
|---|---|---|
| The **Mean** | $\bar{x} = \dfrac{\text{sum of items}}{\text{quantity of items}} = \dfrac{\sum\limits_{i=1}^{n} x_i}{n}$ | $\bar{x} = \dfrac{\text{sum of items}}{\text{quantity of items}} = \dfrac{\sum\limits_{i=1}^{n} f_i x_i}{\sum\limits_{i=1}^{n} f_i}$ |
| | Sometimes called the Average, sometimes called the Arithmetic Mean. | |
| The **Median** | The **middle** value when the items are sequenced in numerical order. If there are two middle values, use their arithmetic mean. | |
| The **Mode** | The **most frequent value**. If more than one value occurs with the maximum frequency, then each of them is a mode. | |

**Examples:**

1. Given the ages:

   15, 15, 15, 16, 17, 18, 20

   find the mean, mode and median.

2. Given the ages:

   14, 15, 15, 15, 16, 16, 16, 17

   find the mean, mode and median.

**Solutions:**

1. mean $= \bar{x} = \dfrac{15+15+15+16+17+18+20}{7}$

   $= \dfrac{116}{7} = 16\frac{4}{7}$

   median: the middle value is 16

   mode $= 15$ the most frequent value

2. mean $= \dfrac{14+15+15+15+16+16+16+17}{8}$

   $= \dfrac{124}{8} = 15\frac{1}{2}$

   median: there are two middle values, 15 and 16

   $\text{median} = \dfrac{15+16}{2} = 15\frac{1}{2}$

   mode $= 15$ and $16$

# Practice Problems 27.1

1. Given the following table, which score is the mode?

| Score | Frequency |
|-------|-----------|
| 98    | 2         |
| 95    | 3         |
| 92    | 2         |
| 87    | 1         |
| 84    | 2         |

   A) 98   B) 95   C) 92   D) 87   E) 84

2. If the mean of four scores is 85, what is the sum of the four scores?

   A) $21\frac{1}{4}$   B) 85   C) 170   D) 340
   E) 500

3. The scores on a test were 75, 75, 85, 90 and 100, which statement about these scores is true?

   A) the mean and the median are the same.
   B) the mode is greater than the median.
   C) the mode is greater than the mean.
   D) the median is greater than the mean.
   E) the mode and median are the same.

4. Express the mean of $(2x+1)$, $(x+1)$ and $(3x-8)$ in terms of $x$.

   A) $6x-6$   B) $2x-2$   C) $3x-3$
   D) $4x-4$   E) $5x-5$

Solutions on page 392

## 27.2 Measures of Dispersion

| | Given sample values $x_1, x_2, x_3, \ldots, x_n$ | Given sample values $x_1, x_2, x_3, \ldots, x_n$ with respective frequencies $f_1, f_2, f_3, \ldots, f_n$ |
|---|---|---|
| | The average of the squares of the deviations from the mean | |
| The **Variance** | $v = \sigma^2 = \dfrac{\sum_{i=1}^{n} (x_i - \bar{x})^2}{n} = \dfrac{1}{n}\sum_{i=1}^{n}(x_i - \bar{x})^2$ | $v = \sigma^2 = \dfrac{1}{n}\sum f_i(x_i - \bar{x})^2$ |
| | The square root of the variance | |
| The **Standard Deviation** | $\sigma = \text{S.D.} = \sqrt{\dfrac{1}{n}\sum_{i=1}^{n}(x_i - \bar{x})^2}$ | $\sigma = \text{S.D.} = \sqrt{\dfrac{1}{n}\sum f_i(x_i - \bar{x})^2}$ |
| | The difference between the highest and lowest values of $x_i$ | |
| The **Range** | $R = x_{max} - x_{min}$ | |

**Examples:**

1. Using the accompanying data

   find: the mean
   the median
   the mode
   the variance
   the standard deviation

| Score | Frequency |
|-------|-----------|
| 70 | 3 |
| 75 | 4 |
| 80 | 7 |
| 85 | 6 |

**Solutions:**

1. The mean $= \dfrac{\sum f_i x_i}{n}$

$$= \frac{70(3) + 75(4) + 80(7) + 85(6)}{20}$$

$$= \frac{210 + 300 + 560 + 510}{20}$$

$$\bar{x} = \frac{1580}{20} = 79$$

There are 20 items, the median is between the 10th and 11th in sequence. Both values are 80.

$\therefore$ The median $= 80$.

The most frequent value is the mode.

The mode $= 80$.

| $x_i$ (score) | $f_i$ (freq) | $\bar{x}$ | $x_i - \bar{x}$ | $(x_i - \bar{x})^2$ | $f_i(x_i - \bar{x})^2$ |
|------|------|------|------|------|------|
| 70 | 3 | 79 | $-9$ | 81 | 243 |
| 75 | 4 | 79 | $-4$ | 16 | 64 |
| 80 | 7 | 79 | 1 | 1 | 7 |
| 85 | 6 | 79 | 6 | 36 | 216 |
| Total | 20 | | | | 530 |

$$v = \frac{1}{n} \sum f_i(x_i - \bar{x})^2 = \frac{1}{20}(530) = 26.5$$

$$\sigma = \sqrt{v} = \sqrt{26.5} \approx 5.1$$

# Practice Problems 27.2

1. Find the standard deviation for the following set of data. (Answer may be left in radical form.)

| Value | Frequency |
|-------|-----------|
| 60 | 1 |
| 75 | 4 |
| 80 | 3 |
| 90 | 2 |

A) 78   B) 660   C) $\sqrt{660}$

D) $\sqrt{66}$   E) $\sqrt{17}$

2. The ages of ten teachers at Jamaica High School are 33, 23, 36, 29, 36, 36, 33, 29, 36 and 29. Find the standard deviation of these ages.

A) $\sqrt{174}$   B) $\sqrt{29}$   C) 27
D) $\sqrt{27}$   E) $\sqrt{17.4}$

3. Find the range for the following data

$$75, 99, 88, 91, 95, 92, 83, 82, 81.$$

A) 99   B) 75   C) 83   D) 24   E) 18

4. Find the variance for the following set of data.

| Score | Frequency |
|-------|-----------|
| 50 | 4 |
| 58 | 4 |
| 62 | 3 |
| 64 | 6 |
| 65 | 2 |
| 68 | 1 |

A) 68   B) 64   C) 60
D) 31.9   E) $\sqrt{31.9}$

Solutions on page 392

## 27.3 Normal Distribution and σ-Limits

The data indicated in the diagram are accurate but approximate.

**Examples:**

1. On a standardized test with a normal distribution, the mean was 75 and the standard deviation was 4.0. Approximately what percentage of scores would fall between the range 71 to 79?

**Solutions:**

1.

$75 - 4 = 71 \quad 75 \quad 75 + 4 = 79$

Answer. 68%

2. What approximate percentage of scores on a normal distribution would be expected to fall within two standard deviations of the mean?

2.

Answer: 95%

## Practice Problems 27.3

1. In the accompanying diagram, the shaded area represents approximately 95% of the scores on a standardized test with a normal distribution. If the scores range from 52 to 96, which could be the standard deviation?

A) 11   B) 22   C) 44   D) 74   E) 76

2. For a standardized test with a normal distribution, the mean was 75 and the standard deviation was 5.8. Which score lies more than two standard deviations from the mean?

A) 90   B) 85   C) 80   D) 70   E) 65

3. A set of measures that follow a bell curve (normal distribution) has a mean of 50 and standard deviation of 5. Approximately what percent of the measures fall between 45 and 55?

A) 13.5   B) 34   C) 68   D) 95   E) 99

4. On a standardized test with a normal distribution, the mean is 76 and the standard deviation is 4. Between which two scores will ~68% of the scores fall?

A) 68 and 84   B) 74 and 84
C) 72 and 80   D) 74 and 78
E) 76 and 80

Solutions on page 393

# 28. Computer Programs

## 28.1 Fundamental Facts

**Computer programs are detailed step by step instructions.** They are followed in sequence unless they direct to do otherwise from within the program.

As an illustration, the program below will be analyzed in detail.

1. **LET** A = 9
2. **LET** B = −1
3. **IF** A = B², **GO TO INSTRUCTION** 6
4. **INCREASE** B by 2
5. **GO TO INSTRUCTION** 3
6. **PRINT** THE VALUE OF B
7. **STOP**

Instructions 1 and 2 provide us with starting values for A and B.

Instruction 3 asks us to test if A = B².

Since A = 9 and B = −1, $9 \neq (-1)^2$.

They are not equal; go on to the next instruction.

Instruction 4 tells us to replace the existing value of B (which is −1) with +1 (−1 + 2 = +1).

Instruction 5 tells us to go back to instruction 3 and test if A = B².

Since A = 9 and B = +1, $9 \neq (+1)^2$.

They are not equal; go on to the next instruction.

Instruction 4 tells us to replace the existing value of B (which is +1) with +3 (+1 + 2 = +3).

Instruction 5 tells us to go back to instruction 3 and test A = B².

Since A = 9 and B = 3, $9 = 3^2$.

They are equal. We are directed to instruction 6, which tells us to print the current value of B.

$$B = +3$$

# Practice Problems 28.1

1. What is the final value of $X$ that will be printed by instruction 6?

   1. LET X = 0
   2. LET F = X² + 8X − 20
   3. IF F = 0, GO TO INSTRUCTION 6
   4. INCREASE X BY 1
   5. GO BACK TO INSTRUCTION 2
   6. PRINT THE VALUE OF X
   7. PRINT THE SENTENCE
      "THIS IS A ROOT OF THE
      EQUATION"
   8. STOP

   A) −10   B) −20   C) 8   D) 2   E) 0

2. What is the final value of A that will be printed by instruction 7?

   1. LET A = 10
   2. LET B = 10
   3. IF $\frac{A}{B} = \frac{2}{3}$ GO TO 7
   4. LET A BECOME A − 1
   5. LET B BECOME B + 1
   6. GO BACK TO INSTRUCTION 3
   7. PRINT THE VALUE OF A
   8. STOP

   A) 11   B) 12   C) 8   D) 9   E) 10

3. What will this program print?

    1. LET A = 5
    2. LET B = A + 7
    3. LET C = B + 2
    4. IF $A^2 + B^2 = C^2$, GO TO INSTRUCTION 7
    5. PRINT "THE VALUES OF A, B, AND C DO NOT REPRESENT THE LENGTHS OF THE SIDES OF A RIGHT TRIANGLE"
    6. STOP
    7. PRINT "THE VALUES OF A, B, AND C REPRESENT THE LENGTHS OF THE SIDES OF A RIGHT TRIANGLE"
    8. STOP
    9. PRINT THE VALUE OF A
    10. PRINT THE VALUE OF B
    11. PRINT THE VALUE OF C
    12. STOP

A) THE VALUE OF A
B) THE VALUE OF B
C) THE VALUE OF C
D) THE VALUES A, B. AND C DO NOT REPRESENT THE LENGTHS OF THE SIDES OF A RIGHT TRIANGLE
E) THE VALUE OF A, B, AND C REPRESENT THE LENGTHS OF THE SIDES OF A RIGHT TRIANGLE

4. A, B and C represent the coefficients of a quadratic equation $Ax^2 + Bx + C = 0$. What will this program print?

    1. LET A = $-2$
    2. LET B = $-4$
    3. LET C = $-2$
    4. IF $B^2 - 4AC > 0$ GO TO INSTRUCTION 7
    5. IF $B^2 - 4AC = 0$ GO TO INSTRUCTION 9
    6. IF $B^2 - 4AC < 0$ GO TO INSTRUCTION 11
    7. PRINT "ROOTS ARE REAL AND UNEQUAL"
    8. STOP
    9. PRINT "ROOTS ARE REAL AND EQUAL"
    10. STOP
    11. PRINT "ROOTS ARE IMAGINARY"
    12. STOP
    13. PRINT "ROOTS ARE RATIONAL"
    14. STOP
    15. PRINT "ROOTS ARE IRRATIONAL"
    16. STOP

A) ROOTS ARE REAL AND UNEQUAL
B) ROOTS ARE REAL AND EQUAL
C) ROOTS ARE IMAGINARY
D) ROOTS ARE RATIONAL
E) ROOTS ARE IRRATIONAL

Solutions on page 394

# Cumulative Review Post-Test

1. The expression $\log r + \log r^2$ is equal to

   A) $\log(r+r^2)$  B) $3 \log r$  C) $\log r$
   D) $r^3$  E) $r$

2. In a plane, the locus of points a distance 6 units from a given line is

   A) a line  B) a circle
   C) two intersecting lines
   D) two parallel lines
   E) 3 points

3. In the accompanying figure, $DE \parallel AB$. If $CD=6$, $CA=18$ and $DE=4$, what is the length of $AB$?

   A) 16  B) 14  C) 12  D) 6  E) 8

4. What is the length of the line which connects the points $J(5,6)$ and $L(2,-1)$?

   A) $\sqrt{34}$  B) $\sqrt{58}$  C) 8  D) $\sqrt{20}$  E) 14

5. If $\log_3 9^{2x+1}=4$, solve for $x$.

   A) 2  B) 1  C) $\frac{1}{2}$  D) $-\frac{1}{2}$  E) $-1$

6. Find the reciprocal of $1+\sqrt{2}$.

   A) $1+\sqrt{2}$  B) $\sqrt{2}-1$  C) $1-\sqrt{2}$
   D) $\frac{\sqrt{2}-1}{3}$  E) $1+\frac{\sqrt{2}}{2}$

7. The sum of the interior angles of a polygon is $1,980°$. How many sides does the polygon have?

   A) 13  B) 14  C) 15  D) 16  E) 17

8. What is the solution set of the equation

   $|5-2x|=7$?

   A) $\{6, -1\}$  B) $\{6\}$  C) $\{1\}$
   D) $\{-1\}$  E) $\{ \}$

9. The expression $x^{-1}+y^{-1}$ is equal to

   A) $(x+y)^{-1}$  B) $\frac{1}{x+y}$  C) $x+y$
   D) $\frac{x+y}{xy}$  E) $\frac{xy}{x+y}$

10. The graph of $x=\sin \theta$, $y=3\csc \theta$ is a(n)

    A) line  B) circle  C) ellipse
    D) parabola  E) hyperbola

11. If $A$ is a positive acute angle, find the smallest value of $A$ that satisfies the equation

    $$\sin(A+40)° = \cos 10°.$$

    A) $-30$  B) $-20$  C) 0  D) 20  E) 40

12. If $f(x)=x+\frac{1}{x}$, then $f\left(\frac{1}{x}\right)=$

    A) $x^2+1$  B) $f(x)$  C) $x^2+x$
    D) $\frac{2}{x}$  E) $\frac{x^2+1}{x^2}$

13. If the roots of a quadratic equation are 3 and 1, find the equation.

    A) $x^2+4x+3=0$  B) $x^2-4x-3=0$
    C) $x+4x-3=0$  D) $x^2+2x+3=0$
    E) $x^2-4x+3=0$

14. In a group of 40 students, 25 applied to **NJL** College and 30 applied to Joshua University. If 3 students applied to neither NJL or Joshua how many applied to both schools?

    A) 6  B) 12  C) 18  D) 19  E) 11

15. In the accompanying figure $AB$ and $AC$ are tangents to circle $O$ from point $A$. If major arc $BC = 210°$, what is the measure of angle $A$?

A) 30   B) 45   C) 90   D) 150   E) 75

16. The graph shown represents which equation?

A) $y = |x + 2|$   B) $y = |x| + 2$
C) $y = |x - 2|$   D) $y = |x| - 2$
E) $y + 2 = |x|$

17. Which is not an identity?

A) $\sin^2 2\theta + \cos^2 2\theta = 1$
B) $\dfrac{\tan 7x}{\sin 7x} = \sec 7x$
C) $\tan 3t + \cot 3t = 1$
D) $\dfrac{1}{\sin 4r} = \csc 4r$
E) $\tan^2 m^2 + 1 = \sec^2 m^2$

18. Simplify the product $(1 + 2\sqrt{7}i)^2$.

A) 29   B) $-27 + 4\sqrt{7}i$   C) $-13$
D) $29 + 4\sqrt{7}i$   E) $15i$

19. The equation $\dfrac{x}{3} + \dfrac{y}{2} = 1$ represents a(n)

A) line   B) parabola   C) circle
D) ellipse   E) hyperbola

20. What is the remainder when $x^5 - 1$ is divided by $x - 1$?

A) 0   B) $-2$   C) 2   D) $-1$   E) 1

21. Multiply $(x + 2)(x^2 - 2x + 4)$.

A) $x^3 - 6x^2 + 2x + 8$   B) $x^3 + 8$
C) $x^3 + 6x^2 - 3x + 4$   D) $x^3 - 8$
E) $x^3 + 2x^2 - 4x + 8$

22. In the accompanying figure, the area of circle $O$ is $36\pi$ and the area of a sector of the circle is $4\pi$. Find the number of degrees in the measure of the central angle of the sector.

A) 40   B) 4   C) 36   D) 144   E) 20

23. What is the negation of $\sim p \vee \sim q$?

A) $p \wedge q$   B) $\sim p \wedge \sim q$   C) $p \vee q$
D) $p \wedge \sim q$   E) $p \vee \sim q$

24. If $\log 2 = .3010$, $\log 2y$ is equal to

A) $.3010y$   B) $.3010 + y$   C) $\dfrac{.3010}{y}$
D) $.3010 + \log y$   E) $.3010 \log y$

25. Solve the following system of equations:

$$x^2 - 3y^2 = 13$$
$$x + 3y = 1$$

A) $(4,1)$ and $(5, -2)$
B) $(-5,2)$ and $(4, -1)$
C) $(3, -1)$ and $(0, \tfrac{1}{3})$
D) $(6, -1)$ and $(5, -2)$
E) $(4,2)$ and $(-5, -1)$

26. Express $\sin(-42°)$ as a function of a positive acute angle.

A) $\sin 42°$   B) $\cos 42°$
C) $-\sin 48°$   D) $-\cos 42°$
E) $-\cos 48°$

27. $2(\cos 30° + i \sin 30°) =$

A) $1 + 2i$   B) $\sqrt{3} + i$   C) $2 + \dfrac{\sqrt{3}}{2}$
D) $\dfrac{1 + 2\sqrt{3}}{2}$   E) $1 + \sqrt{3}i$

28. Find the average of $\frac{1}{4}$, 25% and .25.

    A) .21   B) .22   C) .24   D) .25   E) .26

29. Find the $\sqrt{3}$ to the nearest tenth.

    A) 1.5   B) 1.57   C) 1.6   D) 1.7   E) 1.8

30. Find the reciprocal of $2i$.

    A) $\frac{1}{2}i$   B) $\frac{2}{i}$   C) $-\frac{1}{2}i$   D) $-4$   E) $\frac{1}{2}$

31. In the accompanying diagram $BDC$ is a straight line and $DE \perp AEC$. If $m\angle ADB=80$ and $m\angle EDC=50$, find the $m\angle DAC$.

    A) 45   B) 60   C) 50   D) 40   E) 25

32. The graph of the equation $x^2 + 2x + \dfrac{y^2}{2} = 18$ is a(n)

    A) line   B) parabola
    C) circle   D) ellipse
    E) hyperbola

33. $[\frac{1}{2}(\cos 10° + i \sin 10°)]^3 =$

    A) $\frac{1}{2}(\cos 1000° + i \sin 1000°)$
    B) $\frac{1}{8}(\cos 10° + i \sin 10°)$
    C) $\frac{1}{8}(\cos 30° + i \sin 30°)$
    D) $\frac{1}{2}(\cos 30° + i \sin 30°)$
    E) $\frac{1}{6}(\cos 13° + i \sin 13°)$

34. The number of points which are at a given distance $d$ from a line and also equally distant from two given points on the given line is

    A) 0   B) 1   C) 2   D) 3   E) 4

35. In the accompanying diagram $AB$, $BC$, $CD$ and $DA$ are tangent to the circle at $R$, $V$, $T$ and $S$,

respectively. If $VC=8$ and $SD=10$, find $CD$.

    A) 21   B) 18   C) 14   D) 12   E) 9

36. The expression $\cos 2A$ is equivalent to

    A) $2 \sin^2 A - 1$   B) $\sin^2 A - \cos^2 A$
    C) $1 - 2 \cos^2 A$
    D) $(\cos A + \sin A)(\cos A - \sin A)$
    E) $1 - \sin^2 2A$

37. The period of the graph $y = 3 \cos \frac{1}{2}x$ is

    A) 1   B) 180°   C) 3   D) 720°   E) $\frac{1}{2}$

38. Find the sum of the infinite series

    $2, \frac{4}{3}, \frac{8}{9}, \ldots$ .

    A) $8\frac{3}{4}$   B) $7\frac{1}{5}$   C) 6   D) $5\frac{3}{8}$   E) $5\frac{1}{4}$

39. The sum of the digits of a two-digit number is 11. The number obtained by interchanging the digits exceeds twice the original number by 34. Find the original number.

    A) 47   B) 38   C) 29   D) 56   E) 65

40. If $x$ varies inversely as $t$ and if $x=6$ when $t=2$, find $x$ when $t=4$.

    A) $1\frac{1}{2}$   B) 3   C) 1   D) 12   E) 36

41. If the volume of a right circular cylinder, whose height $= 10$, is $90\pi$, find the lateral area.

    A) 3   B) 60   C) $60\pi$   D) $78\pi$   E) $180\pi$

42. How many irrational roots does the following equation have?

    $$x^4 - 7x^3 + 9x^2 + 13x - 4 = 0$$

    A) 4   B) 3   C) 2   D) 1   E) 0

43. The measure, in degrees, of an angle of 1 radian is

    A) 180   B) $\dfrac{180}{\pi}$   C) 360

    D) $\dfrac{360}{\pi}$   E) $\pi^2$

44. Evaluate $\sum_{k=3}^{5} (k^2-1)$.

    A) 24   B) 12   C) 48   D) 49   E) 47

45. If the probability of throwing a four on a loaded die is $\frac{1}{5}$, what is the probability of throwing exactly 2 fours out of 5 trials.

    A) $_5C_2(\frac{1}{5})(\frac{4}{5})$   B) $_5C_2(\frac{1}{5})^3$
    C) $_5C_2(\frac{1}{5})^2(\frac{4}{5})^3$   D) $_5C_3(\frac{1}{5})^3(\frac{4}{5})^2$
    E) $_5C_3(\frac{1}{5})(\frac{4}{5})^4$

46. Which is the negation of the statement "All squares are parallelograms"?

    A) All squares are not parallelograms.
    B) No squares are parallelograms.
    C) Some squares are parallelograms.
    D) Some squares are not parallelograms.
    E) If it is a square, then it is a parallelogram.

47. The average of $2P$, $3Q$ and a third number is $A$. In terms of $A$, $P$ and $Q$ represent the third number.

    A) $A-2P-3Q$   B) $A-2P+3Q$
    C) $3A-2P+3Q$
    D) $3A-2P-3Q$   E) $A+2P-3Q$

48. What is the value of $C$ that will be printed?

    1. LET A$=1$
    2. LET B$=1$
    3. LET F$=A^2-B$
    4. IF F$\geq 0$ GO TO INSTRUCTION 6
    5. IF F$<0$ GO TO INSTRUCTION 8
    6. PRINT C$=2$
    7. STOP
    8. PRINT C$=7$
    9. STOP

    A) 1   B) 2   C) 0   D) 7   E) 2

49. If $\ln x + \ln(x+2) = \ln 3$, solve for $x$.

    A) 1   B) 0   C) 2   D) $e$   E) $e^2$

50. The numbers $6$, $-\frac{3}{4}$, $.175$, $.171\overline{717}$ are *all* elements of which of the following?

    A) whole numbers   B) integers
    C) fractions   D) rational
    E) irrational

Solutions on page 394

# Diagnostic Chart Post-Test

# Solutions to Problems in Unit Seven

## Practice Problems 25.1

1. **(B)** $x$ is an integer, for example, $-2$ or $1$.

$$(-2)^2 + 2(-2) = 0 \quad \text{True}$$
$$(1)^2 + 2(1) \neq 0$$

There are some $x$'s for which $x^2 + 2x = 0$

$$\therefore \exists_x : x^2 + 2x = 0.$$

2. **(A)**

$$p \wedge q \Rightarrow F \wedge F \Rightarrow F$$
$$p \rightarrow q \Rightarrow F \rightarrow F \Rightarrow T$$
$$\sim p \rightarrow \sim q \Rightarrow T \rightarrow T \Rightarrow T$$
$$p \leftrightarrow q \Rightarrow F \leftrightarrow F \Rightarrow T$$
$$q \rightarrow p \Rightarrow F \rightarrow F \Rightarrow T$$

3. **(E)** The converse of a statement is its backwards statement:

$$p \rightarrow \sim q$$

Converse: $\qquad \sim q \rightarrow p$

4. **(A)** The inverse of a statement is the negation of the hypothesis and the negation of the conclusion
statement: $\qquad \sim p \rightarrow q$
Inverse: $\qquad p \rightarrow \sim q$

5. **(C)** If $x = 11$, then $p$ is true and $q$ is false.

$$\sim p \vee q \Rightarrow \sim T \vee F \Rightarrow F \vee F \Rightarrow F$$
$$p \wedge q \Rightarrow T \wedge F \Rightarrow F$$

$$\boxed{\sim p \rightarrow q \Rightarrow \sim T \rightarrow F \Rightarrow F \rightarrow F \Rightarrow T}$$

$$p \rightarrow q \Rightarrow T \rightarrow F \Rightarrow F$$
$$p \leftrightarrow q \Rightarrow T \leftrightarrow F \Rightarrow F$$

6. **(C)** The contrapositive of a statement is always logically equivalent to the statement.
Statement: If I live, then I love.
Contrapositive: If I don't love, then I don't live.

7. **(C)** The negation of $\exists_x : \sim p$ is $\forall_x : p$.
$\exists_x : \sim p =$ Some students do not do their homework
$\forall_x : p =$ All students do their homework

8. **(A)** $\sim (r \wedge \sim t)$ by De Morgan's Law
$$\overline{\phantom{\sim (r \wedge \sim t)}}$$
$$\sim r \vee t$$

## Solutions to Practice Problems 26.1

1. **(D)** $S$ is a subset of $R$.

$$S \subset R$$

2. **(B)** The set $\{0\}$ contains the element 0, which is not an element of the original set.

3. **(D)** The elements of $P$ are not elements of $R$
$\therefore P$ is not a subset of $R$.

4. **(D)** $A$ is the solution set of

$$|x - 2| = 5$$

$$(x - 2) = -5 \qquad x - 2 = 5$$

$$x - 2 = -5 \qquad\qquad x = 7$$
$$x = -3$$

check $\qquad\qquad$ check
$$|x - 2| = 5 \qquad |x - 2| = 5$$
$$|-3 - 2| = 5 \qquad |7 - 2| = 5$$
$$|-5| = 5 \qquad\qquad 5 = 5 \checkmark$$
$$5 = 5 \checkmark$$

$$A = \{-3, 7\}$$

390

**5. (D)**

$M = \{\text{all numbers} > 0\}$

$x^2 + 5x - 24 = 0$

$(x + 8)(x - 3) = 0$

$x = -8 \quad x = 3$

$\therefore N = \{-8, 3\}$

$M \cap N = \{3\}$

**6. (E)**

$A = \{4 \text{ irrational numbers}\}$

$B = \{\text{all rational numbers}\}$

$A$ and $B$ are disjoint. $A \cap B = \emptyset$ or $\{\ \}$

# Practice Problems 26.2

**1. (D)**

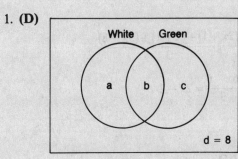

$a + b = 8, \quad b + c = 18, \quad d = 8$

$$\underbrace{\text{Total}}_{28} = \underbrace{a + b + c + \underbrace{d}_{8}}$$

$$28 = 8 + c + 8$$

$$\therefore c = 12$$

Since $b + c = 18$ and $c = 12$

$$\therefore b = 6$$

**2. (C)**

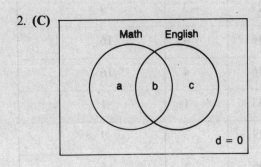

$$d = 0$$

$$a + b = 125$$

$$b + c = 117$$

$$\underbrace{a + b}_{125} + c + \underbrace{d}_{0} = 175$$

$$\therefore c = 50$$

Since $b + c = 117$ and $c = 50$

$$\therefore b = 67$$

**3. (A)**

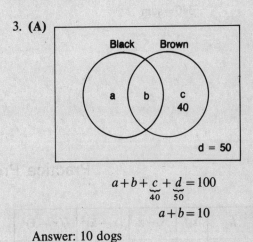

$$a + b + \underbrace{c}_{40} + \underbrace{d}_{50} = 100$$

$$a + b = 10$$

Answer: 10 dogs

**4. (E)** Let $x$ = number of students taking math

$2x$ = number of students taking English

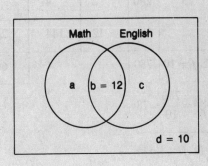

$$b = 12$$

$$a + b = x \Rightarrow a = x - 12$$

$$b + c = 2x \Rightarrow c = 2x - 12$$

$$\underbrace{a + b}_{x} + \underbrace{c}_{2x - 12} + \underbrace{d}_{10} = 40$$

$$x + (2x - 12) + 10 = 40$$

$$3x = 42$$

$$x = 14$$

Math $= a + b = x = 14$

# Practice Problems 27.1

1. **(B)** The mode is the most frequent value.

$$\text{mode} = 95$$

2. **(D)**

$$\text{mean} = \frac{\text{sum}}{\text{qty of items}}$$

$$85 = \frac{\text{sum}}{4}$$

$$340 = \text{sum}$$

3. **(D)**

$$\text{mean} = \frac{75+75+85+90+100}{5}$$

$$= \frac{425}{5} = 85$$

$$\text{median} = \frac{85+90}{2} = 87\tfrac{1}{2}$$

$$\text{mode} = 75$$

4. **(B)**

$$\text{mean} = \frac{(2x+1)+(x+1)+(3x-8)}{3}$$

$$= \frac{6x-6}{3} = 2x-2$$

# Practice Problems 27.2

1. **(D)**

| $x_i$ | $f_i$ | $x_if_i$ | $x_i-\bar{x}$ | $(x_i-\bar{x})^2$ | $f_i(x_i-\bar{x})^2$ |
|---|---|---|---|---|---|
| 60 | 1 | 60 | $-18$ | 324 | 324 |
| 75 | 4 | 300 | $-3$ | 9 | 36 |
| 80 | 3 | 240 | 2 | 4 | 12 |
| 90 | 2 | 180 | 12 | 144 | 288 |
| | $\Sigma f_i = 10$ | 780 | | | 660 |

$$\bar{x} = \frac{\Sigma x_if_i}{\Sigma f_i} = \frac{780}{10} = 78$$

$$\sigma = \sqrt{\frac{1}{n}(\Sigma f_i(x_i-\bar{x})^2}$$

$$= \sqrt{\frac{1}{10}(660)} = \sqrt{66}$$

2. **(E)**

| $x_i$ | $x_i-\bar{x}$ | $(x_i-\bar{x})^2$ |
|---|---|---|
| 33 | 1 | 1 |
| 23 | $-9$ | 81 |
| 36 | 4 | 16 |
| 29 | $-3$ | 9 |
| 36 | 4 | 16 |
| 36 | 4 | 16 |
| 33 | 1 | 1 |
| 29 | $-3$ | 9 |
| 36 | 4 | 16 |
| 29 | $-3$ | 9 |
| $\Sigma x_i = 320$ | | $\Sigma(x_i-\bar{x})^2 = 174$ |

$$\bar{x} = \frac{\Sigma x_i}{n} = \frac{320}{10} = 32$$

$$\sigma = \sqrt{\frac{1}{10}(174)} = \sqrt{17.4}$$

**3. (D)**

$$\text{Range} = x_{max} - x_{min}$$
$$= 99 - 75$$
$$= 24$$

**4. (D)**

| $x_i$ | $f_i$ | $x_i f_i$ | $x_i - \bar{x}$ | $(x_i - \bar{x})^2$ | $f_i(x_i - \bar{x})^2$ |
|---|---|---|---|---|---|
| 50 | 4 | 200 | −10 | 100 | 400 |
| 58 | 4 | 232 | −2 | 4 | 16 |
| 62 | 3 | 186 | 2 | 4 | 12 |
| 64 | 6 | 384 | 4 | 16 | 96 |
| 65 | 2 | 130 | 5 | 25 | 50 |
| 68 | 1 | 68 | 8 | 64 | 64 |
| $\Sigma f_i = n$ $= 20$ | | 1200 | | | 638 |

$$\bar{x} = \frac{\Sigma x_i f_i}{\Sigma f_i} = \frac{1200}{20} = 60$$
$$v = \sigma^2 = \frac{1}{n} \Sigma f_i(x_i - \bar{x})^2$$
$$= \frac{1}{20}(638) = 31.9$$

# Practice Problems 27.3

**1. (A)**

$$\bar{x} = \frac{52 + 96}{2} = 74$$

The 95% region is $\bar{x} \pm 2\sigma$

$$\therefore 96 = 74 + 2\sigma.$$
$$2\sigma = 22$$
$$\sigma = 11$$

**2. (A)**

$$\bar{x} \pm 2\sigma$$
$$75 \pm 2(5.8)$$
$$63.4 \text{ to } 86.6$$

The score of 90 is *not* within 2 standard deviations from the mean.

**3. (C)**

$$\bar{x} - \sigma = 45 \qquad \bar{x} = 50 \qquad \bar{x} + \sigma = 55$$

$$\longleftarrow \quad 68\% \quad \longrightarrow$$

**4. (C)**

$$\bar{x} - \sigma \qquad \bar{x} = 76 \qquad \bar{x} + \sigma$$

$$\longleftarrow \quad 68\% \quad \longrightarrow$$

$$\bar{x} - \sigma = 76 - 4 = 72 \qquad \bar{x} + \sigma = 76 + 4 = 80$$

$$\bar{x} - \sigma = 76 - 4 = 72 \qquad \bar{x} + \sigma = 76 + 4 = 80$$

# Practice Problems 28.1

**1. (D)**

| Values of $x$ | Values of $F = x^2 + 8x - 20$ | Compare |
|---|---|---|
| 0 | $F = 0^2 + 8(0) - 20 = -20$ | $-20 \neq 0$ |
| 1 | $F = 1^2 + 8(1) - 20 = -1$ | $-11 \neq 0$ |
| 2 | $F = 2^2 + 8(2) - 20 = 0$ | $0 = 0$ |

The program will print
$$2$$

This is a root of the equation.

**2. (C)**

| Values of $A$ | Values of $B$ | Values of $\dfrac{A}{B}$ | Compare |
|---|---|---|---|
| 10 | 10 | 1 | $1 \neq \frac{2}{3}$ |
| 9 | 11 | $\frac{9}{11}$ | $\frac{9}{11} \neq \frac{2}{3}$ |
| 8 | 12 | $\frac{8}{12}$ | $\frac{8}{12} = \frac{2}{3}$ |

The program will print
$$8$$

**3. (D)**

| Values of $A$ | Values of $B$ | Values of $C$ | Values of $A^2 + B^2$ |
|---|---|---|---|
| 5 | 12 | 14 | $5^2 + 12^2$ |

Compare vs. $C^2$
$$5^2 + 12^2 \neq 14^2$$

The program will print

The values of $A$, $B$ and $C$ do not represent the lengths of the sides of a right triangle.

**4. (B)**

| Values of $A$ | Values of $B$ | Values of $C$ | Values of $B^2 - 4AC$ |
|---|---|---|---|
| $-2$ | $-4$ | $-2$ | $16 - 16 = 0$ |

The program will print

Roots are real and equal.

# Cumulative Review Post-Test

**1. (B)**

$$\log r + \log r^2 = \log r \cdot r^2 = \log r^3 = 3 \log r$$

**2. (D)**

3. **(C)**

$$\text{line ratio} = \frac{\text{little } \Delta}{\text{big } \Delta} = \frac{6}{18} = \frac{1}{3}$$

$$\frac{6}{18} = \frac{4}{x}$$

$$6x = 4(18)$$

$$x = 12$$

4. **(B)**

$$d = \sqrt{(x_2 - x_1)^2 + (y_2 - y_1)^2}$$

$$= \sqrt{(5-2)^2 + (6-(-1))^2}$$

$$= \sqrt{9 + 49} = \sqrt{58}$$

5. **(C)**

$$3^4 = 9^{2x+1}$$

$$3^4 = (3^2)^{2x+1} = 3^{4x+2}$$

$$4 = 4x + 2$$

$$\tfrac{1}{2} = x$$

6. **(B)**

$$\frac{1}{1+\sqrt{2}}\left(\frac{1-\sqrt{2}}{1-\sqrt{2}}\right) = \frac{1-\sqrt{2}}{1-2}$$

$$= \frac{1-\sqrt{2}}{-1} \text{ or } \sqrt{2}-1$$

7. **(A)**

$$(n-2)(180) = 1980$$

$$n - 2 = 11$$

$$n = 13$$

8. **(A)**

$$|5 - 2x| = 7$$

| | |
|---|---|
| $-5 + 2x = 7$ | $5 - 2x = 7$ |
| $2x = 12$ | $-2x = 2$ |
| $x = 6$ | $x = -1$ |
| check | check |
| $\|5 - 2x\| = 7$ | $\|5 - 2x\| = 7$ |
| $\|5 - 2(6)\| \overset{?}{=} 7$ | $\|5 - 2(-1)\| = 7$ |
| $\|-7\| \overset{?}{=} 7$ | $\|5 + 2\| \overset{?}{=} 7$ |
| $7 = 7 \checkmark$ | $7 = 7 \checkmark$ |

9. **(D)**

$$x^{-1} + y^{-1} = \frac{1}{x} + \frac{1}{y} = \left(\frac{y}{y}\right)\frac{1}{x} + \left(\frac{x}{x}\right)\frac{1}{y}$$

$$= \frac{y + x}{xy}$$

10. **(E)**

$$x = \sin \theta \quad \frac{y}{3} = \csc \theta$$

$$\sin \theta = \frac{1}{\csc \theta} \Rightarrow x = \frac{1}{\frac{y}{3}} = \frac{3}{y}$$

$$xy = 3$$

hyperbola

11. **(E)** sin and cos are cofunction

$$(A + 40) + 10 = 90$$

$$A = 40$$

12. **(B)**

$$f(x) = x + \frac{1}{x}, \quad f\left(\frac{1}{x}\right) = \frac{1}{x} + \frac{1}{\frac{1}{x}}$$

$$= \frac{1}{x} + x = f(x)$$

13. **(E)**
If the roots are $x = 3$ and $x = 1$,
the factors are $(x - 3)$ and $(x - 1)$

$$(x - 3)(x - 1) = 0$$

$$x^2 - 4x + 3 = 0$$

14. **(C)**

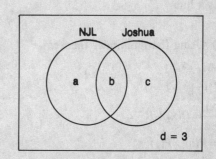

$$a+b+c+\underset{3}{\underbrace{d}}=40$$
$$\therefore a+b+c=37$$
$$a+b=25$$
$$b+c=30$$
$$\overline{\underset{37}{\underbrace{a+b+c}}+b=55}$$
$$\therefore b=18$$
$$b=18, \quad a=7, \quad c=12$$

15. **(A)**

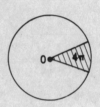

$$\text{major arc } BC = 210°$$
$$\text{minor arc } BC = 360° - 210° = 150°$$
$$m\angle A = \frac{210-150}{2} = 30$$

16. **(C)**

$y=|x-2|$ has critical value at $(x-2)=0$
$\therefore x=2, \quad y=0.$

The legs from left to right have slopes of $-1$ and $1$, respectively.

17. **(C)** $\tan x + \cot x = 1$ is *not* an identity.

18. **(B)**

$$(1+2i\sqrt{7})(1+2i\sqrt{7})$$
$$=1+2i\sqrt{7}+2i\sqrt{7}+4i^2(7)$$
$$=1+4i\sqrt{7}-28$$
$$=-27+4i\sqrt{7}$$

19. **(A)** All the terms are linear; the equation represents a line.

20. **(A)** Using the remainder theorem

$$P(x)=x^5-1$$
$$P(1)=1^5-1=0$$
$$\frac{P(x)}{x-a}=Q(x)+R$$
$$R=P(a) \quad P(1)$$
$$R=0$$

21. **(B)**

$$(x+2)(x^2-2x+4)=x^3-2x^2+4x+2x^2-4x+8$$
$$=x^3+8$$

22. **(A)**

$\dfrac{4\pi}{36\pi}=\dfrac{1}{9}$   The sector is $\frac{1}{9}$ of the circle.

$\frac{1}{9}(360°)=40°$

23. **(A)** By DeMorgan's law

$$\sim(\sim p \vee \sim q)=p \wedge q$$

24. **(D)**

$$\log 2y = \underset{.3010}{\underbrace{\log 2}} + \log y$$
$$=.3010+\log y$$

23. **(B)**

$$x^2-3y^2=13$$
$$x+3y=1 \Rightarrow x=1-3y$$

Substitute $1-3y$ for $x$

$$(1-3y)^2-3y^2=13$$
$$1-6y+9y^2-3y^2=13$$
$$6y^2-6y-12=0$$
$$y^2-y-2=0$$
$$(y-2)(y+1)=0$$

| $y=2$ | $y=-1$ |
|---|---|
| $x+3y=1$ | $x+3y=1$ |
| $x+6=1$ | $x-3=1$ |
| $x=-5$ | $x=4$ |
| | |
| $x=-5$ | $x=4$ |
| $y=2$ | $y=-1$ |
| $(-5, 2)$ | $(4, -1)$ |

26. **(E)**

Notes:     $\sin(-42°) = \sin 318°$

$\begin{array}{r} -42 \\ +360 \\ \hline 318 \end{array}$     $= -\sin 42°$

$= -\cos 58°$

27. **(B)**

$$\cos 30° = \frac{\sqrt{3}}{2}, \ \sin 30° = \tfrac{1}{2}$$

$$2(\cos 30° + i \sin 30°) = 2\left(\frac{\sqrt{3}}{2} + \frac{i}{2}\right)$$

$$= \sqrt{3} + i$$

28. **(D)** $\tfrac{1}{4} = .25$   $25\% = .25$

$$\frac{.25 + .25 + .25}{3} = .25$$

29. **(D)** 1. 7 3

$\begin{array}{r} \overline{3.0000} \\ 1 \\ \hline 27\overline{\smash)200} \\ 189 \\ \hline 343\overline{\smash)1100} \end{array}$

Notes:

$\begin{array}{ccc} 345 & 344 & 343 \\ \times 5 & \times 4 & \times 3 \\ \hline 1725 & 1376 & 1029 \end{array}$

1.7 to the nearest tenth.

30. **(C)**

$$\frac{1}{2i} = \frac{1}{2i}\left(\frac{i}{i}\right) = \frac{i}{2i^2} = \frac{i}{-2} \ \ \text{or} \ \ -\tfrac{1}{2}i$$

31. **(D)**

A straight line contains 180°.

$m\angle y = 180 - (80 + 50) = 50$

The sum of the measures of the angles of a triangle is 180°.

$$m\angle x + \underbrace{m\angle y}_{50} + 90 = 180$$

$$m\angle x = 40$$

32. **(D)** Both variables are squared and have the same sign, but different coefficients

∴ it is an ellipse.

33. **(C)**

$[r(\cos\theta + i\sin\theta)]^n$

$= r^n(\cos n\theta + i\sin n\theta)$

$= (\tfrac{1}{2})^3(\cos 3\cdot10° + i\sin 3\cdot10°)$

$= \tfrac{1}{8}(\cos 30° + i\sin 30°)$

34. **(C)**

2 points

2 points indicated by $x_s$

35. **(B)**

$VC = CT = 8$

$SD = DT = 10$

$CD = CT + TD = 18$

**36. (D)**

$$\cos 2A = \cos^2 A - \sin^2 A$$
$$= (\cos A + \sin A)(\cos A - \sin A)$$

**37. (D)**

$$\text{the period} = \frac{360°}{\text{freq}} = \frac{360°}{\frac{1}{2}} = 720°$$

**38. (C)**

$$S_\infty = \frac{a}{1-r}$$
$$= \frac{2}{1-\frac{2}{3}} = \frac{2}{\frac{1}{3}} = 6 \quad \text{Note: } r = \frac{\frac{4}{3}}{2} = \frac{4}{6} = \frac{2}{3}$$

**39. (C)**

$$t + u = 11$$
$$(10u + t) - 34 = 2(10t + u)$$
$$10u + t - 34 = 20t + 2u$$

Combine like terms and rearrange

$$8u - 19t = 34$$

Multiply $t + u = 11$ by $-8$ and add the two equations

$$\begin{aligned} 8u - 19t &= 34 \\ -8u - 8t &= -88 \\ \hline -27t &= -54 \\ t &= 2 \end{aligned}$$

Since $t + u = 11$ and $t = 2$

$$u = 9$$

The number is 29

**40. (B)**

$$x_1 t_1 = x_2 t_2$$
$$6(2) = x(4)$$
$$3 = x$$

**41. (C)**

$$V = \pi r^2 h = 90\pi$$
$$\pi r^2 (10) = 90\pi$$
$$r = 3$$

$$\begin{aligned} \text{L.A.} &= 2\pi rh \\ &= 2\pi(3)(10) \\ &= 60\pi \end{aligned}$$

**42. (C)**

$$P(x) = x^4 - 7x^3 + 9x^2 + 13x - 4$$

$$\frac{p}{q} = \frac{\pm 4, \ \pm 2, \ \pm 1}{\pm 1}$$

$P(x)$ has 3 sign changes $\therefore$ either 3 or 1 positive root.
$P(-x) = x^4 + 7x^3 + 9x^2 - 13x - 4$ has 1 sign change $\therefore$ only 1 negative root.

$$P(1) = 1 - 7 + 9 + 13 - 4 \neq 0$$
$$P(2) = 16 - 56 + 36 + 26 - 4 \neq 0$$
$$P(3) = 81 - 189 + 81 + 39 - 4 \neq 0$$
$$P(4) = 256 - 448 + 144 + 52 - 4 = 0$$

$x = 4$ is a root, $x - 4$ is a factor

$$\begin{array}{r|rrrrr} 4 & 1 & -7 & 9 & 13 & -4 \\ & & 4 & -12 & -12 & 4 \\ \hline & 1 & -3 & -3 & 1 & \underline{|0|} \end{array}$$

$$P(x) = x^4 - 7x^3 + 9x^2 + 13x - 4$$
$$= (x-4)(x^3 - 3x^2 - 3x + 1)$$

using the depressed equation

$$D_1(x) = x^3 - 3x^2 - 3x + 1$$

try to locate another root

$$D_1(-1) = -1 - 3 + 3 + 1 = 0$$

$\therefore x = -1$ is a root, $x + 1$ is a factor

$$\begin{array}{r|rrrr} -1 & 1 & -3 & -3 & 1 \\ & & -1 & 4 & -1 \\ \hline & 1 & -4 & 1 & \underline{|0|} \end{array}$$

$$P(x) = (x-4)(x^3 - 3x^2 - 3x + 1)$$
$$= (x-4)(x+1)(\underbrace{x^2 - 4x + 1})$$

$$x = \frac{-b \pm \sqrt{b^2 - 4ac}}{2a}$$

$$= \frac{4 \pm \sqrt{(-4)^2 - 4(1)(1)}}{2(1)}$$

$$= \frac{4 \pm \sqrt{12}}{2} = 2 \pm \sqrt{3}$$

2 rational, 2 irrational roots

**43. (B)**

$$\text{radians} \times \frac{180}{\pi} = \text{degrees}$$

$$1 \cdot \frac{180}{\pi} = \frac{180}{\pi} \text{ degrees}$$

**44. (E)**

$$\sum_{k=3}^{5}(k^2-1)=[3^2-1]+[4^2-1]+[5^2-1]$$
$$=8+15+24=47$$

**45. (C)**

$$_5C_2(\tfrac{1}{5})^2(\tfrac{4}{5})^3$$

**46. (D)** The negation of $\forall_x: p$ is $\exists_x: \sim p$.

Some squares are not parallelograms.

**47. (D)**

$$\frac{2P+3Q+x}{3}=A$$

$$x=3A-2P-3Q$$

**48. (E)**

| Value of $A$ | Value of $B$ | Value of $F=A^2-B$ |
|---|---|---|
| 1 | 1 | $1^2-1=0$ |

The program will print
2

**49. (A)**

$$\ln x+\ln(x+2)=\ln 3$$
$$\ln x(x+2)=\ln 3$$
$$x(x+2)=3$$
$$x^2+2x-3=0$$
$$(x+3)(x-1)=0$$
$$x=-3 \mid x=1$$
Reject
$$\therefore x=1$$

**50. (D)** All of the numbers are rational.

# Appendix

## Tables of Natural Trigonometric Functions
(For use with 9th and 10th Year Mathematics Regents Examinations)

| Angle | Sine | Cosine | Tangent | Angle | Sine | Cosine | Tangent |
|-------|------|--------|---------|-------|------|--------|---------|
| 1° | .0175 | .9998 | .0175 | 46° | .7193 | .6947 | 1.0355 |
| 2° | .0349 | .9994 | .0349 | 47° | .7314 | .6820 | 1.0724 |
| 3° | .0523 | .9986 | .0524 | 48° | .7431 | .6691 | 1.1106 |
| 4° | .0698 | .9976 | .0699 | 49° | .7547 | .6561 | 1.1504 |
| 5° | .0872 | .9962 | .0875 | 50° | .7660 | .6428 | 1.1918 |
| 6° | .1045 | .9945 | .1051 | 51° | .7771 | .6293 | 1.2349 |
| 7° | .1219 | .9925 | .1228 | 52° | .7880 | .6157 | 1.2799 |
| 8° | .1392 | .9903 | .1405 | 53° | .7986 | .6018 | 1.3270 |
| 9° | .1564 | .9877 | .1584 | 54° | .8090 | .5878 | 1.3764 |
| 10° | .1736 | .9848 | .1763 | 55° | .8192 | .5736 | 1.4281 |
| 11° | .1908 | .9816 | .1944 | 56° | .8290 | .5592 | 1.4826 |
| 12° | .2079 | .9781 | .2126 | 57° | .8387 | .5446 | 1.5399 |
| 13° | .2250 | .9744 | .2309 | 58° | .8480 | .5299 | 1.6003 |
| 14° | .2419 | .9703 | .2493 | 59° | .8572 | .5150 | 1.6643 |
| 15° | .2588 | .9659 | .2679 | 60° | .8660 | .5000 | 1.7321 |
| 16° | .2756 | .9613 | .2867 | 61° | .8746 | .4848 | 1.8040 |
| 17° | .2924 | .9563 | .3057 | 62° | .8829 | .4695 | 1.8807 |
| 18° | .3090 | .9511 | .3249 | 63° | .8910 | .4540 | 1.9626 |
| 19° | .3256 | .9455 | .3443 | 64° | .8988 | .4384 | 2.0503 |
| 20° | .3420 | .9397 | .3640 | 65° | .9063 | .4226 | 2.1445 |
| 21° | .3584 | .9336 | .3839 | 66° | .9135 | .4067 | 2.2460 |
| 22° | .3746 | .9272 | .4040 | 67° | .9205 | .3907 | 2.3559 |
| 23° | .3907 | .9205 | .4245 | 68° | .9272 | .3746 | 2.4751 |
| 24° | .4067 | .9135 | .4452 | 69° | .9336 | .3584 | 2.6051 |
| 25° | .4226 | .9063 | .4663 | 70° | .9397 | .3420 | 2.7475 |
| 26° | .4384 | .8988 | .4877 | 71° | .9455 | .3256 | 2.9042 |
| 27° | .4540 | .8910 | .5095 | 72° | .9511 | .3090 | 3.0777 |
| 28° | .4695 | .8829 | .5317 | 73° | .9563 | .2924 | 3.2709 |
| 29° | .4848 | .8746 | .5543 | 74° | .9613 | .2756 | 3.4874 |
| 30° | .5000 | .8660 | .5774 | 75° | .9659 | .2588 | 3.7321 |
| 31° | .5150 | .8572 | .6009 | 76° | .9703 | .2419 | 4.0108 |
| 32° | .5299 | .8480 | .6249 | 77° | .9744 | .2250 | 4.3315 |
| 33° | .5446 | .8387 | .6494 | 78° | .9781 | .2079 | 4.7046 |
| 34° | .5592 | .8290 | .6745 | 79° | .9816 | .1908 | 5.1446 |
| 35° | .5736 | .8192 | .7002 | 80° | .9848 | .1736 | 5.6713 |
| 36° | .5878 | .8090 | .7265 | 81° | .9877 | .1564 | 6.3138 |
| 37° | .6018 | .7986 | .7536 | 82° | .9903 | .1392 | 7.1154 |
| 38° | .6157 | .7880 | .7813 | 83° | .9925 | .1219 | 8.1443 |
| 39° | .6293 | .7771 | .8098 | 84° | .9945 | .1045 | 9.5144 |
| 40° | .6428 | .7660 | .8391 | 85° | .9962 | .0872 | 11.4301 |
| 41° | .6561 | .7547 | .8693 | 86° | .9976 | .0698 | 14.3007 |
| 42° | .6691 | .7431 | .9004 | 87° | .9986 | .0523 | 19.0811 |
| 43° | .6820 | .7314 | .9325 | 88° | .9994 | .0349 | 28.6363 |
| 44° | .6947 | .7193 | .9657 | 89° | .9998 | .0175 | 57.2900 |
| 45° | .7071 | .7071 | 1.0000 | 90° | 1.0000 | .0000 | |

## Table A: Common Logarithms of Numbers*

| N | 0 | 1 | 2 | 3 | 4 | 5 | 6 | 7 | 8 | 9 |
|---|---|---|---|---|---|---|---|---|---|---|
| 10 | 0000 | 0043 | 0086 | 0128 | 0170 | 0212 | 0253 | 0294 | 0334 | 0374 |
| 11 | 0414 | 0453 | 0492 | 0531 | 0569 | 0607 | 0645 | 0682 | 0719 | 0755 |
| 12 | 0792 | 0828 | 0864 | 0899 | 0934 | 0969 | 1004 | 1038 | 1072 | 1106 |
| 13 | 1139 | 1173 | 1206 | 1239 | 1271 | 1303 | 1335 | 1367 | 1399 | 1430 |
| 14 | 1461 | 1492 | 1523 | 1553 | 1584 | 1614 | 1644 | 1673 | 1703 | 1732 |
| 15 | 1761 | 1790 | 1818 | 1847 | 1875 | 1903 | 1931 | 1959 | 1987 | 2014 |
| 16 | 2041 | 2068 | 2095 | 2122 | 2148 | 2175 | 2201 | 2227 | 2253 | 2279 |
| 17 | 2304 | 2330 | 2355 | 2380 | 2405 | 2430 | 2455 | 2480 | 2504 | 2529 |
| 18 | 2553 | 2577 | 2601 | 2625 | 2648 | 2672 | 2695 | 2718 | 2742 | 2765 |
| 19 | 2788 | 2810 | 2833 | 2856 | 2878 | 2900 | 2923 | 2945 | 2967 | 2989 |
| 20 | 3010 | 3032 | 3054 | 3075 | 3096 | 3118 | 3139 | 3160 | 3181 | 3201 |
| 21 | 3222 | 3243 | 3263 | 3284 | 3304 | 3324 | 3345 | 3365 | 3385 | 3404 |
| 22 | 3424 | 3444 | 3464 | 3483 | 3502 | 3522 | 3541 | 3560 | 3579 | 3598 |
| 23 | 3617 | 3636 | 3655 | 3674 | 3692 | 3711 | 3729 | 3747 | 3766 | 3784 |
| 24 | 3802 | 3820 | 3838 | 3856 | 3874 | 3892 | 3909 | 3927 | 3945 | 3962 |
| 25 | 3979 | 3997 | 4014 | 4031 | 4048 | 4065 | 4082 | 4099 | 4116 | 4133 |
| 26 | 4150 | 4166 | 4183 | 4200 | 4216 | 4232 | 4249 | 4265 | 4281 | 4298 |
| 27 | 4314 | 4330 | 4346 | 4362 | 4378 | 4393 | 4409 | 4425 | 4440 | 4456 |
| 28 | 4472 | 4487 | 4502 | 4518 | 4533 | 4548 | 4564 | 4579 | 4594 | 4609 |
| 29 | 4624 | 4639 | 4654 | 4669 | 4683 | 4698 | 4713 | 4728 | 4742 | 4757 |
| 30 | 4771 | 4786 | 4800 | 4814 | 4829 | 4843 | 4857 | 4871 | 4886 | 4900 |
| 31 | 4914 | 4928 | 4942 | 4955 | 4969 | 4983 | 4997 | 5011 | 5024 | 5038 |
| 32 | 5051 | 5065 | 5079 | 5092 | 5105 | 5119 | 5132 | 5145 | 5159 | 5172 |
| 33 | 5185 | 5198 | 5211 | 5224 | 5237 | 5250 | 5263 | 5276 | 5289 | 5302 |
| 34 | 5315 | 5328 | 5340 | 5353 | 5366 | 5378 | 5391 | 5403 | 5416 | 5428 |
| 35 | 5441 | 5453 | 5465 | 5478 | 5490 | 5502 | 5514 | 5527 | 5539 | 5551 |
| 36 | 5563 | 5575 | 5587 | 5599 | 5611 | 5623 | 5635 | 5647 | 5658 | 5670 |
| 37 | 5682 | 5694 | 5705 | 5717 | 5729 | 5740 | 5752 | 5763 | 5775 | 5786 |
| 38 | 5798 | 5809 | 5821 | 5832 | 5843 | 5855 | 5866 | 5877 | 5888 | 5899 |
| 39 | 5911 | 5922 | 5933 | 5944 | 5955 | 5966 | 5977 | 5988 | 5999 | 6010 |
| 40 | 6021 | 6031 | 6042 | 6053 | 6064 | 6075 | 6085 | 6096 | 6107 | 6117 |
| 41 | 6128 | 6138 | 6149 | 6160 | 6170 | 6180 | 6191 | 6201 | 6212 | 6222 |
| 42 | 6232 | 6243 | 6253 | 6263 | 6274 | 6284 | 6294 | 6304 | 6314 | 6325 |
| 43 | 6335 | 6345 | 6355 | 6365 | 6375 | 6385 | 6395 | 6405 | 6415 | 6425 |
| 44 | 6435 | 6444 | 6454 | 6464 | 6474 | 6484 | 6493 | 6503 | 6513 | 6522 |
| 45 | 6532 | 6542 | 6551 | 6561 | 6571 | 6580 | 6590 | 6599 | 6609 | 6618 |
| 46 | 6628 | 6637 | 6646 | 6656 | 6665 | 6675 | 6684 | 6693 | 6702 | 6712 |
| 47 | 6721 | 6730 | 6739 | 6749 | 6758 | 6767 | 6776 | 6785 | 6794 | 6803 |
| 48 | 6812 | 6821 | 6830 | 6839 | 6848 | 6857 | 6866 | 6875 | 6884 | 6893 |
| 49 | 6902 | 6911 | 6920 | 6928 | 6937 | 6946 | 6955 | 6964 | 6972 | 6981 |
| 50 | 6990 | 6998 | 7007 | 7016 | 7024 | 7033 | 7042 | 7050 | 7059 | 7067 |
| 51 | 7076 | 7084 | 7093 | 7101 | 7110 | 7118 | 7126 | 7135 | 7143 | 7152 |
| 52 | 7160 | 7168 | 7177 | 7185 | 7193 | 7202 | 7210 | 7218 | 7226 | 7235 |
| 53 | 7243 | 7251 | 7259 | 7267 | 7275 | 7284 | 7292 | 7300 | 7308 | 7316 |
| 54 | 7324 | 7332 | 7340 | 7348 | 7356 | 7364 | 7372 | 7380 | 7388 | 7396 |
| N | 0 | 1 | 2 | 3 | 4 | 5 | 6 | 7 | 8 | 9 |

* This table gives the mantissas of numbers with the decimal point omitted in each case. Characteristics are determined from the numbers by inspection.

## Table A: Common Logarithms of Numbers*

| N | 0 | 1 | 2 | 3 | 4 | 5 | 6 | 7 | 8 | 9 |
|---|---|---|---|---|---|---|---|---|---|---|
| 55 | 7404 | 7412 | 7419 | 7427 | 7435 | 7443 | 7451 | 7459 | 7466 | 7474 |
| 56 | 7482 | 7490 | 7497 | 7505 | 7513 | 7520 | 7528 | 7536 | 7543 | 7551 |
| 57 | 7559 | 7566 | 7574 | 7582 | 7589 | 7597 | 7604 | 7612 | 7619 | 7627 |
| 58 | 7634 | 7642 | 7649 | 7657 | 7664 | 7672 | 7679 | 7686 | 7694 | 7701 |
| 59 | 7709 | 7716 | 7723 | 7731 | 7738 | 7745 | 7752 | 7760 | 7767 | 7774 |
| 60 | 7782 | 7789 | 7796 | 7803 | 7810 | 7818 | 7825 | 7832 | 7839 | 7846 |
| 61 | 7853 | 7860 | 7868 | 7875 | 7882 | 7889 | 7896 | 7903 | 7910 | 7917 |
| 62 | 7924 | 7931 | 7938 | 7945 | 7952 | 7959 | 7966 | 7973 | 7980 | 7987 |
| 63 | 7993 | 8000 | 8007 | 8014 | 8021 | 8028 | 8035 | 8041 | 8048 | 8055 |
| 64 | 8062 | 8069 | 8075 | 8082 | 8089 | 8096 | 8102 | 8109 | 8116 | 8122 |
| 65 | 8129 | 8136 | 8142 | 8149 | 8156 | 8162 | 8169 | 8176 | 8182 | 8189 |
| 66 | 8195 | 8202 | 8209 | 8215 | 8222 | 8228 | 8235 | 8241 | 8248 | 8254 |
| 67 | 8261 | 8267 | 8274 | 8280 | 8287 | 8293 | 8299 | 8306 | 8312 | 8319 |
| 68 | 8325 | 8331 | 8338 | 8344 | 8351 | 8357 | 8363 | 8370 | 8376 | 8382 |
| 69 | 8388 | 8395 | 8401 | 8407 | 8414 | 8420 | 8426 | 8432 | 8439 | 8445 |
| 70 | 8451 | 8457 | 8463 | 8470 | 8476 | 8482 | 8488 | 8494 | 8500 | 8506 |
| 71 | 8513 | 8519 | 8525 | 8531 | 8537 | 8543 | 8549 | 8555 | 8561 | 8567 |
| 72 | 8573 | 8579 | 8585 | 8591 | 8597 | 8603 | 8609 | 8615 | 8621 | 8627 |
| 73 | 8633 | 8639 | 8645 | 8651 | 8657 | 8663 | 8669 | 8675 | 8681 | 8686 |
| 74 | 8692 | 8698 | 8704 | 8710 | 8716 | 8722 | 8727 | 8733 | 8739 | 8745 |
| 75 | 8751 | 8756 | 8762 | 8768 | 8774 | 8779 | 8785 | 8791 | 8797 | 8802 |
| 76 | 8808 | 8814 | 8820 | 8825 | 8831 | 8837 | 8842 | 8848 | 8854 | 8859 |
| 77 | 8865 | 8871 | 8876 | 8882 | 8887 | 8893 | 8899 | 8904 | 8910 | 8915 |
| 78 | 8921 | 8927 | 8932 | 8938 | 8943 | 8949 | 8954 | 8960 | 8965 | 8971 |
| 79 | 8976 | 8982 | 8987 | 8993 | 8998 | 9004 | 9009 | 9015 | 9020 | 9025 |
| 80 | 9031 | 9036 | 9042 | 9047 | 9053 | 9058 | 9063 | 9069 | 9074 | 9079 |
| 81 | 9085 | 9090 | 9096 | 9101 | 9106 | 9112 | 9117 | 9122 | 9128 | 9133 |
| 82 | 9138 | 9143 | 9149 | 9154 | 9159 | 9165 | 9170 | 9175 | 9180 | 9186 |
| 83 | 9191 | 9196 | 9201 | 9206 | 9212 | 9217 | 9222 | 9227 | 9232 | 9238 |
| 84 | 9243 | 9248 | 9253 | 9258 | 9263 | 9269 | 9274 | 9279 | 9284 | 9289 |
| 85 | 9294 | 9299 | 9304 | 9309 | 9315 | 9320 | 9325 | 9330 | 9335 | 9340 |
| 86 | 9345 | 9350 | 9355 | 9360 | 9365 | 9370 | 9375 | 9380 | 9385 | 9390 |
| 87 | 9395 | 9400 | 9405 | 9410 | 9415 | 9420 | 9425 | 9430 | 9435 | 9440 |
| 88 | 9445 | 9450 | 9455 | 9460 | 9465 | 9469 | 9474 | 9479 | 9484 | 9489 |
| 89 | 9494 | 9499 | 9504 | 9509 | 9513 | 9518 | 9523 | 9528 | 9533 | 9538 |
| 90 | 9542 | 9547 | 9552 | 9557 | 9562 | 9566 | 9571 | 9576 | 9581 | 9586 |
| 91 | 9590 | 9595 | 9600 | 9605 | 9609 | 9614 | 9619 | 9624 | 9628 | 9633 |
| 92 | 9638 | 9643 | 9647 | 9652 | 9657 | 9661 | 9666 | 9671 | 9675 | 9680 |
| 93 | 9685 | 9689 | 9694 | 9699 | 9703 | 9708 | 9713 | 9717 | 9722 | 9727 |
| 94 | 9731 | 9736 | 9741 | 9745 | 9750 | 9754 | 9759 | 9763 | 9768 | 9773 |
| 95 | 9777 | 9782 | 9786 | 9791 | 9795 | 9800 | 9805 | 9809 | 9814 | 9818 |
| 96 | 9823 | 9827 | 9832 | 9836 | 9841 | 9845 | 9850 | 9854 | 9859 | 9863 |
| 97 | 9868 | 9872 | 9877 | 9881 | 9886 | 9890 | 9894 | 9899 | 9903 | 9908 |
| 98 | 9912 | 9917 | 9921 | 9926 | 9930 | 9934 | 9939 | 9943 | 9948 | 9952 |
| 99 | 9956 | 9961 | 9965 | 9969 | 9974 | 9978 | 9983 | 9987 | 9991 | 9996 |
| N | 0 | 1 | 2 | 3 | 4 | 5 | 6 | 7 | 8 | 9 |

* This table gives the mantissas of numbers with the decimal point omitted in each case. Characteristics are determined from the numbers by inspection.

## Table B: Values of Trigonometric Functions

| Angle | Sin | Cos | Tan | Cot | |
|---|---|---|---|---|---|
| 0° 00' | .0000 | 1.0000 | .0000 | — | 90° 00' |
| 10 | .0029 | 1.0000 | .0029 | 343.77 | 50 |
| 20 | .0058 | 1.0000 | .0058 | 171.89 | 40 |
| 30 | .0087 | 1.0000 | .0087 | 114.59 | 30 |
| 40 | .0116 | .9999 | .0116 | 85.940 | 20 |
| 50 | .0145 | .9999 | .0145 | 68.750 | 10 |
| 1° 00' | .0175 | .9998 | .0175 | 57.290 | 89° 00' |
| 10 | .0204 | .9998 | .0204 | 49.104 | 50 |
| 20 | .0233 | .9997 | .0233 | 42.964 | 40 |
| 30 | .0262 | .9997 | .0262 | 38.188 | 30 |
| 40 | .0291 | .9996 | .0291 | 34.368 | 20 |
| 50 | .0320 | .9995 | .0320 | 31.242 | 10 |
| 2° 00' | .0349 | .9994 | .0349 | 28.636 | 88° 00' |
| 10 | .0378 | .9993 | .0378 | 26.432 | 50 |
| 20 | .0407 | .9992 | .0407 | 24.542 | 40 |
| 30 | .0436 | .9990 | .0437 | 22.904 | 30 |
| 40 | .0465 | .9989 | .0466 | 21.470 | 20 |
| 50 | .0494 | .9988 | .0495 | 20.206 | 10 |
| 3° 00' | .0523 | .9986 | .0524 | 19.081 | 87° 00' |
| 10 | .0552 | .9985 | .0553 | 18.075 | 50 |
| 20 | .0581 | .9983 | .0582 | 17.169 | 40 |
| 30 | .0610 | .9981 | .0612 | 16.350 | 30 |
| 40 | .0640 | .9980 | .0641 | 15.605 | 20 |
| 50 | .0669 | .9978 | .0670 | 14.924 | 10 |
| 4° 00' | .0698 | .9976 | .0699 | 14.301 | 86° 00' |
| 10 | .0727 | .9974 | .0729 | 13.727 | 50 |
| 20 | .0756 | .9971 | .0758 | 13.197 | 40 |
| 30 | .0785 | .9969 | .0787 | 12.706 | 30 |
| 40 | .0814 | .9967 | .0816 | 12.251 | 20 |
| 50 | .0843 | .9964 | .0846 | 11.826 | 10 |
| 5° 00' | .0872 | .9962 | .0875 | 11.430 | 85° 00' |
| 10 | .0901 | .9959 | .0904 | 11.059 | 50 |
| 20 | .0929 | .9957 | .0934 | 10.712 | 40 |
| 30 | .0958 | .9954 | .0963 | 10.385 | 30 |
| 40 | .0987 | .9951 | .0992 | 10.078 | 20 |
| 50 | .1016 | .9948 | .1022 | 9.7882 | 10 |
| 6° 00' | .1045 | .9945 | .1051 | 9.5144 | 84° 00' |
| 10 | .1074 | .9942 | .1080 | 9.2553 | 50 |
| 20 | .1103 | .9939 | .1110 | 9.0098 | 40 |
| 30 | .1132 | .9936 | .1139 | 8.7769 | 30 |
| 40 | .1161 | .9932 | .1169 | 8.5555 | 20 |
| 50 | .1190 | .9929 | .1198 | 8.3450 | 10 |
| 7° 00' | .1219 | .9925 | .1228 | 8.1443 | 83° 00' |
| 10 | .1248 | .9922 | .1257 | 7.9530 | 50 |
| 20 | .1276 | .9918 | .1287 | 7.7704 | 40 |
| 30 | .1305 | .9914 | .1317 | 7.5958 | 30 |
| 40 | .1334 | .9911 | .1346 | 7.4287 | 20 |
| 50 | .1363 | .9907 | .1376 | 7.2687 | 10 |
| 8° 00' | .1392 | .9903 | .1405 | 7.1154 | 82° 00' |
| 10 | .1421 | .9899 | .1435 | 6.9682 | 50 |
| 20 | .1449 | .9894 | .1465 | 6.8269 | 40 |
| 30 | .1478 | .9890 | .1495 | 6.6912 | 30 |
| 40 | .1507 | .9886 | .1524 | 6.5606 | 20 |
| 50 | .1536 | .9881 | .1554 | 6.4348 | 10 |
| 9° 00' | .1564 | .9877 | .1584 | 6.3138 | 81° 00' |
| 10 | .1593 | .9872 | .1614 | 6.1970 | 50 |
| 20 | .1622 | .9868 | .1644 | 6.0844 | 40 |
| 30 | .1650 | .9863 | .1673 | 5.9758 | 30 |
| 40 | .1679 | .9858 | .1703 | 5.8708 | 20 |
| 50 | .1708 | .9853 | .1733 | 5.7694 | 10 |
| 10° 00' | .1736 | .9848 | .1763 | 5.6713 | 80° 00' |
| 10 | .1765 | .9843 | .1793 | 5.5764 | 50 |
| 20 | .1794 | .9838 | .1823 | 5.4845 | 40 |
| 30 | .1822 | .9833 | .1853 | 5.3955 | 30 |
| 40 | .1851 | .9827 | .1883 | 5.3093 | 20 |
| 50 | .1880 | .9822 | .1914 | 5.2257 | 10 |
| 11° 00' | .1908 | .9816 | .1944 | 5.1446 | 79° 00' |
| 10 | .1937 | .9811 | .1974 | 5.0658 | 50 |
| 20 | .1965 | .9805 | .2004 | 4.9894 | 40 |
| 30 | .1994 | .9799 | .2035 | 4.9152 | 30 |
| 40 | .2022 | .9793 | .2065 | 4.8430 | 20 |
| 50 | .2051 | .9787 | .2095 | 4.7729 | 10 |
| 12° 00' | .2079 | .9781 | .2126 | 4.7046 | 78° 00' |
| | Cos | Sin | Cot | Tan | Angle |

| Angle | Sin | Cos | Tan | Cot | |
|---|---|---|---|---|---|
| 12° 00' | .2079 | .9781 | .2126 | 4.7046 | 78° 00' |
| 10 | .2108 | .9775 | .2156 | 4.6382 | 50 |
| 20 | .2136 | .9769 | .2186 | 4.5736 | 40 |
| 30 | .2164 | .9763 | .2217 | 4.5107 | 30 |
| 40 | .2193 | .9757 | .2247 | 4.4494 | 20 |
| 50 | .2221 | .9750 | .2278 | 4.3897 | 10 |
| 13° 00' | .2250 | .9744 | .2309 | 4.3315 | 77° 00' |
| 10 | .2278 | .9737 | .2339 | 4.2747 | 50 |
| 20 | .2306 | .9730 | .2370 | 4.2193 | 40 |
| 30 | .2334 | .9724 | .2401 | 4.1653 | 30 |
| 40 | .2363 | .9717 | .2432 | 4.1126 | 20 |
| 50 | .2391 | .9710 | .2462 | 4.0611 | 10 |
| 14° 00' | .2419 | .9703 | .2493 | 4.0108 | 76° 00' |
| 10 | .2447 | .9696 | .2524 | 3.9617 | 50 |
| 20 | .2476 | .9689 | .2555 | 3.9136 | 40 |
| 30 | .2504 | .9681 | .2586 | 3.8667 | 30 |
| 40 | .2532 | .9674 | .2617 | 3.8208 | 20 |
| 50 | .2560 | .9667 | .2648 | 3.7760 | 10 |
| 15° 00' | .2588 | .9659 | .2679 | 3.7321 | 75° 00' |
| 10 | .2616 | .9652 | .2711 | 3.6891 | 50 |
| 20 | .2644 | .9644 | .2742 | 3.6470 | 40 |
| 30 | .2672 | .9636 | .2773 | 3.6059 | 30 |
| 40 | .2700 | .9628 | .2805 | 3.5656 | 20 |
| 50 | .2728 | .9621 | .2836 | 3.5261 | 10 |
| 16° 00' | .2756 | .9613 | .2867 | 3.4874 | 74° 00' |
| 10 | .2784 | .9605 | .2899 | 3.4495 | 50 |
| 20 | .2812 | .9596 | .2931 | 3.4124 | 40 |
| 30 | .2840 | .9588 | .2962 | 3.3759 | 30 |
| 40 | .2868 | .9580 | .2994 | 3.3402 | 20 |
| 50 | .2896 | .9572 | .3026 | 3.3052 | 10 |
| 17° 00' | .2924 | .9563 | .3057 | 3.2709 | 73° 00' |
| 10 | .2952 | .9555 | .3089 | 3.2371 | 50 |
| 20 | .2979 | .9546 | .3121 | 3.2041 | 40 |
| 30 | .3007 | .9537 | .3153 | 3.1716 | 30 |
| 40 | .3035 | .9528 | .3185 | 3.1397 | 20 |
| 50 | .3062 | .9520 | .3217 | 3.1084 | 10 |
| 18° 00' | .3090 | .9511 | .3249 | 3.0777 | 72° 00' |
| 10 | .3118 | .9502 | .3281 | 3.0475 | 50 |
| 20 | .3145 | .9492 | .3314 | 3.0178 | 40 |
| 30 | .3173 | .9483 | .3346 | 2.9887 | 30 |
| 40 | .3201 | .9474 | .3378 | 2.9600 | 20 |
| 50 | .3228 | .9465 | .3411 | 2.9319 | 10 |
| 19° 00' | .3256 | .9455 | .3443 | 2.9042 | 71° 00' |
| 10 | .3283 | .9446 | .3476 | 2.8770 | 50 |
| 20 | .3311 | .9436 | .3508 | 2.8502 | 40 |
| 30 | .3338 | .9426 | .3541 | 2.8239 | 30 |
| 40 | .3365 | .9417 | .3574 | 2.7980 | 20 |
| 50 | .3393 | .9407 | .3607 | 2.7725 | 10 |
| 20° 00' | .3420 | .9397 | .3640 | 2.7475 | 70° 00' |
| 10 | .3448 | .9387 | .3673 | 2.7228 | 50 |
| 20 | .3475 | .9377 | .3706 | 2.6985 | 40 |
| 30 | .3502 | .9367 | .3739 | 2.6746 | 30 |
| 40 | .3529 | .9356 | .3772 | 2.6511 | 20 |
| 50 | .3557 | .9346 | .3805 | 2.6279 | 10 |
| 21° 00' | .3584 | .9336 | .3839 | 2.6051 | 69° 00' |
| 10 | .3611 | .9325 | .3872 | 2.5826 | 50 |
| 20 | .3638 | .9315 | .3906 | 2.5605 | 40 |
| 30 | .3665 | .9304 | .3939 | 2.5386 | 30 |
| 40 | .3692 | .9293 | .3973 | 2.5172 | 20 |
| 50 | .3719 | .9283 | .4006 | 2.4960 | 10 |
| 22° 00' | .3746 | .9272 | .4040 | 2.4751 | 68° 00' |
| 10 | .3773 | .9261 | .4074 | 2.4545 | 50 |
| 20 | .3800 | .9250 | .4108 | 2.4342 | 40 |
| 30 | .3827 | .9239 | .4142 | 2.4142 | 30 |
| 40 | .3854 | .9228 | .4176 | 2.3945 | 20 |
| 50 | .3881 | .9216 | .4210 | 2.3750 | 10 |
| 23° 00' | .3907 | .9205 | .4245 | 2.3559 | 67° 00' |
| 10 | .3934 | .9194 | .4279 | 2.3369 | 50 |
| 20 | .3961 | .9182 | .4314 | 2.3183 | 40 |
| 30 | .3987 | .9171 | .4348 | 2.2998 | 30 |
| 40 | .4014 | .9159 | .4383 | 2.2817 | 20 |
| 50 | .4041 | .9147 | .4417 | 2.2637 | 10 |
| 24° 00' | .4067 | .9135 | .4452 | 2.2460 | 66° 00' |
| | Cos | Sin | Cot | Tan | Angle |

## Table B: Values of Trigonometric Functions

| Angle | Sin | Cos | Tan | Cot | |
|---|---|---|---|---|---|
| 24° 00′ | .4067 | .9135 | .4452 | 2.2460 | 66° 00′ |
| 10 | .4094 | .9124 | .4487 | 2.2286 | 50 |
| 20 | .4120 | .9112 | .4522 | 2.2113 | 40 |
| 30 | .4147 | .9100 | .4557 | 2.1943 | 30 |
| 40 | .4173 | .9088 | .4592 | 2.1775 | 20 |
| 50 | .4200 | .9075 | .4628 | 2.1609 | 10 |
| 25° 00′ | .4226 | .9063 | .4663 | 2.1445 | 65° 00′ |
| 10 | .4253 | .9051 | .4699 | 2.1283 | 50 |
| 20 | .4279 | .9038 | .4734 | 2.1123 | 40 |
| 30 | .4305 | .9026 | .4770 | 2.0965 | 30 |
| 40 | .4331 | .9013 | .4806 | 2.0809 | 20 |
| 50 | .4358 | .9001 | .4841 | 2.0655 | 10 |
| 26° 00′ | .4384 | .8988 | .4877 | 2.0503 | 64° 00′ |
| 10 | .4410 | .8975 | .4913 | 2.0353 | 50 |
| 20 | .4436 | .8962 | .4950 | 2.0204 | 40 |
| 30 | .4462 | .8949 | .4986 | 2.0057 | 30 |
| 40 | .4488 | .8936 | .5022 | 1.9912 | 20 |
| 50 | .4514 | .8923 | .5059 | 1.9768 | 10 |
| 27° 00′ | .4540 | .8910 | .5095 | 1.9626 | 63° 00′ |
| 10 | .4566 | .8897 | .5132 | 1.9486 | 50 |
| 20 | .4592 | .8884 | .5169 | 1.9347 | 40 |
| 30 | .4617 | .8870 | .5206 | 1.9210 | 30 |
| 40 | .4643 | .8857 | .5243 | 1.9074 | 20 |
| 50 | .4669 | .8843 | .5280 | 1.8940 | 10 |
| 28° 00′ | .4695 | .8829 | .5317 | 1.8807 | 62° 00′ |
| 10 | .4720 | .8816 | .5354 | 1.8676 | 50 |
| 20 | .4746 | .8802 | .5392 | 1.8546 | 40 |
| 30 | .4772 | .8788 | .5430 | 1.8418 | 30 |
| 40 | .4797 | .8774 | .5467 | 1.8291 | 20 |
| 50 | .4823 | .8760 | .5505 | 1.8165 | 10 |
| 29° 00′ | .4848 | .8746 | .5543 | 1.8040 | 61° 00′ |
| 10 | .4874 | .8732 | .5581 | 1.7917 | 50 |
| 20 | .4899 | .8718 | .5619 | 1.7796 | 40 |
| 30 | .4924 | .8704 | .5658 | 1.7675 | 30 |
| 40 | .4950 | .8689 | .5696 | 1.7556 | 20 |
| 50 | .4975 | .8675 | .5735 | 1.7437 | 10 |
| 30° 00′ | .5000 | .8660 | .5774 | 1.7321 | 60° 00′ |
| 10 | .5025 | .8646 | .5812 | 1.7205 | 50 |
| 20 | .5050 | .8631 | .5851 | 1.7090 | 40 |
| 30 | .5075 | .8616 | .5890 | 1.6977 | 30 |
| 40 | .5100 | .8601 | .5930 | 1.6864 | 20 |
| 50 | .5125 | .8587 | .5969 | 1.6753 | 10 |
| 31° 00′ | .5150 | .8572 | .6009 | 1.6643 | 59° 00′ |
| 10 | .5175 | .8557 | .6048 | 1.6534 | 50 |
| 20 | .5200 | .8542 | .6088 | 1.6426 | 40 |
| 30 | .5225 | .8526 | .6128 | 1.6319 | 30 |
| 40 | .5250 | .8511 | .6168 | 1.6212 | 20 |
| 50 | .5275 | .8496 | .6208 | 1.6107 | 10 |
| 32° 00′ | .5299 | .8480 | .6249 | 1.6003 | 58° 00′ |
| 10 | .5324 | .8465 | .6289 | 1.5900 | 50 |
| 20 | .5348 | .8450 | .6330 | 1.5798 | 40 |
| 30 | .5373 | .8434 | .6371 | 1.5697 | 30 |
| 40 | .5398 | .8418 | .6412 | 1.5597 | 20 |
| 50 | .5422 | .8403 | .6453 | 1.5497 | 10 |
| 33° 00′ | .5446 | .8387 | .6494 | 1.5399 | 57° 00′ |
| 10 | .5471 | .8371 | .6536 | 1.5301 | 50 |
| 20 | .5495 | .8355 | .6577 | 1.5204 | 40 |
| 30 | .5519 | .8339 | .6619 | 1.5108 | 30 |
| 40 | .5544 | .8323 | .6661 | 1.5013 | 20 |
| 50 | .5568 | .8307 | .6703 | 1.4919 | 10 |
| 34° 00′ | .5592 | .8290 | .6745 | 1.4826 | 56° 00′ |
| 10 | .5616 | .8274 | .6787 | 1.4733 | 50 |
| 20 | .5640 | .8258 | .6830 | 1.4641 | 40 |
| 30 | .5664 | .8241 | .6873 | 1.4550 | 30 |
| 40 | .5688 | .8225 | .6916 | 1.4460 | 20 |
| 50 | .5712 | .8208 | .6959 | 1.4370 | 10 |
| 35° 00′ | .5736 | .8192 | .7002 | 1.4281 | 55° 00′ |
| 10 | .5760 | .8175 | .7046 | 1.4193 | 50 |
| 20 | .5783 | .8158 | .7089 | 1.4106 | 40 |
| 30 | .5807 | .8141 | .7133 | 1.4019 | 30 |
| 40 | .5831 | .8124 | .7177 | 1.3934 | 20 |
| 50 | .5854 | .8107 | .7221 | 1.3848 | 10 |
| 36° 00′ | .5878 | .8090 | .7265 | 1.3764 | 54° 00′ |
| | Cos | Sin | Cot | Tan | Angle |

| Angle | Sin | Cos | Tan | Cot | |
|---|---|---|---|---|---|
| 36° 00′ | .5878 | .8090 | .7265 | 1.3764 | 54° 00′ |
| 10 | .5901 | .8073 | .7310 | 1.3680 | 50 |
| 20 | .5925 | .8056 | .7355 | 1.3597 | 40 |
| 30 | .5948 | .8039 | .7400 | 1.3514 | 30 |
| 40 | .5972 | .8021 | .7445 | 1.3432 | 20 |
| 50 | .5995 | .8004 | .7490 | 1.3351 | 10 |
| 37° 00′ | .6018 | .7986 | .7536 | 1.3270 | 53° 00′ |
| 10 | .6041 | .7969 | .7581 | 1.3190 | 50 |
| 20 | .6065 | .7951 | .7627 | 1.3111 | 40 |
| 30 | .6088 | .7934 | .7673 | 1.3032 | 30 |
| 40 | .6111 | .7916 | .7720 | 1.2954 | 20 |
| 50 | .6134 | .7898 | .7766 | 1.2876 | 10 |
| 38° 00′ | .6157 | .7880 | .7813 | 1.2799 | 52° 00′ |
| 10 | .6180 | .7862 | .7860 | 1.2723 | 50 |
| 20 | .6202 | .7844 | .7907 | 1.2647 | 40 |
| 30 | .6225 | .7826 | .7954 | 1.2572 | 30 |
| 40 | .6248 | .7808 | .8002 | 1.2497 | 20 |
| 50 | .6271 | .7790 | .8050 | 1.2423 | 10 |
| 39° 00′ | .6293 | .7771 | .8098 | 1.2349 | 51° 00′ |
| 10 | .6316 | .7753 | .8146 | 1.2276 | 50 |
| 20 | .6338 | .7735 | .8195 | 1.2203 | 40 |
| 30 | .6361 | .7716 | .8243 | 1.2131 | 30 |
| 40 | .6383 | .7698 | .8292 | 1.2059 | 20 |
| 50 | .6406 | .7679 | .8342 | 1.1988 | 10 |
| 40° 00′ | .6428 | .7660 | .8391 | 1.1918 | 50° 00′ |
| 10 | .6450 | .7642 | .8441 | 1.1847 | 50 |
| 20 | .6472 | .7623 | .8491 | 1.1778 | 40 |
| 30 | .6494 | .7604 | .8541 | 1.1708 | 30 |
| 40 | .6517 | .7585 | .8591 | 1.1640 | 20 |
| 50 | .6539 | .7566 | .8642 | 1.1571 | 10 |
| 41° 00′ | .6561 | .7547 | .8693 | 1.1504 | 49° 00′ |
| 10 | .6583 | .7528 | .8744 | 1.1436 | 50 |
| 20 | .6604 | .7509 | .8796 | 1.1369 | 40 |
| 30 | .6626 | .7490 | .8847 | 1.1303 | 30 |
| 40 | .6648 | .7470 | .8899 | 1.1237 | 20 |
| 50 | .6670 | .7451 | .8952 | 1.1171 | 10 |
| 42° 00′ | .6691 | .7431 | .9004 | 1.1106 | 48° 00′ |
| 10 | .6713 | .7412 | .9057 | 1.1041 | 50 |
| 20 | .6734 | .7392 | .9110 | 1.0977 | 40 |
| 30 | .6756 | .7373 | .9163 | 1.0913 | 30 |
| 40 | .6777 | .7353 | .9217 | 1.0850 | 20 |
| 50 | .6799 | .7333 | .9271 | 1.0786 | 10 |
| 43° 00′ | .6820 | .7314 | .9325 | 1.0724 | 47° 00′ |
| 10 | .6841 | .7294 | .9380 | 1.0661 | 50 |
| 20 | .6862 | .7274 | .9435 | 1.0599 | 40 |
| 30 | .6884 | .7254 | .9490 | 1.0538 | 30 |
| 40 | .6905 | .7234 | .9545 | 1.0477 | 20 |
| 50 | .6926 | .7214 | .9601 | 1.0416 | 10 |
| 44° 00′ | .6947 | .7193 | .9657 | 1.0355 | 46° 00′ |
| 10 | .6967 | .7173 | .9713 | 1.0295 | 50 |
| 20 | .6988 | .7153 | .9770 | 1.0235 | 40 |
| 30 | .7009 | .7133 | .9827 | 1.0176 | 30 |
| 40 | .7030 | .7112 | .9884 | 1.0117 | 20 |
| 50 | .7050 | .7092 | .9942 | 1.0058 | 10 |
| 45° 00′ | .7071 | .7071 | 1.0000 | 1.0000 | 45° 00′ |
| | Cos | Sin | Cot | Tan | Angle |

## Table C: Logarithms of Trigonometric Functions*

| Angle | L Sin | L Cos | L Tan | L Cot | |
|---|---|---|---|---|---|
| 0° 00' | — | 10.0000 | — | — | 90° 00' |
| 10 | 7.4637 | 10.0000 | 7.4637 | 12.5363 | 50 |
| 20 | 7.7648 | 10.0000 | 7.7648 | 12.2352 | 40 |
| 30 | 7.9408 | 10.0000 | 7.9409 | 12.0591 | 30 |
| 40 | 8.0658 | 10.0000 | 8.0658 | 11.9342 | 20 |
| 50 | 8.1627 | 10.0000 | 8.1627 | 11.8373 | 10 |
| 1° 00' | 8.2419 | 9.9999 | 8.2419 | 11.7581 | 89° 00' |
| 10 | 8.3088 | 9.9999 | 8.3089 | 11.6911 | 50 |
| 20 | 8.3668 | 9.9999 | 8.3669 | 11.6331 | 40 |
| 30 | 8.4179 | 9.9999 | 8.4181 | 11.5819 | 30 |
| 40 | 8.4637 | 9.9998 | 8.4638 | 11.5362 | 20 |
| 50 | 8.5050 | 9.9998 | 8.5053 | 11.4947 | 10 |
| 2° 00' | 8.5428 | 9.9997 | 8.5431 | 11.4569 | 88° 00' |
| 10 | 8.5776 | 9.9997 | 8.5779 | 11.4221 | 50 |
| 20 | 8.6097 | 9.9996 | 8.6101 | 11.3899 | 40 |
| 30 | 8.6397 | 9.9996 | 8.6401 | 11.3599 | 30 |
| 40 | 8.6677 | 9.9995 | 8.6682 | 11.3318 | 20 |
| 50 | 8.6940 | 9.9995 | 8.6945 | 11.3055 | 10 |
| 3° 00' | 8.7188 | 9.9994 | 8.7194 | 11.2806 | 87° 00' |
| 10 | 8.7423 | 9.9993 | 8.7429 | 11.2571 | 50 |
| 20 | 8.7645 | 9.9993 | 8.7652 | 11.2348 | 40 |
| 30 | 8.7857 | 9.9992 | 8.7865 | 11.2135 | 30 |
| 40 | 8.8059 | 9.9991 | 8.8067 | 11.1933 | 20 |
| 50 | 8.8251 | 9.9990 | 8.8261 | 11.1739 | 10 |
| 4° 00' | 8.8436 | 9.9989 | 8.8446 | 11.1554 | 86° 00' |
| 10 | 8.8613 | 9.9989 | 8.8624 | 11.1376 | 50 |
| 20 | 8.8783 | 9.9988 | 8.8795 | 11.1205 | 40 |
| 30 | 8.8946 | 9.9987 | 8.8960 | 11.1040 | 30 |
| 40 | 8.9104 | 9.9986 | 8.9118 | 11.0882 | 20 |
| 50 | 8.9256 | 9.9985 | 8.9272 | 11.0728 | 10 |
| 5° 00' | 8.9403 | 9.9983 | 8.9420 | 11.0580 | 85° 00' |
| 10 | 8.9545 | 9.9982 | 8.9563 | 11.0437 | 50 |
| 20 | 8.9682 | 9.9981 | 8.9701 | 11.0299 | 40 |
| 30 | 8.9816 | 9.9980 | 8.9836 | 11.0164 | 30 |
| 40 | 8.9945 | 9.9979 | 8.9966 | 11.0034 | 20 |
| 50 | 9.0070 | 9.9977 | 9.0093 | 10.9907 | 10 |
| 6° 00' | 9.0192 | 9.9976 | 9.0216 | 10.9784 | 84° 00' |
| 10 | 9.0311 | 9.9975 | 9.0336 | 10.9664 | 50 |
| 20 | 9.0426 | 9.9973 | 9.0453 | 10.9547 | 40 |
| 30 | 9.0539 | 9.9972 | 9.0567 | 10.9433 | 30 |
| 40 | 9.0648 | 9.9971 | 9.0678 | 10.9322 | 20 |
| 50 | 9.0755 | 9.9969 | 9.0786 | 10.9214 | 10 |
| 7° 00' | 9.0859 | 9.9968 | 9.0891 | 10.9109 | 83° 00' |
| 10 | 9.0961 | 9.9966 | 9.0995 | 10.9005 | 50 |
| 20 | 9.1060 | 9.9964 | 9.1096 | 10.8904 | 40 |
| 30 | 9.1157 | 9.9963 | 9.1194 | 10.8806 | 30 |
| 40 | 9.1252 | 9.9961 | 9.1291 | 10.8709 | 20 |
| 50 | 9.1345 | 9.9959 | 9.1385 | 10.8615 | 10 |
| 8° 00' | 9.1436 | 9.9958 | 9.1478 | 10.8522 | 82° 00' |
| 10 | 9.1525 | 9.9956 | 9.1569 | 10.8431 | 50 |
| 20 | 9.1612 | 9.9954 | 9.1658 | 10.8342 | 40 |
| 30 | 9.1697 | 9.9952 | 9.1745 | 10.8255 | 30 |
| 40 | 9.1781 | 9.9950 | 9.1831 | 10.8169 | 20 |
| 50 | 9.1863 | 9.9948 | 9.1915 | 10.8085 | 10 |
| 9° 00' | 9.1943 | 9.9946 | 9.1997 | 10.8003 | 81° 00' |
| 10 | 9.2022 | 9.9944 | 9.2078 | 10.7922 | 50 |
| 20 | 9.2100 | 9.9942 | 9.2158 | 10.7842 | 40 |
| 30 | 9.2176 | 9.9940 | 9.2236 | 10.7764 | 30 |
| 40 | 9.2251 | 9.9938 | 9.2313 | 10.7687 | 20 |
| 50 | 9.2324 | 9.9936 | 9.2389 | 10.7611 | 10 |
| 10° 00' | 9.2397 | 9.9934 | 9.2463 | 10.7537 | 80° 00' |
| 10 | 9.2468 | 9.9931 | 9.2536 | 10.7464 | 50 |
| 20 | 9.2538 | 9.9929 | 9.2609 | 10.7391 | 40 |
| 30 | 9.2606 | 9.9927 | 9.2680 | 10.7320 | 30 |
| 40 | 9.2674 | 9.9924 | 9.2750 | 10.7250 | 20 |
| 50 | 9.2740 | 9.9922 | 9.2819 | 10.7181 | 10 |
| 11° 00' | 9.2806 | 9.9919 | 9.2887 | 10.7113 | 79° 00' |
| 10 | 9.2870 | 9.9917 | 9.2953 | 10.7047 | 50 |
| 20 | 9.2934 | 9.9914 | 9.3020 | 10.6980 | 40 |
| 30 | 9.2997 | 9.9912 | 9.3085 | 10.6915 | 30 |
| 40 | 9.3058 | 9.9909 | 9.3149 | 10.6851 | 20 |
| 50 | 9.3119 | 9.9907 | 9.3212 | 10.6788 | 10 |
| 12° 00' | 9.3179 | 9.9904 | 9.3275 | 10.6725 | 78° 00' |
| | L Cos | L Sin | L Cot | L Tan | Angle |

| Angle | L Sin | L Cos | L Tan | L Cot | |
|---|---|---|---|---|---|
| 12° 00' | 9.3179 | 9.9904 | 9.3275 | 10.6725 | 78° 00' |
| 10 | 9.3238 | 9.9901 | 9.3336 | 10.6664 | 50 |
| 20 | 9.3296 | 9.9899 | 9.3397 | 10.6603 | 40 |
| 30 | 9.3353 | 9.9896 | 9.3458 | 10.6542 | 30 |
| 40 | 9.3410 | 9.9893 | 9.3517 | 10.6483 | 20 |
| 50 | 9.3466 | 9.9890 | 9.3576 | 10.6424 | 10 |
| 13° 00' | 9.3521 | 9.9887 | 9.3634 | 10.6366 | 77° 00' |
| 10 | 9.3575 | 9.9884 | 9.3691 | 10.6309 | 50 |
| 20 | 9.3629 | 9.9881 | 9.3748 | 10.6252 | 40 |
| 30 | 9.3682 | 9.9878 | 9.3804 | 10.6196 | 30 |
| 40 | 9.3734 | 9.9875 | 9.3859 | 10.6141 | 20 |
| 50 | 9.3786 | 9.9872 | 9.3914 | 10.6086 | 10 |
| 14° 00' | 9.3837 | 9.9869 | 9.3968 | 10.6032 | 76° 00' |
| 10 | 9.3887 | 9.9866 | 9.4021 | 10.5979 | 50 |
| 20 | 9.3937 | 9.9863 | 9.4074 | 10.5926 | 40 |
| 30 | 9.3986 | 9.9859 | 9.4127 | 10.5873 | 30 |
| 40 | 9.4035 | 9.9856 | 9.4178 | 10.5822 | 20 |
| 50 | 9.4083 | 9.9853 | 9.4230 | 10.5770 | 10 |
| 15° 00' | 9.4130 | 9.9849 | 9.4281 | 10.5719 | 75° 00' |
| 10 | 9.4177 | 9.9846 | 9.4331 | 10.5669 | 50 |
| 20 | 9.4223 | 9.9843 | 9.4381 | 10.5619 | 40 |
| 30 | 9.4269 | 9.9839 | 9.4430 | 10.5570 | 30 |
| 40 | 9.4314 | 9.9836 | 9.4479 | 10.5521 | 20 |
| 50 | 9.4359 | 9.9832 | 9.4527 | 10.5473 | 10 |
| 16° 00' | 9.4403 | 9.9828 | 9.4575 | 10.5425 | 74° 00' |
| 10 | 9.4447 | 9.9825 | 9.4622 | 10.5378 | 50 |
| 20 | 9.4491 | 9.9821 | 9.4669 | 10.5331 | 40 |
| 30 | 9.4533 | 9.9817 | 9.4716 | 10.5284 | 30 |
| 40 | 9.4576 | 9.9814 | 9.4762 | 10.5238 | 20 |
| 50 | 9.4618 | 9.9810 | 9.4808 | 10.5192 | 10 |
| 17° 00' | 9.4659 | 9.9806 | 9.4853 | 10.5147 | 73° 00' |
| 10 | 9.4700 | 9.9802 | 9.4898 | 10.5102 | 50 |
| 20 | 9.4741 | 9.9798 | 9.4943 | 10.5057 | 40 |
| 30 | 9.4781 | 9.9794 | 9.4987 | 10.5013 | 30 |
| 40 | 9.4821 | 9.9790 | 9.5031 | 10.4969 | 20 |
| 50 | 9.4861 | 9.9786 | 9.5075 | 10.4925 | 10 |
| 18° 00' | 9.4900 | 9.9782 | 9.5118 | 10.4882 | 72° 00' |
| 10 | 9.4939 | 9.9778 | 9.5161 | 10.4839 | 50 |
| 20 | 9.4977 | 9.9774 | 9.5203 | 10.4797 | 40 |
| 30 | 9.5015 | 9.9770 | 9.5245 | 10.4755 | 30 |
| 40 | 9.5052 | 9.9765 | 9.5287 | 10.4713 | 20 |
| 50 | 9.5090 | 9.9761 | 9.5329 | 10.4671 | 10 |
| 19° 00' | 9.5126 | 9.9757 | 9.5370 | 10.4630 | 71° 00' |
| 10 | 9.5163 | 9.9752 | 9.5411 | 10.4589 | 50 |
| 20 | 9.5199 | 9.9748 | 9.5451 | 10.4549 | 40 |
| 30 | 9.5235 | 9.9743 | 9.5491 | 10.4509 | 30 |
| 40 | 9.5270 | 9.9739 | 9.5531 | 10.4469 | 20 |
| 50 | 9.5306 | 9.9734 | 9.5571 | 10.4429 | 10 |
| 20° 00' | 9.5341 | 9.9730 | 9.5611 | 10.4389 | 70° 00' |
| 10 | 9.5375 | 9.9725 | 9.5650 | 10.4350 | 50 |
| 20 | 9.5409 | 9.9721 | 9.5689 | 10.4311 | 40 |
| 30 | 9.5443 | 9.9716 | 9.5727 | 10.4273 | 30 |
| 40 | 9.5477 | 9.9711 | 9.5766 | 10.4234 | 20 |
| 50 | 9.5510 | 9.9706 | 9.5804 | 10.4196 | 10 |
| 21° 00' | 9.5543 | 9.9702 | 9.5842 | 10.4158 | 69° 00' |
| 10 | 9.5576 | 9.9697 | 9.5879 | 10.4121 | 50 |
| 20 | 9.5609 | 9.9692 | 9.5917 | 10.4083 | 40 |
| 30 | 9.5641 | 9.9687 | 9.5954 | 10.4046 | 30 |
| 40 | 9.5673 | 9.9682 | 9.5991 | 10.4009 | 20 |
| 50 | 9.5704 | 9.9677 | 9.6028 | 10.3972 | 10 |
| 22° 00' | 9.5736 | 9.9672 | 9.6064 | 10.3936 | 68° 00' |
| 10 | 9.5767 | 9.9667 | 9.6100 | 10.3900 | 50 |
| 20 | 9.5798 | 9.9661 | 9.6136 | 10.3864 | 40 |
| 30 | 9.5828 | 9.9656 | 9.6172 | 10.3828 | 30 |
| 40 | 9.5859 | 9.9651 | 9.6208 | 10.3792 | 20 |
| 50 | 9.5889 | 9.9646 | 9.6243 | 10.3757 | 10 |
| 23° 00' | 9.5919 | 9.9640 | 9.6279 | 10.3721 | 67° 00' |
| 10 | 9.5948 | 9.9635 | 9.6314 | 10.3686 | 50 |
| 20 | 9.5978 | 9.9629 | 9.6348 | 10.3652 | 40 |
| 30 | 9.6007 | 9.9624 | 9.6383 | 10.3617 | 30 |
| 40 | 9.6036 | 9.9618 | 9.6417 | 10.3583 | 20 |
| 50 | 9.6065 | 9.9613 | 9.6452 | 10.3548 | 10 |
| 24° 00' | 9.6093 | 9.9607 | 9.6486 | 10.3514 | 66° 00' |
| | L Cos | L Sin | L Cot | L Tan | Angle |

* These tables give the logarithms increased by 10. Hence in each case 10 should be subtracted.